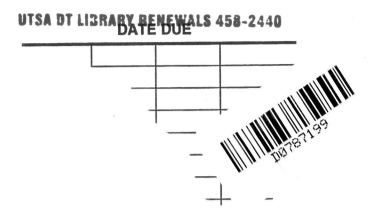

# ADVANCES IN URBAN ECOLOGY

# ADVANCES IN URBAN ECOLOGY
## Integrating Humans and Ecological Processes in Urban Ecosystems

by

**Marina Alberti**
University of Washington
Seattle, Washington, USA

Springer

Marina Alberti
University of Washington
Seattle, Washington, USA

Cover Design: Eric Knapstein

ADVANCES IN URBAN ECOLOGY:
Integrating Humans and Ecological Processes in Urban Ecosystems

Library of Congress Control Number: 2007936241

ISBN-13: 978-0-387-75509-0          e-ISBN-13: 978-0-387-75510-6

Printed on acid-free paper.

9 8 7 6 5 4 3 2 1

springer.com

*To Antonio, Leda
and Matteo*

# CONTENTS

# PREFACE

Natural and social scientists face a great challenge in the coming decades: to understand the role that humans play in ecosystems, particularly urban ecosystems. Cities and urbanizing regions are complex coupled human-natural systems in which people are the dominant agents. As humans transform natural landscapes into highly human-dominated environments, they create a new set of ecological conditions by changing ecosystem processes and dynamics. Urbanization changes natural habitats and species composition, alters hydrological systems, and modifies energy flows and nutrient cycles. Although the impacts of urban development on ecosystems occur locally, they cause environmental changes at larger scales. Environmental changes resulting from urbanization influence human behaviors and dynamics and affect human health and well-being.

Remarkable progress has been made in studying the impact of urban development on ecosystem functions (McDonnell and Pickett 1993, McDonnell et al. 1997, Grimm et al. 2000, Pickett et al. 2001, Alberti et al. 2003), yet the interactions and feedback between human processes and ecosystem dynamics in urbanizing regions are still poorly understood. In this book I argue that new syntheses across the natural and social sciences are necessary if urban and ecological dynamics are to be successfully integrated into a common framework to advance urban ecology research. If we remain within the traditional disciplinary boundaries, we will not make progress towards a theory of urban ecosystems as coupled human-ecological systems, because no single discipline can provide an unbiased and integrated perspective. Questions and methods of inquiry specific to disciplinary domains yield partial views that reflect different epistemologies and understandings of the world.

It is critical that we develop an integrated approach at a time when urbanizing regions are faced with rapid environmental change. Planners and managers worldwide face unprecedented challenges in supporting urban populations and improving their well-being while simultaneously maintaining ecosystem functions. Agencies must devise policies to guide urban development and make decisions about investing in infrastructure that is both economically viable and ecologically sustainable. An integrated framework is required to assess the environmental implications of alternative urban development patterns and to develop policies to manage urban areas in the face of change. In particular, strategies for urban growth management will require such integrated knowledge to maintain ecological resilience by preventing development pressure on the urban fringe, reducing resource use and emissions of pollutants, and minimizing impacts on aquatic and terrestrial ecosystems.

Scholars of urban ecology have started to recognize the importance of explicitly linking human and ecological processes in studying the dynamics of urban ecosystems. Not only are human decisions the main driving force behind urban ecosystems; changes in environmental conditions also control some important human decisions. Integrated studies of coupled human-natural systems have started to uncover new and complex mechanisms that are not visible to either social or natural scientists who study human and natural systems separately (Liu et al. 2007, Collins et al. 2000, Alberti et al. 2003). Simply linking scientific diciplines is not enough to achieve the level of synthesis required to see the urban ecosystem as a whole. Yet virtually no plan exists for synthesizing these processes into one coherent research framework.

The idea that humans are an integral part of ecosystems and that cities cannot be fully understood outside of their ecological context is hardly new. The evolution of cities as part of nature dates back at least to Geddes (1915) if not much earlier. Anne Spirn (1985) noted that an understanding of the interdependence between cities and nature was already present in the writings of Hippocrates (ca. $5^{th}$ century BCE), Vitruvius (ca. $1^{st}$ century BCE) and Leon Battista Alberti (1485). During the last century, the idea took form and evolved in initial areas of study in various disciplines including sociology (Park et al. 1925, Duncan 1960), geography (Berry 1964, Johnston 1982, Williams 1973, Zimmerer 1994), ecology (Odum 1953, Wolman 1971, Sukopp 1990, McDonnel et al. 1993), anthropology (Rappaport 1968, Kemp 1969, Thomas 1973), history (Cronon 1991), and urban design and planning (McHarg 1969, Spirn 1984, Lynch 1961), only to mention some of the earlier scholars. More recently, new attempts at interdisciplinary studies have emerged (McDonnell et al. 1997, Grimm et al. 2000, Pickett et al. 2001, Alberti et al. 2003).

What is new today is the acknowledgment that the sciences of ecology and of cities have pretty much ignored each other until very recently. The theoretical perspectives developed to explain or predict urban development and ecosystem dynamics have been created in isolation; neither perspective fully recognizes their interdependence. Ecologists have primarily studied the dynamics of species populations, communities, and ecosystems in non-urban environments. They have intentionally avoided or vastly simplified human processes and institutions. Landscape ecology is, perhaps, the first consistent effort to study how human action (i.e., changing spatial patterns) influences ecological processes (e.g., fluxes of organisms and materials) in urbanizing environments. Social scientists, on the other hand, have only primitive ways to represent ecological processes. Neoclassical economics, for example, uses the theory of land rent to explain the behaviors of households, businesses, and governments that lead to patterns of urban development, completely disregarding the dynamic interactions between land development and environmental change.

In studying the ways that humans and ecological processes interact, we must consider that many factors work simultaneously at various levels. If we simply link traditional disciplinary models of human and ecological systems, we may misrepresent system dynamics because system interactions may occur at levels that our models fail to consider. This is particularly true in urban ecosystems, since urban development controls ecosystem structure and function in complex ways. Furthermore, these interactions are spatially determined. The dynamics of land development and resource uses and their ecological impacts depend on the spatial patterns of human activities and their interactions with biophysical processes at various scales. Humans generate spatial heterogeneity as they transform land, extract resources, introduce exotic species, and modify natural agents of disturbance. In turn, spatial heterogeneity, both natural and human-induced, affects resource fluxes and ecological processes in urbanizing ecosystems.

In this book I seek to bring together—systematically—a wide range of theories, models, and findings by scholars of urban ecosystems in both the natural and social sciences.[1] What sets this work apart from other efforts to assess the human role in ecosystems is my specific focus on urban areas. Although interest in urban ecosystems is growing, no single theory incorporates the different processes and approaches. A major obstacle to integration is the absence of a consistent understanding of related concepts and a common language (Tress et al. 2004). I address several disciplinary perspectives—ecology, economics, geography, landscape ecology, and planning—each with its own assumptions, methods of analysis, and standards of validation. Without the previous work of scholars from these many disciplines I could not have possibly covered all the areas of research or touched on the complex scientific problems emerging in these fields. Many aspects remain outside the scope of this book, since an attempt to address them all would have made the task impossible. My goal is to explore the opportunities for a synthesis and provide a framework that can stimulate scholars in these disciplines to generate theories and hypotheses, and identify areas for future research. Using the Puget Sound as a case example I present a range of theoretical issues and methodological implications.

When I started writing this book I thought I could synthesize the challenges that the study of urban ecosystems poses to both social and natural

---

[1] I focus primarily on North America and only in part on the European schools. There are important contributions in many parts of the world that are not included in this book—not because they are not relevant to the study of urban ecology—but simply because they are outside the scope of this book.

scientists; I aimed at the "consilience," or unity of knowledge across fields, that Wilson (1998) argues has eluded science. As this work proceeded, it became clear that many syntheses are possible—at least one for each team of scientists and practitioners that comes together to study urban ecology. All are potentially accurate accounts, but all are incomplete views of urban ecology. In this book, I attempt to provide one of these possible syntheses, building on the collective work and thoughts of the Urban Ecology team of faculty and students at the University of Washington in Seattle. I propose that cities are hybrid phenomena—driven simultaneously by human and bio-physical processes. We cannot understand them fully just by studying their component parts separately. Thus urban ecology is the study of the ways that human and ecological systems evolve together in urbanizing regions.

To fully integrate humans into ecosystems, ecology must deal with the complexity and diversity of human cultures, values, and perceptions, and their evolution over time. Culture and values play a key role in the ways that humans build cities and shape the built environment. As Lynch (1961,79) put it in *A Theory of Good City Form*, "we must learn what is desirable so as study what is possible." In this book I do not explicitly address culture and values, not because I do not consider them essential to an understanding of how urban ecosystems work, but because I could not possibly do justice to the complexity of relationships that culture and values bring to the study of urban ecosystems. I leave this task to those scholars in anthropology, sociology, planning, and political science who study culture and values and can more effectively and thoroughly build a bridge with the perspective proposed here. I hope that by bringing humans into the study of the eco-systems this book will lead the way in efforts to fully integrate them.

The book starts with a review of urban ecological theory and the evolving concept of the urban ecosystem. Chapter 1 examines existing approaches for integrating human and ecological systems and articulates the rationale for a new synthesis, based on the fact that humans are driving the dynamics of urban ecosystems. Chapter 2 explores the role of humans and societal pro-cesses, and identifies human-induced stresses and disturbances. Major human driving forces are demographics, socioeconomic organization, political structures, and technology. Human behaviors—the underlying rationales for the actions that give rise to these forces—directly influence the use of land, as well as the demand for and supply of resources. In urban areas these forces combine to affect the spatial distribution of activities, and ultimately affect the spatial heterogeneity of ecological processes and disturbances. Chapter 3 focuses on how urban patterns affect ecosystem dynamics. I summarize what we do and do not know about the relationships between urban patterns and ecosystem functions.

To study urban ecosystems and test hypotheses about mechanisms that govern their dynamics, we need to detect and accurately quantify the urban landscape pattern and its change over time. In Chapter 4, I propose that hybrid landscapes in urbanizing regions have distinctive signatures and propose an approach to quantify them. Chapters 5 through 8 examine the impacts of urban patterns on the biophysical environment and the resulting effects on ecosystem dynamics. Throughout these chapters I explore the connections between human and ecological processes and their implications for integrated research. Chapter 9 addresses the complexity and uncertainty in modeling urban ecosystems, their variability and dynamics, and the causes and effects of heterogeneity on ecological and economic processes at various scales. Many important ecological processes are sensitive to spatial heterogeneity and its effects on fluxes of organisms, materials, and energy. Spatial heterogeneity also affects the fluxes of economic resources that ultimately control the underlying urban pattern. Scale is a critical factor in understanding the interactions between human and natural disturbances, since spatial heterogeneity may affect the outcome of changes in driving forces only at certain scales. I discuss the challenges of predicting future dynamics of coupled human-ecological systems. Then, I explore scenario planning as an approach to adaptive planning and management. In Chapter 10 I provide a synthesis and a research agenda for the future.

An integrated knowledge of the processes and mechanisms that govern urban ecosystem dynamics will be crucial if we are to advance ecological research, and to help new generations of planners and managers solve complex urban environmental problems. This is a book aimed at ecologists, economists, geographers, engineers, political scientists, and planners interested in understanding the dynamic of coupled human-natural systems in urbanizing regions and the resilience of urban ecosystems under alternative scenarios of urban development and environmental change.

Marina Alberti
Seattle, Washington
August 2007

# ACKNOWLEDGMENTS

Many people have contributed to this project in crucial ways. It would have been impossible without the students and faculty of the Urban Ecology Program at the University of Washington (UW) in Seattle. John Marzluff, Gordon Bradley, Clare Ryan, Craig Zumbrunnen, and Eric Shulenberger have been instrumental in the development and evolution of the ideas presented here. I am grateful to them for an exciting and intellectually stimulating collaboration that has led to a research framework for urban ecology and the emergence of a school of thought.

The ideas and work that I present in the book are the product of many scholars involved or affiliated with the Urban Ecology Research Lab (UERL). Indeed, without the lab team, I would have had very little to write about. Stefan Coe, Jeff Hepinstall, Daniele Spirandelli, Michal Russo, Karis Puruncajas, Yan Jiang, Bekkah Couburn, Marcie Bidwell, Camille Russell, Debashis Mondal, Erik Botsford, and Alex Cohen have all contributed to the empirical research conducted in Puget Sound and presented in this book. Jeff Hepinstall has been instrumental to the development of the Land Cover Change Model. Lucy Hutyra and Steven Walters have contributed to discussions of aspects of earth sciences and landscape ecology. Several of my Ph.D. students have contributed to discussions on theoretical questions posed here: Vivek Shandas, Adrienne Greve, Yan Jiang, Karis Puruncajas, Daniele Spirandelli, Michelle Kondo, and David Hsu. My collaborations with John Marzluff (ecology), Paul Waddell (modeling), Derek Booth (hydrology), Robin Weeks (remote sensing), and Hilda Blanco (planning) have been instrumental to many aspects of this book.

The team involved in the Biocomplexity project BE/CNH (Urban Landscape Patterns: Complex Dynamics and Emergent Properties) has inspired several of the key ideas contained in this book. The project is a joint effort by the UW Urban Ecology Research Lab and the Arizona State University Global Institute of Sustainability. The project team includes Jianguo Wu, Charles Redman, John Marzluff, Mark Handcock, Marty Anderies, Paul Waddell, Dieter Fox, Henry Kautz, and Jeff Hepinstall. I am also grateful to several federal and state agencies that have supported the research presented here: the National Science Foundation, the US Environmental Protection Agency, the Washington State Department of Ecology, the Puget Sound Action Team, and King County.

I am indebted to many scientists whose work in urban ecology has made this project possible. I am particularly grateful to Stewart Pickett, Mark McDonnell, and Nancy Grimm for their pioneering work and the leadership they have provided to the field of urban ecology in the United States and to Herbert Sukopp for his pioneering work in Europe. Several other outstanding thinkers have influenced my thinking in urban ecology, especially on complex coupled human-natural systems: Buzz Hollings, Stuart Kaufman, Per Back, and Steve Carpenter. Kevin Lynch has influenced my view of urban design and planning. Three people have taught me to challenge my assumptions about how human and natural systems work: Virginio Bettini, Larry Susskind, and Paul Ehrlich.

Finally, I am indebted to several people for vital contributions to the production process. Michal Russo produced all the graphics and illustrations for this book, translating complex concepts and data into effective visual representations. Japhet Koteen has conducted literature research in the initial stages of this book. John Marzluff, Steven Walters, Lucy Hutyra, and Michal Russo provided invaluable input on content, and feedback on editorial style, in many chapters. Steven also contributed to the literature search. I am grateful to five anonymous referees for their constructive input at an initial stage of this project. Helen Snively edited the book carefully and thoroughly; her excellent comments and critical eye have substantially improved the writing style and readability. Sue Letsinger provided additional editing and created a camera-ready manuscript—and with great dedication and patience. Additional editorial comments were provided by Sue Blake. I also thank Melinda Paul, my editor at Springer-Verlag, for her support and great patience.

This book is dedicated to three important people in my life. My father, Antonio, and my mother, Leda, who have taught me to think critically and openly across many aspects of science and human history. This book might not have existed at all, without one very young boy: my son. I did not know him when I was writing this book. He has motivated me to complete this project before his arrival and inspired my work, because he will be the one living in the cities of the future.

# Chapter 1

# THE URBAN ECOSYSTEM

Cities are complex ecological systems dominated by humans. The human elements make them different from natural ecosystems in many ways. From an ecological perspective, urban ecosystems differ from natural ones in several respects: in their climate, soil, hydrology, species composition, population dynamics, and flows of energy and matter (Rebele 1994, Collins et al. 2000, Pickett et al. 2001). Humans create distinctive ecological patterns, processes, disturbances, and subtle effects (McDonnel et al. 1993). Planners must consider all these factors in order to effectively plan cities that will be ecologically resilient. Managing these systems requires an understanding of the mechanisms that link human and ecological processes and control their dynamics and evolution. Because change is an inherent property of ecological systems, the capacity of urban ecosystems to respond and adapt to these changes is an important factor in making cities sustainable over the long term (Alberti and Marzluff 2004).

Ecology has provided increasing evidence that humans are dramatically changing Earth's ecosystems by increasing landscape heterogeneity and transforming their energy and material cycles (Vitousek et al. 1997). We know a great deal about the processes through which human activities affect the material and energy budgets that cause heterogeneity. For example, we appropriate natural resources, convert land surfaces, modify land forms, burn fossil fuels, and build artificial drainage networks. Human action has transformed 30% to 50% of the world's land surface and humans use more than half of the accessible fresh water. More nitrogen is now fixed synthetically than naturally in terrestrial ecosystems (Vitousek et al. 1986). According to the most recent global ecosystem assessment, humans have changed ecosystems more rapidly during the past 50 years than in any other time in human history, and as a consequence have irreversibly modified biodiversity (Figure 1.1, Turner et al. 1990, MEA 2005).

It is becoming quite evident that Earth's ecosystems are increasingly influenced by both the pace and patterns of urban growth, and to a great extent the future of ecosystems will depend upon how we will be able to make urban regions sustainable. The remarkable change urbanization has made across the globe can be observed from space (Figure 1.2, NASA 2000). Cities have grown remarkably in the last few decades and are

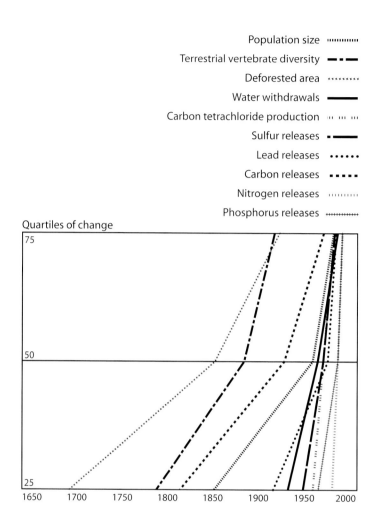

Figure 1.1. Trends in selected human-induced drivers of environmental change (Turner et al. 1990, p. 7).

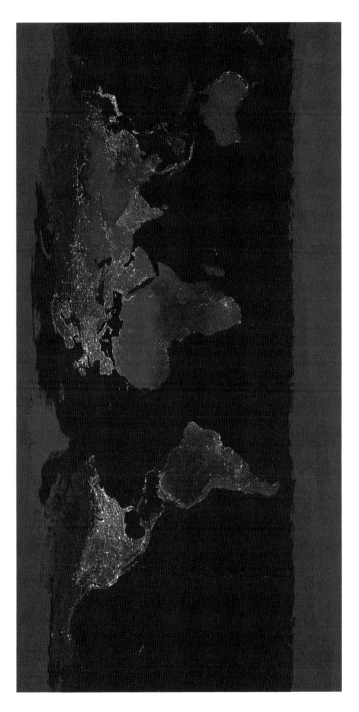

Figure 1.2. Earth lights (National Aeronautics and Space Administration 2000).

growing rapidly worldwide with a total of 20 cities now boasting populations of over 20 million, compared to just two in 1950 (Figure 1.3). All of the population growth expected in the next 25 years (2000 and 2030, approximately two billion people) will be concentrated in urban areas. The world urban population will reach more than 60 percent (4.9 billion) by the year 2030. This is three times the total population of the planet 100 years ago (1.7 billion people) (UN 2005, Figure 1.4).

During the last half century we have learned much about the impact of urbanization on ecosystems (McDonnell and Pickett 1993). Early descriptions of urban ecosystems have focused on both the "ecology in cities" (which primarily focuses on the study of habitats or organisms within cities) and an "ecology of cities" (which studies urban areas from an ecological systems perspective) (Grimm et al. 2000). In terms of energy metabolism, cities are "hot spots" on the biosphere's surface (Odum 1963, 1997). Human activities directly affect land cover, which controls primary productivity and biotic diversity (Sukopp 1990). Urbanization also influences biogeochemical processes, and modifies microclimates and air quality by altering the nature of the land surface and generating heat (Oke 1987). As urbanization also increases the impervious surface area, it affects both geomorphological and hydrological processes and changes fluxes of water, nutrients, and sediment (Leopold 1968, Arnold and Gibbons 1996). But the mechanisms through which urbanization patterns affect ecosystem processes are still virtually unknown. Nor do we know how biophysical patterns and processes and their dynamic changes affect human choices regarding their spatial arrangement on the landscape. We do not know how urban ecosystems evolve through the interactions between human and ecological processes, nor do we know what factors control their dynamics.

Although a substantial body of urban research has focused on the dynamics of urban systems—their sociology, economics, ecology, and policies—these diverse dimensions have yet to be synthesized into one coherent theoretical framework. Models of urban systems designed to explain or predict urban dynamics are limited in their ability to simultaneously represent human and ecological processes. Ecological models of urban ecosystems vastly simplify human processes. Even though ecologists have studied urban areas for quite some time, only recently have they realized that we cannot study urban ecosystems unless we also understand how humans and their organizations function in them. Social scientists, on the other hand, have only recently started to recognize that people are biological organisms and that the natural environment may be a key factor in explaining many of the choices people make. Simply linking existing approaches in an "additive" fashion may not adequately address the processes and behaviors that couple human and natural systems, because human and ecological processes may interact at levels that are not represented in each separate disciplinary framework (Pickett et al. 1994).

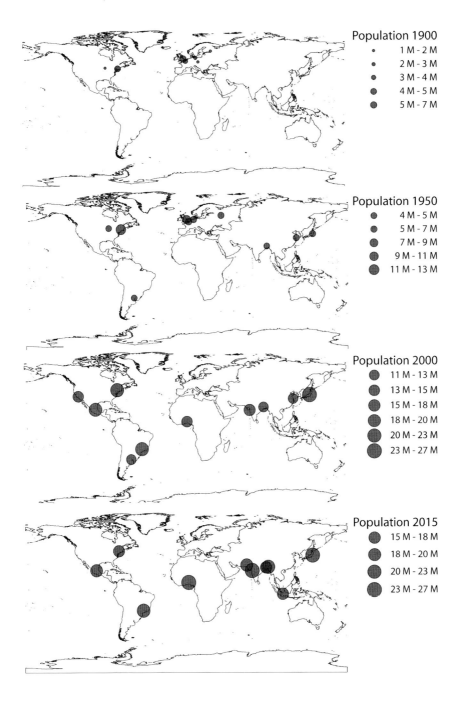

Figure 1.3. Population of the top ten largest cities in the world (UN 2006).

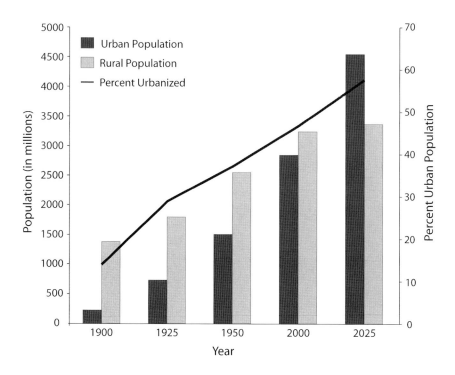

Figure 1.4. World urbanized and rural population growth, actual and projected 1900–2025 (UN 2006).

In this book I argue that we can achieve a new understanding of the relationships between cities and the natural environment by looking at cities as hybrid phenomena that emerge from the interactions between human and ecological processes. Complex systems theory provides the conceptual basis and methodology for studying urban ecosystems to decode "emergent" phenomena, such as urban sprawl, and devise effective policies to minimize their effects on ecosystem function. Complex structures can evolve from multiple agents operating according to simple decision rules (Resnick 1994, Nicolis and Prigogine 1989). Some fundamental attributes of complex human and ecological adaptive systems—multiple interacting agents, emergent structures, decentralized control, and adapting behavior—can help researchers to understand how urbanizing landscapes work, and to study urban ecosystems as integrated human-ecological phenomena.

Complex coupled human-natural systems in metropolitan areas challenge both ecological and current planning and management paradigms. The study of urban ecosystems requires a radical change in the way scholars frame questions about urban ecology. Instead of asking: "How do humans affect ecological systems?" the question should be: "How do humans interacting with their biophysical environment generate emergent collective behaviors (of humans, other species, and the systems themselves) in urbanizing landscapes?" Theories about complex adaptive systems provide the conceptual foundations we can use to analyze how the landscape structure and processes emerge in urbanizing regions, how they are maintained, and how they evolve through the local interactions of processes that occur at smaller scales among social, economic, ecological, and physical agents—in other words, how self-organization occurs in urban landscapes.

Scholars in several disciplines have started to recognize the importance of explicitly addressing human and ecological interactions when studying urban dynamics. Furthermore, both the social and natural sciences have made fundamental changes in the assumptions underlying their theories. They no longer regard social and natural systems as closed, self-regulating entities that "mature" to reach equilibrium. Instead, they recognize that they are multiequilibria systems—open, dynamic, and highly unpredictable. In such systems, change is a frequent intrinsic characteristic (Pickett et al. 1992). Such change has multiple causes, can follow multiple pathways, and is highly dependent on the environmental and historical context. In the newer, nonequilibrium paradigm, systems are driven by processes (rather than toward end points) and are often regulated by external forces. Sharp shifts in behaviors are natural for these systems. The challenge for both ecological and urban scholars is to apply this perspective to study coupled human-ecological systems in urbanizing landscapes.

Several questions inspire this book. First, I ask how ecosystem patterns in urbanizing regions emerge from the complex interactions of multiple agents and processes. These interactions occur among and between human and non-human agents and between agents and their environment (e.g., species and their habitat, housing and its neighborhood context). Interactions occur between human and ecological processes. With this question, I aim to understand how the local interaction of multiple agents and their environments affects the global composition, configuration, and dynamics of whole metropolitan regions.

Second, I ask how and at what scale, the spatial structure of the ecological, physical, and socioeconomic factors in urban landscapes affects ecosystem function. While several scholars have extensively described the urban landscape as a mosaic of biological and physical patches within a matrix of infrastructure, social institutions, cycles, and order (Machlis et al.

1997), few have explored the mechanisms that link spatial heterogeneity to human and ecological processes and the influence those mechanisms have on ecosystem dynamics (Pickett et al. 1997, Wu and David 2002, Alberti and Marzluff 2004).

A third question is critical for understanding the dynamics of urban ecosystems: How the fluxes of energy and matter in urban ecosystems compare in magnitude to non-human-dominated ecosystems, and how humans drive and control them. Although many natural ecosystems have been studied from this perspective, we do not know much about the inputs and outputs of key energetic and material fluxes and their linkages to human processes in urban areas.

Finally, I ask how complex interactions between human and ecosystem functions over multiple scales affect resilience in urban ecosystems. That is, how much alteration can urban ecosystems tolerate before they reorganize around a new set of structures and processes (Holling 1973, 1996)? Can urban ecosystem evolution be predicted given the complexity and uncertainty of such interactions? And how can we plan in light of such complexity and uncertainty?

In this chapter I explore how different disciplinary perspectives—ecology, economics, geography, landscape ecology, and planning—each with a distinctive approach to the study of cities, have contributed to define the dynamics of urban systems. Building on complex system theory, I suggest that cities are hybrid human and natural phenomena, and that in their separate domains none of these disciplines can completely explain how urban ecosystems emerge and evolve. I then articulate my rationale for a new synthesis and discuss an integrated approach for studying the relationships between natural and human systems in the metropolis.

## 1.1 The Dynamics of Urban (Eco)Systems

The processes that contribute to urban development and ecology are extraordinarily complex, and many scholars have adopted diverse theoretical approaches to explain or predict them. To some extent, researchers within separate disciplines can detect and study distinct urban ecological processes, but because of the processes' complex interactions, the emergence of the city and its evolution are phenomena that cannot be studied within separate domains. Scholars in various disciplines have tried to make sense of urban phenomena through a specific set of lenses, revealing different aspects of what the city is. Pioneer urban thinkers started to draw the connections among the fields. Patrick Geddes (1915) was one of the first to apply concepts from biology and evolutionary theory to the study of cities and their evolution. Inspired by Geddes, Lewis Mumford (1925) expanded the notion of ecological regionalism. Kevin Lynch (1961) classified theories of urban genesis and function according to the dominant images that theorists used to conceive of the city and understand how it works. These functional theories adequately represent the way cities have been conceptualized in different disciplinary domains. The city can be described as a unique historical process, a human ecosystem, a space for producing and consuming goods, a field of forces, a system of linked decisions, or an arena of conflicts (Lynch 1961). However, these theories do not add up to a theory of urban (eco)systems and its dynamics.

The first conceptualization of an "urban ecology" that attempted to use ecosystem ecology to understand urban patterns is the Chicago School of Sociology in the 1920s (Burgess 1925, Park et al. 1925). Park et al. (1925) applied ecosystems ecology to explore how cities work; they posited that cities are governed by many of the same driving forces and mechanisms that govern ecosystems. They suggested that competition for scarce urban resources (i.e., land) drives people to organize the urban space into distinctive ecological niches or "natural areas" where they share similar social characteristics and ecological pressures. Competition for land and resources ultimately leads people to spatially differentiate urban space into zones, with more desirable areas attracting higher rents. Paraphrasing plant ecologists, Park and Burgess described the change in land use as succession. Their model, known as concentric zone theory and first published in *The City* (1925), identified five concentric rings, with areas of social and physical deterioration concentrated near the city center and more prosperous areas located near the city's edge. The Chicago School certainly made the earliest systematic attempt to apply ecological theory in urban studies. Its goal, however, was not to study ecological relationships but to understand urban systems, building on ecological analogies.

In ecology, early efforts to understand the interactions between urban development and environmental change led to the conceptual model of cities as urban ecosystems (Odum 1963, 1997, Duvigneaud 1974, Stern and Montag 1974, Boyden et al. 1981, Douglas 1983). Ecologists have described the city as a heterotrophic ecosystem—highly dependent on large inputs of energy and materials and a vast capacity to absorb emissions and waste (Odum 1963, Duvigneaud 1974, Boyden et al. 1981). Wolman (1965) applied an "urban metabolism" approach to quantify the flows of energy and materials into and out of a hypothetical American city. Systems ecologists provided formal equations to describe the energy balance and the cycling of materials (Douglas 1983).

Parallel to the efforts to conceptualize the city as urban ecosystems are the empirical studies conducted by many scholars in various disciplines including biology (Sukopp and Werner 1982, Trepl 1995, Rebele 1994, Gill and Bonnett 1973), hydrology (Dunne and Leopold 1978, Shaeffer et al. 1982), atmospheric science (Oke 1973, Chandler 1976, Landsberg 1981) and the work of many pioneer landscape architects and urban designers in translating this knowledge into strategies for city design (McHarg 1969, Hough 1984, Spirn 1984, Lyle 1985). Since then, urban and ecological scholars have made important progress in understanding how urban ecosystems operate and how they differ from pristine natural ecosystems.

Early studies of urban ecosystems have provided a rich basis on which to develop a science of urban ecology. However, scholars were working in isolation within each discipline, so these studies often simplified either the human or the ecological dimension. Urban scholars were rightly skeptical about the attempts to integrate biological and socioeconomic concepts into system dynamics models. None of these models could explicitly represent the processes through which humans affect or are affected by the urban environment. At best, human behavior was reduced to a few differential equations. Ecological scholars, on the other hand, have been skeptical about the idea that adding human processes would generate any useful insight for ecological research. Only recently have scholars in both the natural and social sciences started to acknowledge that coupled human-natural systems require us to revise our disciplinary assumptions so we can study the complex interactions and subtle feedback of urban ecosystems (McDonnell et al. 1993).

More recently, complex systems theory has been applied to study human-natural systems. Gunderson and Holling (2002) conceptualize coupled human-natural systems by focusing on the temporal and spatial scale at which each system component operates. They describe coupled human-ecological systems as a hierarchy or a nested set of adaptive cycles. According to the theory of adaptive cycle, dynamic systems do not tend toward some stable condition. Instead they follow four stages: rapid growth, conservation, collapse, and renewal and reorganization (Gunderson et al. 1995, Carpenter et al. 2001, Holling et al. 2002a, 2002b). The traditional view of ecosystem succession is replaced by a new model of dynamic change that is regulated by three properties of ecosystems: the potential for change, the degree of connectedness, and the system resilience. Resilience—the amount of disturbance a system can tolerate before moving into another domain of attraction—determines how vulnerable the system is to unexpected change and surprises (Holling 2001, Holling and Gunderson 2002). This perspective has influenced the more recent studies in urban ecology.

Scholars of urban ecosystems in Phoenix, Baltimore, and Seattle have started to articulate conceptual models in an effort to integrate multiple perspectives, so we can better understand coupled human-ecological interactions in urban ecosystems (Grimm et al. 2000, Pickett et al. 2001, Pickett and Cadenasso 2002, Alberti et al. 2003). These different schools of thought have developed different models to test hypotheses regarding the relationships between urbanization and ecosystem function (Collins et al. 2000, Grimm et al. 2000, Pickett et al. 2001). While all three models are concerned with the relationships between human and ecological patterns and their relationship with human and ecological function, the degree to which these approaches are integrated vary with the specific composition of the teams constructing them.

In Baltimore, Pickett et al. (2001) have developed a framework for the urban Long Term Ecological Research (LTER) site that aims to articulate the different sub-systems that constitute a human ecosystem and link them through a series of direct mechanisms and feedback loops (Figure 1.5). In addition to traditional human dimensions, Pickett et al. (2001) characterize the human social system in terms of social institutions, social cycles, and social order. The link between the human system and the biophysical resources (patterns and processes) is mediated by the resource system, which in turn includes both cultural and socioeconomic mechanisms. They apply a patch dynamic approach to represent the spatially explicit structure of ecological systems, but because urban areas are coupled human-biophysical systems, they propose a hybrid patch dynamic approach to integrate biological, physical, and social patches.

The Phoenix LTER, led by Grimm et al. (2000), proposes a more integrated approach. The team is articulating the mechanisms that link biophysical and socioeconomic drivers to ecosystem dynamics through both human activities and ecosystem processes and patterns (Figure 1.6). Grimm et al. (2000) build on a systems ecology perspective to study the relationships between patterns of human activities and the patterns and processes of ecosystems driven by flows of energy and information, and the cycling of matter. These relationships are mediated by social institutions, culture, behavior, and their interactions. Using this approach, they have shown the interaction among physical, ecological, engineering, social, and management variables and drivers in the new Tempe Town Lake in Arizona (Grimm et al. 2000). The constraints that drive the land-use decision (to establish the lake) are both biophysical (i.e., existence of an alluvial channel with no surface water flow) and societal (i.e., economic cost of the project). When it is filled, the lake is likely to have high levels of nutrients, algal production, and infiltration, and flooding is very likely. These factors may lead to changes in ecological conditions such as eutrophication and loss of water to the groundwater system, and may help establish a robust mosquito population. Societal responses can be direct, such as adding chemical control agents, or indirect, such as diverting upstream inputs of point source nutrients away from the water supply, and thus affecting the underlying ecological patterns and processes that produce the problem (Grimm et al. 2000).

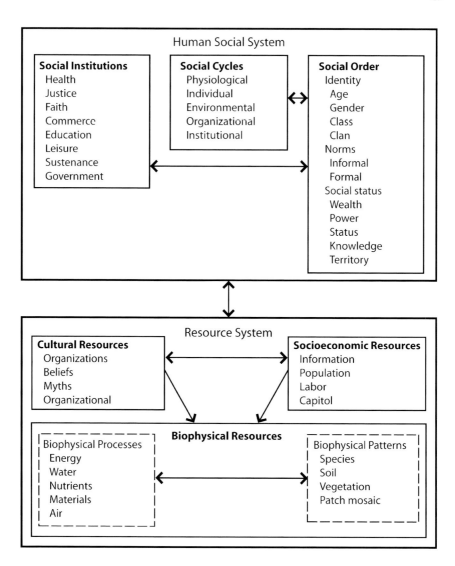

Figure 1.5. Human ecosystem framework. The model represents the relevant interactions and feedbacks of coupled human-biophysical systems in urban ecosystems and constitutes the conceptual framework for the Baltimore Long-Term Ecological Research (BES LTER) site (Pickett et al. 2001, p. 118).

In developing an integrated urban ecology framework, my colleagues on the Seattle urban ecology team and I place the emphasis on the unique interactions between patterns and processes and their human and ecological function. We propose a conceptual framework that does not distinguish

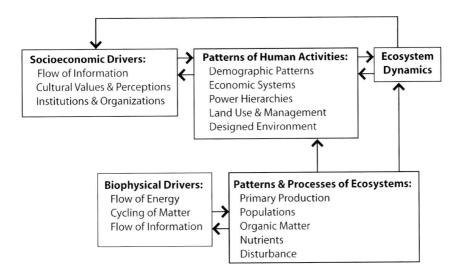

Figure 1.6. Ecosystem dynamics conceptual framework. The model represents the relevant interactions and feedbacks of coupled human-biophysical systems in urban ecosystems and constitutes the conceptual framework for the Central Arizona-Phoenix Long-Term Ecological Research (Grimm et al. 2000, p. 574, © American Institute of Biological Sciences).

between human and ecological patterns and human and ecological processes (Figure 1.7). Instead, our approach recognizes that patterns in urban landscapes are created by micro-scale interactions between human and ecological processes, and that urban ecosystem functions are affected and maintained simultaneously by human and ecological patterns.

Whether we look at the biological, physical, chemical, social, economic, or other constituents, urban landscape patterns are hybrid phenomena emerging from the interplay of human and ecological processes acting on multiple temporal and spatial scales. Land cover in cities is the result of climate and weather—which can vary across a day, a year, or a decade—as well as land clearing and development. The system of urban water flow results from stormwater runoff, which in turn is a product of many other factors: geology, rainfall, topography, land cover, basin size, and the routing of runoff (Dunne and Leopold 1978). In an urban setting, humans alter the basin shape, size, and the movement of water by building infrastructure (i.e., sewers) and changing dispersal vectors and patterns as people and products move throughout the landscape. Species diversity is affected by changes in habitat, predation, and food availability. Natural disturbances, such as fire and flooding, are modified in urban landscapes in terms of their magnitude, intensity, and frequency. Other disturbances (e.g., the concentration of air

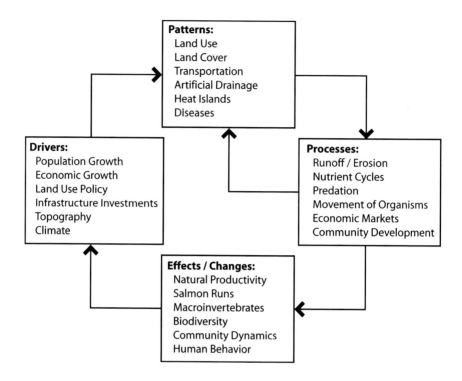

Figure 1.7. Urban ecology conceptual framework. The model represents the relevant interactions and feedbacks of coupled human-biophysical systems in urban ecosystems and constitutes the conceptual framework for the Seattle Urban Ecology Research site (Alberti et al. 2003, p. 1175, © American Institute of Biological Sciences).

pollution from intense traffic) that are unique to the urban environment are introduced as a consequence of human activities.

My team and I see urban ecosystems as complex, adaptive, dynamic systems (Alberti et al. 2003). Cities evolve as the outcome of myriad inter-actions between the individual choices and actions of many human agents (e.g., households, businesses, developers, and governments) and biophysical agents, such as local geomorphology, climate, and natural disturbance regimes. These choices produce different patterns of development, land use, and infrastructure density. They affect ecosystem processes both directly (in and near the city) and remotely through land conversion, use of resources, and generation of emissions and waste. Those changes, in turn, affect human health and well-being (Alberti and Waddell 2000). All three teams conceptualize urban ecosystems as complex and dynamic, but only recently have scholars started to articulate urban ecosystems in a systems dynamic perspective. Parallel work at the three urban sites has led to different perspectives to characterize urban ecological dynamics. Pickett and Cadenasso (2006) propose a patch dynamic approach. Wu and David (2002) are developing a hierarchical modeling approach. Building on Holling's (2002) concept of adaptive cycles, Marzluff (2004) and I propose that urban ecosystems are complex adaptive systems with multiple equilibria. Their resilience is governed by the balance between human and ecological function.

## 1.2 Cities as Human Systems

In urban ecosystems, human dynamics are dominant driving forces through demographics, economics, socioeconomic organizations, political structures, and technology. Some authors describe cities as production systems primarily driven by market forces (Thompson 1975). Others see it as a consumption system (Hallsworth 1978), or a system of both production and consumption. Various attempts to describe the city as an economic, social, and political system have clearly depicted the economic, social, or political dynamics, but have failed to understand the interactions between those dynamics and the ecological dimensions.

Human behaviors—the underlying rationales for the actions that give rise to these forces—directly influence the use of land and the demand for, and supply of, resources (Turner 1989). In urban areas these forces combine to affect the way that activities are distributed over space. Both social (Openshaw 1995) and natural scientists (Pickett et al. 1994) are increasingly observing that it is absurd to model urban ecosystems without explicitly representing humans in them. Would ecologists exclude other species from models of natural ecosystems? However, as Pickett et al. (1997) point out, it is not enough to simply add humans to ecosystems without representing the functions or the mechanisms that link humans to ecosystems.

Most operational urban models rooted in urban economic theory rest on the assumption that both landowners and households seek to maximize their economic return. These models originate with the theory of land rent and land market clearing. Given the location and physical qualities of any parcel of land, people will use it in the way that earns the highest rent. Both Wingo (1961) and Alonso (1964) describe the urban spatial structure within the framework of equilibrium theory and use bid-rent functions to model the way land is distributed to its users. Both models aim to describe the effects of the residential land market on location; they assume that households will aim to maximize their utility and select their residential location by trading off housing prices and transportation costs. These urban economic models are cross-sectional, have a general equilibrium, and assume a mono-centric pattern structure. During the last fifty years, urban models have evolved towards much more sophisticated representations of urban dynamics, but as I discuss in Chapter 2 they are still predominantly developed within an equilibrium framework which makes it difficult to integrate them with dynamic ecological models.

Representing human actors and their institutions in models of urban ecosystems will be an important step towards more realistically representing the human dimension of environmental change. Many of the human impacts on the physical environment are mediated through social, economic, and political institutions that control and order human activities (Kates et al. 1990). Also, humans consciously act to mitigate these impacts and build institutional settings that promote such mitigation. Furthermore, humans adapt by learning both individually and collectively. How can we represent these dimensions? Lynch (1981) suggested that "a learning ecology might be more appropriate for human settlement since some of its actors at least are conscious, and capable of modifying themselves and thus changing the rules of the game" (p. 115), for example, by restructuring materials and switching the path of energy flows. Humans, like other species, respond to environmental change but in a more complex way.

## 1.3 Cities as Ecological Systems

Environmental forces—such as climate, topography, hydrology, land cover, and human-induced changes in environmental quality—are also important drivers of urban systems. Moreover, natural hazards—such as hurricanes, floods, and landslides—can cause significant perturbations in human systems. Most models of human systems, however, simply ignore these forces. Even the best models for conceptualizing urban growth do not integrate biophysical processes directly, but include them either as exogenous variables or constants. This is a severe limitation because human decisions are directly related to environmental conditions and changes. To urban modelers, such human-driven factors as the behavior of the job market or the degradation of housing stock are endogenous elements of their models and cannot be removed, but the modelers can and do represent the dynamics of urban systems without considering such environmental factors as the degradation of the environment and the depletion of natural resources.

Just as we cannot simply add humans to ecological models, if we want to represent biophysical processes in urban models we will have to do more than simply add environmental variables to existing urban models.

Researchers have extended the design of a number of current models to include changes in environmental variables such as air quality and noise (Wegener 1995). However, linking urban models to a simplified representation of environmental systems is not sufficient to detect ecological responses and feedback. While we generally understand the effects that urban development may have on ecological processes, we have little consensus on the details: Does increase in urban development have a linear effect on ecological processes? Or can we detect thresholds, and/or differentiate among types of development patterns? Scholars generally agree that changes in land cover associated with urbanization affect biotic diversity, primary productivity, soil quality, runoff, and sedimentation rates. By altering the availability of nutrients and water, urbanization also affects populations, communities, and ecosystem dynamics. Urbanized areas modify the microclimate and air quality by altering the nature of the surface and generating large amounts of heat (i.e., urban heat islands; Oke 1973).

Since ecological processes are tightly linked with the landscape, scholars hypothesize that alternative urban patterns have distinct implications for ecosystem dynamics. Some of our research in Seattle and elsewhere has shown that the transformation of land cover favors organisms that are more capable of rapid colonization, better adapted to the new conditions, and more tolerant of people. As a result, urbanizing areas often have novel combinations of organisms living in unique communities. Diversity may peak at intermediate levels of urbanization, where many native and nonnative species thrive, but this diversity typically declines as urbanization intensifies. Rearranging the pattern of land cover also changes the composition of communities, typically increasing the population of edge species (those inhabiting the interfaces among vegetation types and ecotones), and decreasing the population of interior species, those that rarely occur within a few hundred meters of such interfaces (Alberti et al. 2003).

## 1.4 Cities as Hybrid Ecosystems

In this book I propose that urban ecosystems exhibit unique properties, patterns, and behaviors that arise from a complex coupling of humans and ecological processes. Urban ecosystems are not different from other ecosystems simply because of the magnitude of the impact humans impose on ecosystem processes, nor are they so removed from nature that ecosystem processes become only a social construct in them. If we conceptualize such systems purely in ecological or human systems terms, we limit our ability to fully understand their functioning and dynamics (Collins et al. 2001, Alberti et al. 2003).

Urban sprawl illustrates the complexity of interactions and feedback mechanisms between human decisions and ecological processes in urban ecosystems (Alberti et al. 2003).[1] Sprawl manifests as a rapid development of scattered (fragmented), low-density, built-up areas; Ewing (1994) calls it "leapfrogging." Between 1950 and 1990, US metropolitan areas grew from 538,720 square kilometers ($km^2$) and 84 million people to 1,515,150 $km^2$ and 193 million people. Land development due to urbanization has grown 50% faster than the population (Rusk 1999). Sprawl is driven by demographics (e.g., increases in numbers of households), socioeconomic trends (e.g., housing preferences, industrial restructuring), and biophysical factors (e.g., geomorphological patterns and processes). It is also reinforced by choices about infrastructure investments (e.g., development of highway systems; Ewing 1994). Sprawl is strongly encouraged by the land and real estate markets (Ottensmann 1977) and is now a highly preferred urban living arrangement (Audirac et al. 1990).

The phenomenon of sprawl shows how we miss out on some important mechanisms that drive human-dominated ecosystems if we consider interactions between humans and ecological processes only in the aggregate. Human decisions are the primary driving force behind environmental conditions in urban ecosystems, but we cannot explain these conditions by looking separately at the behavior of the individual agents (e.g., households, businesses, developers) competing in each market (e.g., job market, land and real estate market). Households, which are themselves complex entities, compete simultaneously in the job and real estate markets when people decide where to live. Furthermore, they have preferences and make trade-offs that are highly dependent on biophysical factors. Decisions about land development and infrastructure are strongly influenced by biophysical constraints (e.g., topography) and environmental amenities (e.g., "natural" habitats). Metropolitan patterns eventually emerge from the local interactions among these agents; in turn these patterns affect both human and biophysical processes. Resulting changes in environmental conditions then strongly influence some important human decisions. Furthermore, in these systems uncertainty is important, since any departure from past trends can affect how a system evolves.

Sprawl has important economic, social, and environmental costs (Burchell et al. 2002). It fragments forests, removes native vegetation, degrades water quality, lowers fish populations, and demands high mobility and an intensive transportation infrastructure. Such environmental changes may eventually make

---

[1] This example is presented earlier in Alberti, M., J. Marzluff, E. Shulenberger, G. Bradley, C. Ryan, and C. Zumbrunnen, 2003. Integrating humans into ecology: Opportunities and challenges for studying urban ecosystems. BioScience 53(12):1169–1179.

suburban sprawl areas less desirable for people and may trigger further development at increasingly remote locations. Environmental regulation or urban growth control may emerge from these trends. But such feedback from urban dwellers or planning agencies can take many forms and is often phase-lagged by decades — consider, for example, how long the deliberations over highway development can take. Municipalities in some cases may be responsible for promoting sprawl. For example, they often subsidize sprawl by providing public services (schools, waste disposal, utilities) whose prices do not reflect their real cost and distance from central facilities (Ewing 1997). Usually residents in the sprawled periphery do not pay the full costs of the services they get (Ottensmann 1977), which must be borne by the wider society, either now or in the future.

Take the example of salmon in Puget Sound. By the 1990s, after a long history of diminishing runs, Puget Sound salmon populations were on the verge of collapse and several species were listed as threatened under the Endangered Species Act. The decline of salmon can be attributed to a range of urban pressures on stream ecosystems: changes in flow regime, habitat, food sources, water quality, and biotic interactions (Karr 1991). But salmon are not only an endangered species. As an icon connecting the people of the Puget Sound to their natural environment throughout history, it synthesizes the challenges that arise when humans coexist with other species in the same habitat. Sustainable Seattle (1999) chose it as an indicator of human and ecological health, because people in the region see threats to salmon as threats to themselves. The "Four H's" (hydropower, habitat, hatcheries, and harvest), considered to be the major factors contributing to salmon decline in the Puget Sound region, are inextricably linked with human activities. At the same time, these factors reflect the interdependencies between the human functions that depend on these activities and the biophysical processes that support ecological function.

To address the diverse human and ecological factors that have led to the decrease in salmon populations, public agencies have developed a variety of instruments and policies. The state of Washington passed its Growth Management Act in 1990, which required that local municipalities and counties create and implement "critical areas ordinances," partially to address the growing concern over salmon declines. In addition, counties are required to designate "urban growth areas" where the majority of new development will be concentrated. Development outside of these boundaries is severely limited to prevent sprawl and preserve the rural character of these areas. A new state agency, the Puget Sound Partnership, was established in 2007 to lead efforts to protect and restore Puget Sound by 2020. A wide variety of regulatory tools are being invoked to satisfy the Endangered Species Act's (ESA) required recovery plan at all levels of government. However, our scientific understanding of coupled human-ecological systems is still at its infancy.

## 1.5 Complexity, Emergent Properties, and Self-Organization

The "complex systems" paradigm provides a powerful approach for studying cities as emergent phenomena. Building on examples from current research, I propose that urban ecosystems are hybrid, multi-equilibria, hierarchical systems, in which patterns at higher levels emerge from the local dynamics of multiple agents interacting among themselves and with their environment. They are prototypical complex adaptive systems, which are open, nonlinear, and highly unpredictable (Hartvigsen et al. 1998, Levin 1998, Portugali 2000, Folke et al. 2002, Gunderson and Holling 2002). Disturbance is a frequent intrinsic characteristic (Cook 2000). Change has multiple causes, can follow multiple pathways, and is highly dependent on historical context; that is, it is path-dependent (Allen and Sanglier 1978, 1979, McDonnell and Pickett 1993). Agents are autonomous and adaptive, and they change their rules of action based upon new information.

Complex structures emerge from the amplification and limiting actions of multiple agents who follow a few simple decision rules. (Resnick 1994, Nicolis and Prigogine 1989). One of the least understood aspects of urban development are the ways that local interactions among multiple agents affect the global composition and dynamics of whole metropolitan regions. Consideration of some fundamental attributes of complex human and ecological adaptive systems—multiple interacting agents, emergent structures, decentralized control, and adapting behavior—can help scholars study and manage urban sprawl as an integrated human-ecological phenomenon. The emerging urban landscape structure can be described as a cumulative and aggregate order that results from many locally-made decisions involving many intelligent and adaptive agents. Complex metropolitan systems cannot be managed by a single set of top-down governmental policies (Innes and Booher 1999); instead, they require that multiple independent players coordinate their activities under locally diverse biophysical conditions and constraints, constantly adjusting their behavior to maintain an optimal balance between human and ecological functions.

Urban ecosystems are dynamic complex systems of biophysical and human interactions that evolve through feedback loops, non-linear dynamics, and self-organization (Nicolis and Prigogine 1977). Within a complex system, interactions generate emergent behaviors and structures. The system's self-organization drives it towards either order or chaos. Different theories of self-organization have different implications for the way systems evolve (Patten 1995, Jorgenson 1997, Phillips 1999). As Phillips (1999) suggests in his review of self-organization theories, the question is not whether systems are chaotic or ordered, and divergent or convergent. In fact, there may be as many instances of stable, nonchaotic, convergent phenomena in urban ecosystems as there are unstable and chaotic ones. Perhaps a more important question is how divergent self-organization and patterns are linked to instability and chaos and how they affect system evolution.

A key property of self-organized systems is criticality—a state between stability and instability. An urban ecosystem at the "edge of chaos" is not chaotic. Instead it has reached a critical threshold, a state in which perturbations are not dampened or amplified, but are propagated over long temporal or spatial scales (Bak 1996). Bak provides the example of a sandpile, where local interactions result in frequent small avalanches and infrequent large ones. The concept of criticality is particularly relevant when thinking about system evolution, environmental change, and adaptation of urban ecosystems. According to Kauffman (1993), criticality facilitates the emergence of complex aggregated behaviors that reach an optimal balance between stability and adaptability. Critical systems maximize the ability of the system to use the information about its past to respond to future conditions.

On the other hand, most ecological design for urban development is based on a myth: It assumes that we should aim to produce policies and developments that result in stable social, economic, and environmental behaviors. Urban designers often assume that ecological systems are predictable and behave in a linear way, that their behavior is consistent over time and space and invariant to scale, and that change is continuous. Before we can model these interactions, we need to explicitly recognize the properties of ecosystem organization and the behavior that governs them.

Holling (1996) has pointed out four key characteristics of ecological systems:

1. Ecological systems are complex, dynamic, open, and non-equilibrium. This has implications for defining the multiple states and evaluating the effects of urban patterns in terms of a given system's ability to maintain human and ecological function over the long term. Feedback mechanisms can amplify or regulate a given effect. This has also implications for understanding the dynamic interactions between development patterns and ecological and human functions. Instead of aiming to achieve a specific condition (e.g., fixed density or distance of a development from a stream, as set by critical area ordinances), perhaps development patterns should aim at maintaining characteristics of the system that support the ecosystem and human function (i.e., resilience).
2. Change is neither continuous and gradual nor consistently chaotic. Rather it is episodic, with periods of slow accumulation punctuated by sudden reorganization. These events can shape trajectories far into the future. Critical processes function at very different rates, but they cluster

around dominant frequencies. Episodic behaviors are caused by the interactions between variables that have immediate versus delayed effects.

3. Spatial attributes are neither uniform nor scale-invariant. Instead, they are patchy and discontinuous. Therefore, scaling up is not a simple process of aggregation since nonlinear processes determine how the shift occurs from one scale to another. This has implications for understanding the effects of spatial interactions between human and ecological systems at multiple scales.

4. Ecosystems are moving targets. Knowledge is incomplete and surprise is inevitable. This has implications for the type of strategies we adopt. Instead of fixed policies, perhaps we need to think about flexible mechanisms, and governing institutions that can learn effectively and deal with change.

## 1.6 Resilience in Urban Ecosystems

Ongoing research in Seattle and Phoenix has shown that complex dynamics occur in urban landscapes (Wu and David 2002, Alberti and Marzluff 2004). An essential aspect of complex systems is nonlinearity, which means that their dynamics can lead to multiple possible outcomes (Levin 1998). Urbanizing regions have multiple steady and unstable states. Using a system dynamics framework, John Marzluff and I have described these alternative states (Alberti and Marzluff 2004, Figure 1.8). In urbanizing regions, urban sprawl can cause shifts in the quality of natural land cover, from a natural steady state of abundant and well-connected natural land cover to a second steady state of greatly reduced and highly fragmented natural land cover. The natural "steady" state depends on natural disturbance regimes. The sprawl state is a forced equilibrium that results when agents in the system do not have complete information about the full ecological costs of providing human services to low-density development (Alberti and Marzluff 2004).

Marzluff and I hypothesize that resilience in urban ecosystems is defined by the system's ability to maintain human and ecosystem functions simultaneously. In urbanizing regions, ecological and human functions are interdependent. As urbanization increases, the system moves away from the natural vegetation attractor toward the sprawl attractor and beyond, until increasingly urbanized ecological systems become unable to support the human population. As we replace ecological functions with human functions in urbanizing regions, the processes supporting the ecosystem may reach a threshold and drive the system to collapse. This process drives the system

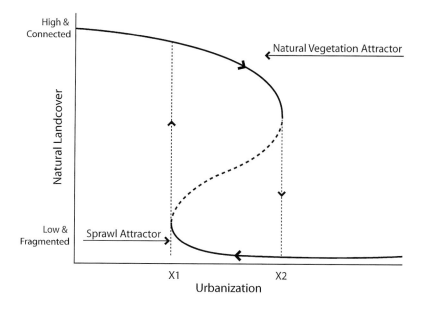

Figure 1.8. Impact of urbanization on resilience. From a system dynamics perspective, we can indicate that the state of an urban region is likely driven between natural and human function states by the amount of urbanization. As urbanization increases, natural vegetation decreases. The system moves along the upper solid line (Natural vegetation attractor) until a point (X2) is reached where natural vegetation is too degraded and fragmented to perform vital ecological functions and the system becomes unstable (dashed portion of curve). As urbanization reduces ecosystem function the system flips into a sprawl state (the lower solid line, sprawl attractor) where human functions replace ecosystem functions. Eventually, ecosystem function is degraded to a point that cannot support human function, urbanization declines and the system becomes unstable again (X1). The system eventually returns to the natural vegetation state. (Alberti and Marzluff 2004, p. 244).

back toward the natural vegetation attractor if ecosystem collapse reduces the system's ability to support human settlement to the point that substantial natural vegetation can re-grow. Our hypothesis assumes a time scale of centuries or millennia. In the past, many human settlements have collapsed possibly because ecological conditions or carrying capacity have changed in response to human pressure or large scale climatic shifts; for example the Mayas, the Anasazi, the Incas, and the Egyptians (Alberti and Marzluff 2004).

In response to the human, ecological, and economic costs of sprawling development, urban planners have attempted to stabilize inherently unstable states—that is, to balance the conversion of natural land cover with the development needed to support human functions (Figure 1.9). The assumption

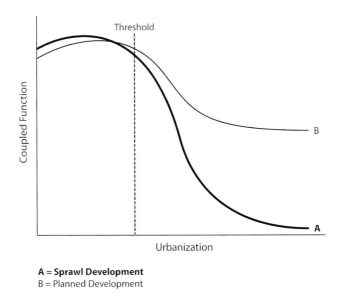

A = **Sprawl Development**
B = Planned Development

Figure 1.9. Relationships between urbanization patterns and ecosystem and human functions. Alternative urbanization patterns have different levels of resilience measured as their capacity to simultaneously support ecological and human functions. Sprawling development (A) leads to a decline in coupled system function associated with the sprawl attractor state represented by the lower left corner of the diagram in Figure 1.8. Planned development (B) is an urban development pattern that simultaneously supports ecological and human functions allowing a greater resilience of the coupled urban ecosystem (Alberti and Marzluff 2004, p. 250).

of planned development is that the development pattern affects ecological conditions and the maintenance of ecosystem and human functions. In the phase of reorganization and renewal, humans have a chance to change the trajectory of urban ecosystems, allowing them to develop self-organizing processes of interacting ecological and socioeconomic functions. But this forced equilibrium is inherently unstable, as it requires balancing the tension between providing human and ecosystem services. The trajectory the system will take depends strongly on information flows, knowledge transfers, and system learning. In urban ecosystems that emphasize foresight, communication, and technology (Holling et al. 2002a), information flows can be used more effectively to control system dynamics and feedback mechanisms, and humans have more opportunities to respond innovatively to ecological crises. It is possible that under these conditions a planned equilibrium can be made more resilient.

The resilience of alternative urban development patterns is a key element of such information. Can planned development simultaneously maintain ecosystem and human functions over the long term? If we are to assess the resilience of urban ecosystems, we must understand how interactions between humans and ecological processes affect the resilience of inherently unstable equilibrium points between the natural vegetation attractor and the sprawl attractors. The challenge for ecological scholars and urban planners is to address a key question: How can we best balance human function and ecosystem function in urban ecosystems?

## 1.7 Rationale for a Synthesis

Studies of urban systems and of ecological systems have evolved in separate knowledge domains (Alberti 1999a). Although urban scholars have focused extensively on the dynamics of urban systems and their ecology, these diverse urban processes have yet to be synthesized into one coherent modeling framework that allows us to study the resilience of such systems. Disciplinary approaches have not adequately addressed the processes and variables that couple human and natural systems. Urban models designed to explain or predict urban development are still extremely limited in their ability to represent ecological processes. On the other hand, ecological models vastly simplify human processes. Only as we have begun to pay more attention to the important role of human activities in environmental change have we seen the need to develop an integrated framework for studying the interactions between biophysical and socioeconomic processes.

A new inter-disciplinary synthesis is necessary if urban and ecological dynamics are to be integrated successfully. Such a synthesis will allow us to take at least six important steps:

1. Develop a shared understanding of coupled human-ecological systems in urbanizing regions and a common definition of urban drivers, patterns, processes, and functions. Such an understanding would help scholars across disciplines to work together to develop and test hypotheses that can provide insights into the dynamics of urban ecosystems.
2. Define a useful set of indicators of human and ecological function and well-being in urbanizing regions to monitor trends and assess the impacts and tradeoffs among alternative patterns.

3. Determine what we know with a reasonable level of confidence about how these patterns affect and maintain human and ecological function and well-being. What are the major uncertainties and data gaps? By identifying gaps in knowledge we can help set the agenda for research in urban ecology.

4. Provide insights into the relative importance of different linkages between urban development and ecological dynamics. What are the slow and fast variables that affect system dynamics? What are the major feedback mechanisms? What are the relative strengths of different mechanisms?

5. Lay out the plausible alternative future scenarios over multiple time scales (i.e., effects of climate change, technological progression, natural resource consumption patterns, etc.).

6. Link scientific research to relevant policy questions regarding urban ecological problems in order to assist decision making. This will help us communicate the complexity of assessing development patterns and help shape smarter growth policies, changing popular ideas to reflect an evolutionary and adaptive perspective.

# Chapter 2

# HUMANS AS A COMPONENT
# OF ECOSYSTEMS

One of the greatest challenges for natural and social scientists over the next few decades will be to understand how urbanizing regions evolve through the extraordinarily complex interactions between humans and biophysical processes. The challenge is to incorporate this complexity in studying urban ecosystems and their evolution. For more than a century, urban theorists have struggled to understand urban systems and their dynamics. During the second half of the last century, ecological scholars started to recognize the subtle human-natural interplay governing the ecology of urbanizing regions. Both social and natural scientists concur that assessing future urban scenarios will be crucial in order to make decisions about urban development, land use, and infrastructure so we can minimize their eco-logical impact. But to fully understand the interactions between urban systems and ecology, we will have to redefine the role of humans in ecosystems and the relationships between urban planning and ecology (Alberti et al. 2003).

That the city evolves as part of both its natural and social history was already clear to Geddes (1915) a century ago. Social ecologists such Mumford (1956), Dubos (1968) and Bookchin (1980) each conceived of the idea of evolutionary humanism from very different perspectives, but they all referred to the continuity between the human and natural world. Pioneer scholars in the social sciences (Park et al. 1925, Burgess 1925, Duncan 1960) and natural sciences (Odum 1963, Sukopp and Werner 1982, Holling and Orians 1971, Dunne and Leopold 1978) as well as in the fields of urban design and planning (McHarg 1969, Hough 1984, Spirn 1984, Lyle 1985) have produced important knowledge in the study of urban ecosystems within their disciplines. More recently in *The Nature of Economies* Jane Jacobs (2000) emphasized that humans exist wholly within nature as part of the natural order in every respect. This view is shared by many historians (Cronon 1991, 1996, Melosi 2000, Klingle 2001), geographers (Zimmerer 1994, Robbins and Sharp 2003, Kaika 2005), anthropologists (Abel and Stepp 2003, Wali et al. 2003, Redman 2005), and planners (Beatley and

Manning 1977, Steiner 2002, Alberti et al. 2003) who are engaged in efforts to understand human dominated ecosystems.

The idea that humans and cities are key agents in Earth's ecosystems has been around for some time. But only recently have scholars in diverse disciplines started to acknowledge that it is not enough to simply expand the study of cities to include humans and ecological processes within each separate disciplinary domain: the assumptions of disciplinary frameworks need to be substantially revised to fully overcome the human-nature dichotomy and address the complexity of hybrid urban ecosystems. Economists, ecologists, and urban planners have built models of urbanizing regions that simplify human and ecological agents and processes to such an extent that their perspectives have proven to be too limited and hence insufficient for understanding the complex interdependence between humans and ecosystems. Building an integrated approach is critical to advancing such understanding.

Over the last three decades, complexity theory has had an important influence on both the natural and social sciences. It provided a new basis for understanding the complex dynamics of coupled human-natural systems: how can a myriad of local interactions among multiple agents generate some simple behavioral patterns and ordered structures? Cities are non-equilibrium systems. Random events produce system shifts, discontinuities and bifurcations (Krugman 1993, 1998, Batty 2005). Patterns emerge from complex interactions that take place at the local scale, suggesting that urban development self-organizes (Batten 2001). Emergent patterns are often scale-invariant and fractal, suggesting that urban morphology is derived from similar processes operating at the local scale (Batty and Longley 1994, Allen et al. 1997).

Instead of asking how emergent patterns of human settlements and activities affect ecological processes, the question we should ask is how humans, interacting with their biophysical environment, generate emergent phenomena in urbanizing ecosystems. And how do these patterns selectively amplify or dampen human and ecological processes and functions? Cities are coupled human-natural systems in which people are the dominant agents. Although extensive urban research has focused on the dynamics of urban systems and their ecology, efforts to understand urban systems have proceeded separately in different disciplinary domains and disciplinary approaches have not adequately addressed the processes and variables that couple human and ecological functions.

Scholars of both urban development and ecology have begun to recognize the importance of explicitly considering human and ecological processes in studying urban systems (Alberti et al. 2003). The challenge for these scholars is to revise fundamental assumptions in the disciplines regarding the nature and functioning of urban systems and their evolving spatial structure. Integrated studies of coupled human-natural systems reveal new and complex patterns and processes that are not evident when either social or natural scientists study them separately (Liu et al. 2007). Simply linking existing models without modifying them will not uncover the behavior of coupled systems, because interactions between human and ecosystem processes may occur at levels of integration that the current models cannot represent. In this chapter I discuss some characteristics of coupled urban ecosystems that influence our ability to understand how settlement patterns emerge and evolve. I then examine different models of urban development in various disciplines. Building on current work on cities and biocomplexity at the University of Washington's Urban Ecology Research Lab, I conclude by providing some directions for modeling urban ecosystems.

## 2.1 Emergence and Evolution of Settlement Patterns

Urban theorists have been interested in how cities form and evolve for a long time (Geddes 1915, Mumford 1961, Lynch 1961). To understand how cities emerge and evolve—how they shift in their spatial structure—we need to examine how both pattern and function co-evolve over space and time. This requires a shift from predominantly static theories of urban form to an evolutionary theoretical perspective on the emergence and change of human settlements (Fujita 1989, Pumain 2000). Several questions emerge: How do spaces that provide a setting for centers linked by flows of people, goods, and services form? What interactions and feedback take place between population, economics, built infrastructure, technological development, land and natural resources over space and time? The interplay among these factors, together with evolving social preferences, has shaped the spatial structure and evolution of modern cities. An historical account of the way urban structures have changed over time points primarily to transportation and communication systems as the key determinants that govern where cities are built and how settlement patterns evolved (Glaab and Brown 1967). But to fully understand the complex relationships that control the evolution of urban form and explain the change in urban densities, the shift from monocentric to polycentric structures, the emergence of the edge cities, and the phenomenon of sprawl, a more complex set of elements, processes, and interactions in space and time need to be taken into account. We need to understand the evolution of cities as part of nature and develop a theory of hybrid urban landscapes.

## Emergent patterns

During the last century urban structures in the US have changed dramatically (Kim 2007). Cities have become more scattered and dispersed, and generally less dense. Their structure has also shifted, from monocentric to polycentric, because of suburban development. In some cases large concentrations of offices and commercial development at the nodes of major highways have emerged into "edge cities" (Garreau 1991, Henderson and Slade 1993, Small and Song 1994). Urban scholars have found that a variety of factors contribute to suburbanization; these factors range from economics to demographics, to land use and transportation dynamics, and from consumer preferences to the unintended consequences of planning (Baldassare 1992).

Many early land use models of urban development have rarely included elements of the biophysical environment except to represent constraints to urban development (Alberti 1999a). Although more recent models have started to integrate some elements of the biophysical environment more explicitly, none of the explanations of suburbanization recognizes that possible interactions between human and ecological functions may govern such patterns (Alberti and Waddell 2000). Nor have they explored the possible interactions between human and biophysical process that may affect such patterns. For example, we do not know the extent to which urban concentrations obliterate the biophysical processes that support human functions such as the water supply, or how natural amenities attract people away from city centers.

## Human and natural drivers

Urbanizing regions provide an excellent laboratory to test hypotheses on emergent human-ecological phenomena (Alberti et al. 2003). A complex set of social, political, economic, institutional, and biophysical factors drives urbanization and affects when, where, how, and at what rate urban development proceeds. In studying interactions between human and ecological processes, researchers need to address explicitly the complexities of many factors working simultaneously on scales ranging from the individual to the regional and global, and from the hourly to decades and centuries (Cumming et al. 2006). Consideration solely of aggregated interactions cannot help understand emergent patterns or explain important feedbacks and outcomes. Institutional settings and policies governing urbanization mediate the complex relationships between humans and ecological processes (Folke et al. 2005). Lag times between human decisions and their environmental effects further complicate understanding of these interactions. For instance, in urban ecosystems, land-use decisions affect species composition directly (e.g., introduction and removal of species) and indirectly (e.g., modification of natural disturbance agents like fire and flood). If ecological productivity controls the regional economy, interactions between local decisions and local-scale ecological processes can cause large-scale environmental changes (Alberti 1999a). All these factors need to be explicitly taken into account in studying emergent patterns of urban ecosystem dynamics. Both economics (Andereis et al. 2006, Alexandridis and Pijanowski 2007) and political ecology (Peterson 2000) that take into account resilience have much to offer to address an integrated approach.

## Competing functions

Although the tendency for people and firms to form clusters across space has long been observed, the regularity of such trends has been described only during the last century (Auerbach 1913, Zipf 1949). Zipf's law states that the distribution of cities is highly skewed and follows a simple "rank-size rule." More recently several authors have advanced the possibility that nonlinear dynamics, discontinuity, and disequilibria, instead of continuity, characterize the relationship between city rank and size (Dendrino 1992, Krugman 1996). For Krugman (1996), these urban macrodynamics reflect the emergence of employment centers from self-organizing forces. He sees no reason to expect that the hierarchical parameters that determine a system of urban clusters will be constant.

At the metropolitan scale, the emergence of polycentric urban structures can be explained by the balance between the forces of agglomeration and the forces of dispersion; this balance determines the level of employment concentration in space (Anas and Kim 1996). Several theoretical models have been developed to explain this phenomenon, ranging from traditional neo-classical economics to complex systems theories (Anas et al. 1998). The spatial clustering of people and firms is one of the central explanations for human settlements. Interaction between the forces of agglomeration and dispersion can be thought of in a stable-equilibrium framework based on scale economies in production (Starrett 1974), retailing (Anas and Kim 1996), externalities (Fujita and Ogawa 1982), and trade (Krugman 1993). Spatial clustering can also be generated by self-organizing forces, in which feedback can amplify or dampen random fluctuations. Krugman (1998) suggests that the interaction among agents can be subject to bifurcations leading to sudden changes and qualitatively different urban structures. However, none of these explanations explicitly address the interactions between human and natural systems and the potential system shifts driven by existing thresholds in such systems.

## Heterogeneous agents

Cities are built by a wide variety of agents: households, businesses, developers, financial agencies, governments, and services. Agents make a myriad of interconnected decisions that influence each other and ultimately the urban form. Households make residential choices associated with a variety of demographic and socioeconomic characteristics (i.e., household size, income, etc.) as well as preferences for space, place, and proximity (Fillion et al. 1999). Businesses make decisions about location, production, and consumption. Business externalities and spillovers are key factors in spatial geography and economics (Clark et al. 2000). Developers make decisions about development and redevelopment by trading off risks and benefits; those in turn are influenced by both demand for housing and the requirements of financial institutions and regulations. Governments determine policies and investments in infrastructure, which may mitigate or exacerbate emerging trends.

While most models assume that agents behave in a homogenous and rational way, increasing evidence suggests that agents are highly hetero-geneous (Benenson and Torrens 2004). Heterogeneity is important in explaining how agents decide about residential location or land develop-ment, and it affects the outcome of models (Anas 1986). Agents make decisions based on their demographic, social, and economic characteristics and their preferences with respect to knowledge and values. Moreover, their decisions are highly affected through their perception, cognition, and evaluation of the landscape (Nassauer 1995).

## Spatial and temporal dynamics

Agents interact with each other and with their environments at various levels across time and space. At the landscape level, humans in urbanizing regions affect and are affected by natural systems through several mech-anisms: (a) they convert land and transform habitats; (b) they extract and deplete natural resources; and (c) they release emissions and wastes. Since the earth's ecosystems (d) provide important goods and services to urban systems, environmental changes occurring at the regional and global scales—such as the contamination of watersheds, loss of biodiversity, and change in climate—(e) affect the quality of the urban environment in the long term and ultimately human health and well-being. All of these factors must be considered in any attempt to develop models of urban ecosystems.

Feedback mechanisms can amplify or dampen changes in urban struc-tures (Krugman 1996). Some interactions lead to positive feedback processes—e.g., urban dispersal causes more dispersal. Other interactions could trigger negative feedback processes—e.g., urban dispersal causes forest fragmentation which leads to degraded environmental conditions, which in turn leads to changes in growth management regimes that would allow ecological conditions to recover. Scale mismatches between human and natural systems processes need to be taken into account to study interactions and feedbacks in such systems. (Cumming et al. 2006). Other important aspects of spatial and temporal dynamics of urbanization include discontinuities, tipping points, phase transitions, and system shifts. Both the spatial and time scales of these interactions may introduce important scale dependence as the urban spatial structure evolves.

## Disequilibrium and criticality

The emerging spatial structures of metropolitan regions can be explained in terms of self-organized criticality (Krugman 1996, Batty 1998). Bak et al. (1989) illustrate self-organized criticality using the example of a pile of sand. As grains of sand are dropped and a sand pile grows, the structure of the pile reaches a critical state such that one additional dropped grain can trigger a large-scale avalanche that restructures the pile to a new state of self-organized criticality. This phenomenon follows a power law: the distribution of avalanche sizes exhibits greater variance than the distribution of sand dropping. Qualitative changes in the urban landscape can be explained by the emergence and growth of urban centers until a threshold is reached. The urban structure can "flip" into an entirely different state indicating the existence of a phase transition. In Bak's example, the sand pile reaches self-organized criticality when local interactions between separate grains of sand are replaced by global communication throughout the whole sand pile.

Although this concept was advanced in physics more than thirty years ago, the empirical evidence of its applications in ecology and in economics has started to emerge only much more recently. Scholars in ecology and economics have shown macroscopic behaviors of forests and systems of cities do display the spatial and temporal scale-invariance characteristic of such systems. While the concept has been applied in various disciplines to explain the behavior of different systems, only recently have scientists posited self-organized criticality as a means of understanding coupled human-natural systems and explaining how humans adapt to environmental change (Holling and Gunderson 2002).

### Urban ecosystems and human well-being

In urban ecosystems, human and ecological functions and well-being are inextricably linked. To articulate this statement into a set of testable hypotheses, our research team in Seattle has developed a conceptual framework linking drivers, patterns, processes and effects. Key human drivers of change are demographics, economics, technology, social organization, and political and governmental structures. Human choices about their location and their consumption behaviors directly influence the way people use land and demand resources. These drivers also interact with biophysical drivers (i.e., climate, topography, etc.). In combination, these forces affect the land cover and ultimately the heterogeneity of the landscape and its natural processes and disturbances. For example, urban development affects ecosystems by fragmenting natural habitat (e.g., land conversion), modifying biophysical processes (e.g., artificial drainage), imposing barriers (e.g., roads), and homogenizing natural patterns (e.g., sprawl often occurs as a regular grid that disregards natural vegetation and topography).

## 2.2 Modeling Urban Development and Ecology

Current models of the interactions of urban development and ecology are inadequate. We must develop coupled human-natural models of urban ecosystems if we are to develop urban development management strategies that will result in more efficient urban land use and resource consumption patterns. Current urban models still exclude biophysical drivers, and ecological models often simply represent the human component as drivers of ecosystem change without incorporating any feedback to the human system. We need models that can explicitly represent the complexity and heterogeneity of both the human and the biophysical agents that drive urban ecological systems and their dynamics.

### Urban models[1]

Although extensive research has focused on the dynamics of urban systems, the hypotheses about their functioning have mostly been tested by developing formal models. Models generally focus on specific subsystems,

---

[1] This review of urban models builds on a previous review I present in Alberti, M. 1999a. Modeling the urban ecosystem: A conceptual framework. Environment and Planning B 26:605-630. Modified paragraph extracts are reprinted with permission.

such as housing, employment, land use, and transportation, where a limited set of elements influences the dynamics. Early models of the spatial distribution of activities were based on simple mechanisms of spatial interaction and economic axioms. But while models have grown in sophistication, they are still very limited in their representation of bio-physical drivers or impacts. Recently, a few modelers have started to address the direct impacts that human activities have on the environment, such as air pollution and noise. But, as we can see quite clearly in the idealized urban model proposed by Wegener (1994), such models only involve unidirectional links between urban systems and the environment.

Today, a vast literature synthesizes the theoretical and methodological foundations of urban simulation models (Wilson et al. 1977, 1981, Batty and Hutchinson 1983, Putman 1983, 1994, Mackett 1985, 1994, Wegener 1994, 1995, Harris 1994). In this section I draw on this literature to explore how modelers consider environmental variables and represent environmental processes. Operational urban models can be classified according to the approach they use to predict the generation and spatial allocation of activities or according to the solution they propose to a variety of model design questions.

### Gravity and maximum entropy models

Lowry (1964) was the first to conceptualize the ways that population, employment, service, and land use are distributed spatially. Lowry's model is based on the simple hypothesis that residences locate in a gravity-like way around employment locations. Two schools of research have provided a statistical basis for the gravity model, guided by Wilson (the entropy-maximizing principle, 1967) and McFadden (utility maximization, 1975). Later analysis showed that the two methods obtained equivalent results (Anas 1983). The models most often used by planning agencies in the US— the Disaggregated Residential Allocation Model (DRAM) and the Employment Allocation Model (EMPAL)—are derived from Lowry's model and use a formulation that assumes maximum entropy. Developed by Putman (1979), and incrementally improved since the early 1970s, DRAM and EMPAL are currently in use in most US metropolitan areas (Putman 1995). The Integrated Transportation Land Use Package (ITLUP), also developed by Putman (1983), provides a feedback mechanism to integrate DRAM, EMPAL and various components of the Urban Transportation Planning System (UTPS) models that have been implemented in most metropolitan areas.

### Economic market-based models

A second urban modeling approach is based on the work of Wingo (1961) and Alonso (1964), who introduced the notion of land rent and land market clearing. Wingo was the first to describe the urban spatial structure in the framework of equilibrium theory. Given the location of employment centers, a particular transportation technology, and a set of households, his model determines the spatial distribution, value, and extent of residential land requirements, assuming that both land owners and households aim to maximize their return. While Wingo uses demand, Alonso uses bid-rent functions to predict how the land will be distributed to its users: his model describes how much each household is willing to pay to live at each location. Both models aim to describe the effects that location has on the residential land market. Those using this approach assume that households will maximize their utility: that is, that households select the best possible residential location by trading off housing prices and transportation costs. The trade-offs are represented in a functional form, based on either demand or bid-rent.

### Discrete-choice models

A more flexible modeling approach, now being used more widely, is the class of discrete-choice models. This approach, first proposed by McFadden (1978), uses random-utility theory to model consumer choices among discrete location alternatives based on the utility each alternative provides. Ellickson (1981) was the first to develop a logit model based on a bid-rent function rather than the utility function. His approach focuses on the landowner's problem of selling to the highest bidder instead of the consumer's problem of choosing among properties based on maximizing their utility function. Anas (1987) developed a general equilibrium model based on discrete choice modeling, extending the traditional urban economic model.

Martinez (1992) built on this work, fully integrating the bid-rent theory with the discrete-choice random utility theory by showing that the approaches are consistent. He developed a "bid-choice" land use model that uses a logit formulation to deal simultaneously with both land supplier and consumer perspectives. His approach is also based on equilibrium assumptions.

Two models that use this approach are UrbanSim, developed by Waddell (1995), and California Urban Future II (CUFF II) developed by Landis and Zhang (1998a, 1998b). Both models are based on random utility theory and use logit models to implement key components. But they differ in one substantial way. UrbanSim models the key decision makers—households, businesses, and developers—and simulates their choices that impact urban development. It also simulates the land market as the interaction of demand and supply, with prices adjusting to clear the market. UrbanSim simulates urban development as a dynamic process as opposed to a cross-sectional or equilibrium approach. CUFF II models the probabilities of land use transitions, based on a set of site and community characteristics such as population and employment growth, accessibility, and the original use of the site and surrounding sites.

## Mathematical programming-based models

A third approach to describing the allocation of urban activity is based on optimization theory. Using mathematical programming, these models design spatial interaction problems in order to optimize an objective function that includes the costs of transportation and of establishing activities. The Herbert-Stevens (Herbert and Stevens 1960) model uses linear programming to simulate the market mechanisms that affect location, based on the economic theory of trading transportation costs and time for space and other amenities in suburban areas, as Wingo and Alonso described earlier. Harris (1962) and Wheaton (1974) developed an optimization model by using a non-linear programming model.

More recently Boyce et al. (1993) have explored the options for integrating spatial interactions of residential, employment, and travel choices within a single optimized modeling framework. The Projective Optimization Land use System (POLIS) developed by Prastacos (1986) is one of the few such models used in planning practice, and has been implemented in the San Francisco Bay Area. It seeks to maximize both the location surplus and the spatial agglomeration benefits of basic employment sectors. Like the previous models, only land availability is included as a determinant of employment allocation to zones.

## Input-Output models

Another important contribution from economic theory to urban modeling is the spatial disaggregated intersectoral Input-Output (I/O) approach, initially developed by Leontief (1967). This approach provides a framework for disaggregating economic activities by sector and integrating them into the models of urban spatial interaction. This approach transforms the basic structure of an input-output table, allowing the modeler to estimate the direct and indirect impacts that exogenous change will have in the economy on a spatially disaggregated scale.

Several operational urban models use this approach, including MEPLAN, TRANUS and the models developed by Kim (1989). MEPLAN includes three modules: a land use model (LUS), a flow model which converts production and consumption into flows of goods and services (FRED), and a transportation model that allocates the transport of goods and passengers to travel modes and routes (TAS). MEPLAN's land use component is based on a process of spatially disaggregating production and consumption factors that include goods, services, and labor. It estimates total consumption using a modified I/O framework that is later converted into trips.

MEPLAN, TRANUS and Kim's models use input-output tables to generate interregional flows of goods, but not to directly model economic-ecological interactions. MEPLAN uses the results of the I/O framework to evaluate environmental impacts. While input-output models have been extended to include environmental variables and to incorporate pollution multipliers, no urban models have attempted to implement this approach to describe economic-ecological interactions. Those who have tried to apply these approaches regionally have encountered various difficulties, particularly in determining how to specify the ecological interprocess matrix and in the assumptions of the fixed coefficients.

### Microsimulation

One major limitation in the way most urban models represent the behavior of households and businesses is that they are aggregated and static. Individuals behave in ways that are influenced by their characteristics and the opportunities from which they choose. Without explicitly representing these individuals it is impossible to predict the tradeoffs they make between jobs, residential locations or travel modes. A distinct approach to modeling individuals' behaviors is micro-analytic simulation, which explicitly represents individuals and their progress through a series of processes (Mackett 1990).

As a modeling technique, microsimulation is particularly suitable for systems in which decisions are made at the individual level and the interactions within the system are complex. In such systems, the outcomes produced by altering the system can vary widely for different groups and are often difficult to predict. Researchers using microsimulation can define the relationships between the various outcomes of decision processes and the characteristics of the decision-maker using a set of rules or a "Monte Carlo" process. They can also use it to simulate the actions of a population through time and to incorporate the dynamics of demographic change. An example is the Micro-Analytical Simulation of Transport, Employment and Residence (MASTER) developed by Mackett (1990). The model simulates the choices of a given population through a set of processes. The outcome of each process is a function of the characteristics of the household or business, the set of available choices, and a set of constraints. Wegener (1983) created a less extensive application of this approach in the Dortmund model.

The UrbanSim model, described in detail by Waddell (1995, 2000, 2002), integrates and extends elements of the approach taken by Martinez's (1992) 'bid-choice' land use model that simultaneously deals with both land supplier and consumer perspectives, a real estate stock adjustment model (DiPasquale and Wheaton 1996), and a dynamic mobility and location choice modeling approach. The theoretical basis of the model draws on random utility theory and the urban economics of location behavior of businesses and households. This is implemented through a simulation modeling framework that explicitly addresses land market clearing, land development, and cumulative change at the metropolitan scale (Waddell 1998). The model predicts the location of businesses and households, developer choices to develop real estate on vacant land or to redevelop existing parcels, and the price of land and buildings. Households, businesses, developers, and government respectively make decisions about location, production, consumption and investments. They dynamically interact in the land and real estate markets and generate physical development and relocation. The model is interfaced with travel demand models to account for the feedback relationships between land use and transportation.

While these models do not include biophysical processes, their improved ability to disaggregate the actors and behaviors has enormous advantages for modeling consumer behavior and environmental impacts. As I indicate later in this chapter, UrbanSim provides the microsimulation of urban agents and behaviors to build a coupled Land Use Land Cover Model.

## Landscape ecology models[2]

Ecologists have primarily modeled the dynamics of species populations, communities, and ecosystems in non-urban environments. Only in the last decade have they turned their attention to the study of urban ecosystems. Also, they have mostly been concerned with describing the processes that create patterns observed in the environment; only more recently have they become interested in studying the effects that the patterns have on the processes. Ecologists have started to develop studies to answer the following questions: What are the fluxes of energy and matter in urban ecosystems? And, how does the spatial structure of ecological, physical, and socioeconomic factors in the metropolis affect ecosystem function? Landscape ecology is perhaps the first consistent effort to study the reciprocal effects of spatial patterns (e.g., patch composition) on ecological processes (e.g., fluxes of organisms and materials).

During the last few decades, landscape models have evolved from indices of climatic variables, towards more sophisticated models of species demography and growth. Landscape ecologists originally extrapolated vegetation

---

[2] This review of landscape models builds on a previous review I published in collaboration with Paul Waddell, in Alberti M. and P. Waddell 2000. An integrated urban development and ecological simulation model. Integrated Assessment 1:215-227. Modified paragraph extracts are reprinted with permission.

cover from climatic models, such as the Holdridge life zone classification system. Recently, these approaches have been replaced by more sophisticated simulations of the biological dynamics of vegetation and the way it interacts with abiotic factors such as soil and topography. Models of species demographics and growth have the advantage of introducing more realistic representations of multiple species and their interactions, but they are data-intensive and difficult to implement on a large area. They have recently been replaced by more commonly used transition probability models.

Three general classes of ecological models are used to predict changes in landscape structure:

–   Individual-based models combine the properties of individual organisms and the mechanisms by which they interact within the environment.

–   Process-based landscape models, focusing on mass balance, predict flows of water and nutrients across the landscape and biotic responses in order to predict changes in spatial landscape patterns.

–   Stochastic landscape models, utilizing gridded representations of a landscape, predict changes in spatial patterns based on the characteristics of a given cell, the structural configuration of the patch to which the cell belongs, and the probability of a transition in the cell's state.

### Spatial stochastic models

Spatially explicit stochastic and process-based simulation models have been applied to various landscapes and biophysical processes (Turner 1989, Turner and Gardner 1991). The land-use change analysis system (LUCAS), an example of a spatially explicit stochastic model (Berry and Minser 1997), is structured around three modules linked by a common database. Its socio-economic models are used to derive transition probabilities associated with change in land cover, and produce a transition probability matrix (TPM) that serves as input for the landscape-change model. Through the LUCAS simulation, the landscape condition labels in the input are matched with equivalent landscape condition values in the TPM. The impact models utilize the landscape-change output to estimate impacts to selected environmental and resource-supply variables.

### Process-based models

However useful they may be, models of landscape transition probability cannot represent the biophysical processes that drive landscape change. Spatially explicit process-based models are considered more realistic since they represent biological and physical processes and can relate causes and effects in simulating real landscapes (Sklar and Costanza 1991). Within process-based models, the landscape is modeled by compartments representing different sectors and by flows between compartments representing transfers of materials and energy dissipation.

One example of a process-based spatial simulation model is the Patuexent landscape model (PLM), which can simulate the succession of complex ecological systems across the landscape. The PLM builds on the coastal ecological landscape spatial simulation (CELSS) model developed by Costanza et al. (1995, 2002). At the core of the PLM is a general ecosystem model (GEM) that simulates the dynamics of various ecosystem types. The PLM divides the study area into 6,000 spatial cells in which a GEM with 21 state variables is replicated to simulate landscape change at the regional scale.

## Patch dynamic models

Recent theoretical developments in landscape ecology have emphasized the importance of spatial heterogeneity in understanding the relationship between pattern and process (Turner 1989). Various approaches for predicting changes in the landscape structure are based on the premise that certain characteristics of the landscape are linked to the structural and functional characteristics of ecosystems. And researchers are now paying more attention to ecosystems as hierarchical mosaics of patches (Wu and Loucks 1995); they are developing hierarchical patch dynamic models that can incorporate the effects of spatial heterogeneity on ecosystem dynamics. The Wu and Levin (1994, 1997) model, for example, combines two models: a spatially explicit, age/size-structured patch demographic model and a multi-specific population model. While most spatially explicit landscape models are grid-based, spatially explicit patch dynamics models emphasize the importance of representing dynamics at the level of discrete, relatively homogeneous patches across the landscape. These models are useful for representing the spatial structure of relatively discrete patches (assuming some level of homogeneity within patches), but gridded/process-based models are often more realistic in terms of capturing gradient patterns and processes, and in some instances can "reproduce" patch dynamics as emergent phenomena.

## Complexity and self-organization

Perhaps the least understood aspect of urban development and ecosystem dynamics is the way that local interactions affect the global composition and dynamics of entire metropolitan regions. Urban ecosystems exhibit some fundamental features of complex and self-organizing systems (Couclelis 1985, 1997, Batty and Xie 1994, Batty 1997, White and Engelen 1997, White et al. 1997). The urban spatial structure can be described as a cumulative and aggregate order that results from many locally-made decisions involving many intelligent and adaptive agents. These agents may well change their rules of action based on new information. The local behaviors of multiple decision-makers eventually can lead to qualitatively different global patterns. Furthermore, in these disequilibrium systems, uncertainty is important since any change that departs from past trends can affect the path of system evolution (Wu 1998a).

## Cellular automata

Several researchers have proposed using cellular automata (CA) to model complex spatially explicit urban dynamics (Couclelis 1985, Batty and Xie 1994, White et al. 1997, Wu 1998a, b). CA are cells arranged in a regular grid that change their state according to specific transition rules. These rules define the new state of the cells as a function of their original state and local neighborhood. CA models have been used successfully to simulate a wide range of environmental systems, including the spread of fires (Green et al. 1990) and diseases (Green 1993), episodes of starfish overpopulation (Hogeweg and Hesper 1990), and forest dynamics (Green 1989). Green (1994) applied a CA approach to model species distribution patterns and the dynamics of ecosystems. Recently, the interest in CA has spread to modelers of urban and regional development (Couclelis 1985, 1997, Batty and Xie 1994, White and Engelen 1997, White et al. 1997, Wu 1998a, Wu and Webster 1998). CA has several advantages for those studying urban phenomena: they are intrinsically geographic (Tobler 1979, Couclelis 1985), relatively simple, and can mirror the way urban systems work (Batty and Xie 1994, Batty 1997).

Several modelers have stressed the need to more realistically represent the spatial behavior of urban actors (White and Engelen 1993). CA can be used to simulate changes in land use and land cover through the iteration of rules. Land use state at time $t + 1$ is determined by the state of the land and development in its neighborhood at time $t$ in accordance with a set of transition rules. Various modelers have also adopted transition rules ranging from simple deterministic rules (White and Engelen 1997), to stochastic rules (White et al. 1997), to self-modification (Clarke et al. 1997), to utility maximization (Wu 1998a). Finally, Wu (1998a) has developed a CA probabilistic simulation based on discrete utility theory.

Urban modelers have managed to generalize CA by relaxing some of the standard assumptions of the approach that do not fit the representation of urban systems. Among these generalizations are the ideas that space is irregular, that states are not uniform, that neighborhoods do not remain stationary, that transition rules are not universal, that time is not regular, and that systems are open (Couclelis 1985, 1997). White and Engelen (1997), for example, applied CA to simulate urban land uses; they defined heterogeneous cell-states across the cell-space. They developed a cellular automaton that is open to outside demographic, macroeconomic, and environmental forcing.

## Agent-based models

The emergence of agent-based models (ABM) provides a potentially effective way to address complex behaviors and heterogeneity in coupled human-natural systems. ABMs model the system as a collection of autonomous agents that make decisions by interacting among themselves and with their environment. Some key characteristics make ABM especially valuable to building models of coupled human-natural systems (Bonabeau 2002). First is their ability to simulate nonlinear behaviors and discontinuity. Agents and their interactions can be treated as heterogeneous. Agents have memory and their behavior can reflect path-dependency. Agents also exhibit complex behavior, including learning and adaptation.

ABMs are currently applied to model very diverse phenomena in a variety of disciplines from ecology to economics (Liebrand et al. 1988, Epstein and Axtell 1996, Janssen and Jager 2000). ABMs have been primarily applied to model emergent phenomena such as road traffic, stock markets, ant colonies, or immune systems. More recently, interdisciplinary teams of modelers have started to apply ABMs to build coupled human-natural models (Berger and Ringler 2002, Hoffmann et al. 2002). A number of scholars have reviewed applications of agent-based land use and land cover change models (Parker et al. 2001, Berger and Parker 2002, Parker et al. 2003). Evans and Kelley (2004) examine how scale may affect the design of an ABM in modeling land use and land cover change.

## 2.3 An Agent-Based Hierarchical Model

Modelers can use a number of human and biophysical variables to represent an urban ecosystem. They can describe its human dimensions from the perspective of its actors, resources, markets, and institutions. Major components are population, economic activities, land, buildings (residential and non-residential), infrastructure (transportation, energy, water supply, waste water), and natural resources (water, forests, and ecosystems). Alternative landscape states can be hypothesized to emerge from behavior of multiple agents and their interactions with each other and with their environment across time and space (e.g., Batty and Torrens 2001, Torrens 2003, Parker et al. 2003). With emergent phenomena, a small number of rules or laws, interacting at the local scale, can generate complex systems.

Detectable patterns emerge from local-scale interactions among variables such as human preferences for residential location, individual mobility patterns, transportation infrastructure, real estate markets, and topography (Torrens and Alberti 2000). Urban landscapes are also organized hierarchically. Higher levels operate on a larger spatio-temporal scale and define the boundary conditions in which the system functions, while the lower levels exhibit much faster processes in space and time and act as initiating conditions (Figure 2.1).

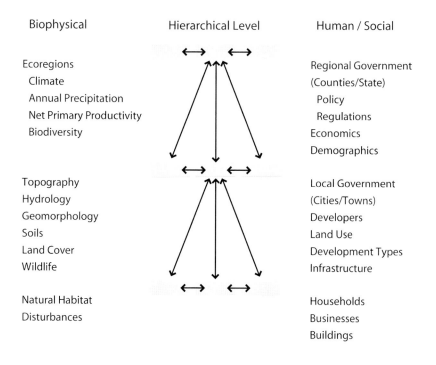

Figure 2.1. Hierarchical relationships of human and ecological systems.

## Complex systems theory and complex adaptive systems

Complex adaptive systems (CAS: Levin 1998, 1999, Gunderson and Holling 2002) provide the theoretical framework that modelers can use to link local interactions between human and social dynamics and biophysical processes to the overall structure and dynamics of urban landscapes. Building on complex systems theory and the theory of patch dynamics, my research team in Seattle is focusing on the ways that the structure, function and dynamics of spatially heterogeneous systems lead to emergent properties (Levin and Paine 1974, Wu and Levin 1994, 1997, Pickett et al. 1997). In this work, we hypothesize that while the dynamics of urban landscapes are primarily driven by bottom-up processes, top-down constraints and hierarchical structures are also important for predicting those dynamics (Wu and Loucks 1995, Wu 1999). Drawing on hierarchy theory, we conceptualize the landscape as nested hierarchies by explicitly representing both its vertical structures (as levels) and its horizontal structures (as holons) (O'Neill et al. 1986, Wu 1999, Wu and David 2002). An agent-based approach is the basis for integrating these approaches: we model the agents, the environment through which the agents interact, the rules that define the relationship between agents and their environment, and finally the rules that determine the sequence of their actions (Parker et al. 2003). Although our model is still in its development, a discussion of its design well summarizes years of experimentation, conceptual refinement, and initial operationalization.

## Modeling the emergent properties of urban landscapes

Urban landscapes can be described as nearly decomposable, nested spatial hierarchies, in which hierarchical levels correspond to structural and functional units operating at distinct spatial and temporal scales (Reynolds and Wu 1999, Wu and David 2002). Near-decomposability is a key tenet of hierarchy theory that allows for diversity, flexibility, and higher efficiency in representing complex dynamic systems. In applying hierarchy theory to urban landscapes, holons (horizontal structure) can be represented by patches (the ecological unit) and parcels (the economic unit). Patches and parcels interact with other patches and parcels at the same and at higher and lower levels of organization through loose horizontal and vertical coupling (Wu 1999).

In the urban landscape, the lowest hierarchical level and the smallest spatial unit of the landscape vary with socioeconomic and biophysical processes; from households and buildings to habitat patches or remnant ecosystems. At a coarser spatial scale parcels and patches interact with each other to create new functional levels and units such as neighborhoods or sub-basins. In turn, neighborhoods and sub-basins initiate and are constrained by regional economic and biophysical processes. Since landscapes are nonlinear systems, they can simultaneously exhibit instability at lower levels and complex meta-stability at broader scales (Wu 1999, Burnett and Blaschke 2003).

The approach proposed by Wu and David (2002) to model the landscape as a hierarchical mosaic of patches (Allen and Starr 1982) can be effectively linked to an agent-based approach to break down the complexity of landscape dynamics. Such a modeling approach provides a more tractable representation of complex urban systems by explicitly representing the hierarchies of interactions between human and biophysical agents and identifying multiple-scale patterns and processes, including top-down constraints and bottom-up mechanisms.

## Coupled land development and ecology model

Using the framework described above, we are working towards the development of a hierarchial agent-based model that links human and ecological processes in urbanizing regions. This model is part of a Biocomplexity project currently being developed at the University of Washington Urban Ecology Research Lab in collaboration with the Arizona State University Global Institute of Sustainability. We build on a micro-simulation of human behaviors represented by UrbanSim (Waddell 2002) and link it to a spatially explicit Land Cover Change Model (LCCM) (Alberti and Hepinstall forthcoming) to predict three types of human-induced environmental stressors: land conversion, resource use, and emissions. Figure 2.2 represents the urban ecological dynamics that the envisioned fully coupled model should address (Alberti and Waddell 2000). Our initial focus has been on modeling changes in land use and land cover. Instead of linking the urban and ecological components sequentially, we propose to integrate them at a functional level.

Our strategy aimed to extend the object properties and methods implemented in the UrbanSim model developed by Waddell (1995) to develop a coupled land use land cover change model for the central Puget Sound region[3] UrbanSim predicts the location behaviors of households,

---

[3] An initial description of the integrated urban development and ecology modeling strategy discussed here and in Section 2.4 is in Alberti and Waddell (2000). An integrated urban development and ecological simulation model. Integrated Assessment 1:215-227. Modified paragraphs extracts are reprinted with permission.

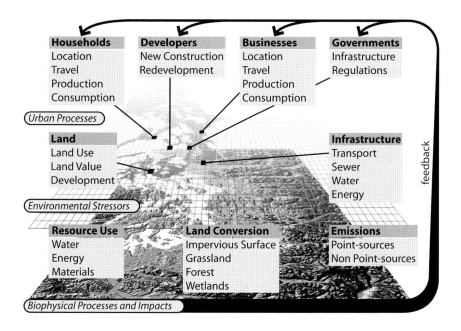

Figure 2.2. Integrated ecological and urban development model (PRISM 2000, revised).

businesses, and developers, and consequent changes in land uses and physical development (Waddell 2002). These are among the required inputs to predict the changes in land cover and ecological impacts. We intend to add the production and consumption behaviors of households and businesses, and link these through an explicit representation of land use and land cover to infrastructure and biophysical processes. The structure of the integrated model identifies the principal objects as households, businesses, buildings, land, infrastructure, natural resources, and various biophysical components. Coupled with the human models are a series of biophysical and resource models. Resource demand and supply models can be integrated to provide predictions of water, energy, and materials usage, which will then be linked to our coupled model on the basis of consumption and infrastructure capacity. Similarly, mass-balance models will simulate pollution loads into the atmosphere, water, and soil, and estimate relative contributions from the various sources.

## Agent behavior

Our ability to realistically represent the behavior of key actors depends on the level of aggregation at which we will represent these elements in the model. Real decision-makers are a diffuse and often diversified group of people who make a series of relevant decisions and trade-offs over a period of time. Their decisions depend on a broad range of characteristics: of their households or businesses, of the buildings they occupy, and of other aspects of their location. Furthermore, in the real world these decision-makers place different values on the costs and benefits of alternative decisions, and distribute them differently. And—importantly—these actors also learn through time. For all these reasons, we will need a highly disaggregated representation of households and businesses in order to account for the diversity of production, consumption, and location behaviors of various actors.

## Space and time

If we are to integrate urban and ecological processes, then we must focus on ways to more realistically represent the spatial and temporal dynamic behavior of urban ecosystems. We will treat space explicitly, in two steps. First we will develop and implement a grid of variable resolution to geo-reference all spatially located objects in the model, a process that can be done much more efficiently within a raster data structure. Using a variable grid will also give us the flexibility to vary spatial resolutions for different processes, and to test the effects of spatial scale on model predictions. In the second step we will focus explicitly on spatial processes across the area. The grid provides a foundation for linking urban spatial processes to processes that occur in the natural environment.

We also improve the treatment of time. Most urban models assume a static, cross-sectional approach. The current UrbanSim model uses annual time steps to simulate household choices, real estate development and redevelopment, and market clearing and price adjustment processes within the market (Waddell 1998). Every 10 years (or after significant changes in the transportation system), the travel accessibility data is updated by running a travel model simulation. UrbanSim can be improved by implementing a more flexible treatment of time for different behaviors in the model, such as location choices and real estate development.

## Feedback mechanisms

The model framework accounts for the interactions between the ecological impacts and urban processes. Ecological changes will feed back on the choices of both households and business locations, and availability of land and resources. For example, the amount, distribution, and health of the vegetation canopy cover in urban areas have both social and ecological functions. Among the most obvious social functions are the attractiveness of the area and therefore its economic value. Vegetation cover has several important ecological functions: it removes air pollutants, mitigates micro-climate—thus saving energy in buildings—and absorbs some rainfall, reducing urban stormwater runoff.

All these factors improve urban environmental quality and provide ecological services to urban residents. These economic benefits are not currently reflected in land values but eventually will affect long-term urban development. To capture these values, we will define a set of parcel—and cell-based environmental quality indices (e.g., air quality, water quality, noise, etc.) and potential risk indices (e.g., floods, landslides etc.) that influence location choices and the profitability of development.

## 2.4 Modeling Changes in Land Use and Land Cover

The dynamics of land use and land cover are at the core of urban eco-system change. They are distinct elements driven by coupled human-biophysical processes. Changes in land cover are driven by climate, hydrology and vegetation dynamics, and involve conversion and modifi-cation (Riebsame et al. 1994). Land conversion is a change from one cover type to another; land modification is a change in conditions within the same cover type. Here we present our specific strategy, developed as part of the Biocomplexity Project, for building the land use-land cover component as a fully coupled model (Alberti et al. 2006a). To model land use and land cover dynamics we build on various modeling approaches rooted in economics, landscape ecology, and complex system science. Instead of separately simulating urban growth and its impacts on biodiversity, we develop a framework to simulate metropolitan areas as they evolve through the dynamic interactions between urban development and ecological processes and link them through a spatially explicit representation of the urban landscape (Figure 2.3). Our hybrid model structure combines four ele-ments: (a) an agent-based structure that allows us to explicitly represent agent choices (location, housing, travel, production consumption, and land development); (b) a spatially-explicit, patch dynamic model structure; (c) a process-based model of physical and ecological dynamics (hydrology, nutrient cycling, primary production, and consumer dynamics); and (d) a hierarchical organization to model different levels of interactions in the coupled system.

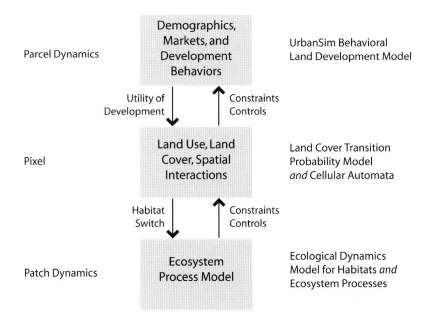

Figure 2.3. Biocomplexity model.

Urban development is modeled using UrbanSim that simulates demographic, market, and real estate development behaviors at the parcel level. Demographic and economic processes drive the demand for built space for various activities, as households and businesses make decisions about locations. In UrbanSim, these are integrated with a market-clearing component. The ecological process models, currently limited to the dynamics of bird species richness, are based on habitat patches and defined by the species under consideration. The land cover change model (LCCM) is the link in our system between urban development events and land use changes to the biophysical model, providing bi-directional feedback and interactions between socioeconomic and biophysical processes. To fully couple the human and biophysical components we have developed a hybrid spatially-explicit microsimulation structure by combining the current UrbanSim behavioral approach, which explicitly represents the land development process, with a spatially explicit approach that represents local spatial dynamics and feedback between land use and land cover change. To this end we are in the process of expanding LCCM and explicitly linking to the Distributed Hydrology Soil Vegetation Model (DHSVM, modified for urbanizing environments) and the MM5 regional climate model.

## Demographic and economic processes

Economic and demographic processes within metropolitan economies can be modeled using diverse approaches ranging from input/output models to structural equations, or a hybrid combination. Demographic processes are often modeled as a function of several factors including the ageing of an age-sex population pyramid; fertility and death probabilities are applied to the population counts in each age-sex cohort, with age- and sex-specific net migration rates predicted as a function of employment opportunities. Conway (1990) uses an hybrid approach for the Washington Projection and Simulation model providing a robust approach to link the evolution of regional- or state-level economic structure to broader national and global economic trends and influences.

One of the major challenges faced in developing a full microsimulation model as proposed in UrbanSim is to reconcile the microsimulation of land demand and supply processes, with macroeconomic processes. In UrbanSim, a Demographic Transition Model simulates births and deaths in the population of households. Externally imposed population control totals determine overall target population values and can be specified in more detail by distribution of income groups, age, size, and presence or absence of children. The Economic Transition Model is responsible for modeling job creation and loss. Employment control totals determine employment targets and can be specified by business sector distributions. A substantial degree of research must still be conducted before we can effectively implement a microsimulation of the macro-behavior of a regional economy, in a way that captures not only the interactions within the region, but also the interactions with economic processes in the nation and the world.

## Household and business demand

The integrated model draws on specifications of multiple agents represented in the household, business location choice, and development model components within the UrbanSim model system (Waddell 1998, Waddell et al. 2003) and links them to biophysical agents. UrbanSim represents human agents and dynamics based on microsimulation of the behavior of households, businesses, and developers and their spatially explicit interactions between agents of the same (i.e., households) or different types (households and developers), and between agents and their spatially-explicit environment (i.e., development types) (Waddell et al. 2003). By linking buildings to a location grid and developing a model infrastructure for spatial query and analysis, urbanism provides a platform to model interactions and feedback between spatially-explicit land use and land cover attributes of location choices. We expect characteristics of the natural and built environment, such as the street pattern, the availability of open space, and access to shopping and services, to be some of the micro-location attributes relevant to demand for residential location. For business location, a spatially explicit representation of the context surrounding available sites should allow for better integrating the demand behavior with access to resources and services and environmental quality of location. A set of spatial metrics used in the land cover change model may be adapted to describe spatial patterns of land use and environmental characteristics that could inform the demand functions of different types of households. This last step will make it possible to establish feedback linkages from biophysical processes and land cover change to the demand for residential locations. We propose implementing a microsimulation of the demand components of the model following the approach developed by Wegener et al. (1999).

## Market clearing and price adjustment

UrbanSim addresses market clearing and the adjustment of land and building prices by matches moving households and businesses to vacancies in the housing or nonresidential building stock based on the consumer surplus of the match. Consumer surplus measures the degree to which the willingness to pay for an alternative exceeds its market cost (Ellickson 1981, Martinez 1992, Waddell 2000). UrbanSim matches active consumers and available vacancies, and then adjusts prices in the real estate submarkets according to the relationship between the current vacancy rate in a submarket and the structural vacancy rate, following an approach described by DiPasquale and Wheaton (1996). Vacancy rates below the structural vacancy rates, push prices. Exceptionally high vacancy rates pull prices downward (Waddell 2002).

## Land development and redevelopment

Models of land use change have generally taken one of two approaches to represent change: (1) as the developer's decision to build a given project on a given parcel; or (2) as the transition of a given land parcel from its original use to a new use. The first formulation requires a microsimulation of the behavior of developers who calculate the profitability of converting each land parcel to alternative development projects. Land use and environmental policies and development fees can all directly influence the behavior of the developer. The second approach emphasizes the probability of transition based on local dynamics and the spatial self-organization of urban land uses. Thus the approach provides a framework in which to address the evolutionary and nonlinear nature of land use change (Alberti and Waddell 2000).

Our hybrid model strategy combines the two approaches. We treat land development as a stochastic formulation of the profitability of a given development project to be realized in a given parcel or group of parcels. UrbanSim uses a a multinomial logit formulation to simulate the profit-maximization behavior of developers. It predicts changes in land use based on expected benefits and on the costs of developing alternative parcels into allowable developed uses. Developers tend to develop all available parcels until they have satisfied the aggregated demand for built space. Based on the predicted land use, the model allocates specific buildings and associated infrastructure to individual cells of high resolution. In order to predict the amount of additional built space, the model identifies specific development types and allocates building and infrastructure attributes to individual cells.

To fully couple land use and land cover dynamics, feedback mechanisms between land cover and land use need to be explicitly represented. Our strategy involves incorporating characteristics of land use and cover patterns of developable and re-developable parcels in the behavioral model component. Agents decisions take into account the patterns and dynamics of both land use and land cover at each location.

## Changes in land cover

Our land cover change model (LCCM) consists of a set of spatially explicit multinomial logit models of transitions in site-based land cover (Alberti and Hepinstall forthcoming). We build on previous efforts to model land cover change, in order to simulate the change as influenced by spatially explicit dynamic interactions between socioeconomic and biophysical processes (Turner et al. 1996, Berry and Minser 1997, Wear and Bolstad 1998, Wear et al. 1998). We estimate the transition probability equations empirically as a function of a set of independent variables comparing land cover data every two years from 1985 to 2002. We considered three types of variables: 1) biophysical; 2) land use; and 3) change variables. We also considered three different spatial effects: 1) attributes of the site; 2) site location along various gradients, including proximity to the most recent land conversion and and most recent development event; and 3) landscape patterns—both landscape composition and configuration of neighboring cells—at multiple spatial scales. We are expanding on this LCCM framework by explicitly representing the urban landscape as hierarchical, dynamic mosaics of patches generated and maintained by processes of patch formation, development, and disappearance (Wu and Levin 1994, 1997, Wu and David 2002). Human decisions, particularly changes in land use, are important drivers of these patch dynamics (e.g., fragmentation). The model component that predicts land cover change is influenced by both biophysical and human agents, each operating through different mechanisms measured through different proximate variables (Alberti and Hepintsall forthcoming). These decisions together with environmental heterogeneity create patchiness in ecological system in time and space. We will first identify "landscape signatures" that describe and quantify the emergent landscape patterns in the Seattle and Phoenix metropolitan regions by extracting spatial characteristics of urban landscape change using data spanning two decades. Different types of urban growth are distinguished in infill, expansion, or outlying which can be characterized by spontaneous, diffusive, cluster and linear branching (Clarke et al. 1997). These various forms of urban growth emerge from the interactions between human agents and biophysical constraints.

## 2.5 Changes in Land Use and Land Cover in Central Puget Sound

The health of the Puget Sound ecosystem, along with its ability to support a growing human population, is highly influenced by complex interactions between human and ecological processes (Alberti et al. 2007). The drivers of ecosystem change include such diverse factors as patterns of human population growth, local and global markets and economies, and climate change. These factors both interact with and affect different components of the ecosystem simultaneously—influencing the ways humans plan their communities and the range of other species that are supported in those areas, for example. It is critical to characterize these complex interactions if we are to assess the possible impacts of human activities and the ecosystem's responses to alternative management approaches.

Although we have yet to develop a systematic account of the impacts of human and natural drivers, their trajectories, and potential interactions in Puget Sound, several studies have started to establish the linkages between human activities and ecological change in this region.

### Patterns of human population and economic activities

Over the last century, population growth has driven dramatic changes in the Puget Sound landscape (Figure 2.4). The human population in this region has increased particularly rapidly over the last two decades: in 2005

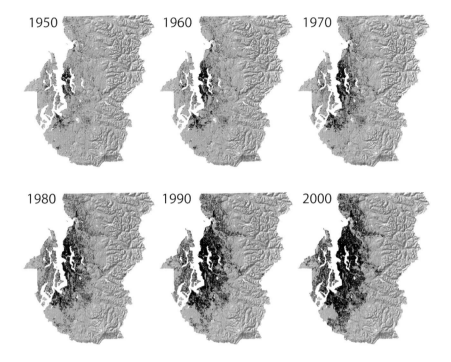

Figure 2.4. Urban development in Central Puget Sound 1950-2000 (UERL 2005a).

it housed approximately 4.4 million people, a 25% increase from 1991. Although estimates vary depending on the area encompassed, the Washington State Office of Financial Management (OFM) predicts that the population will grow to between 4.7 million and 6.1 million residents by 2025. The inter-mediate projection estimated by the state suggests we will see an additional two million people in the basin within the next 20 years (OFM 2005).

Most of the population growth in the Puget Sound region has con-centrated in the Central Puget Sound region (King, Kitsap, Pierce, and Snohomish counties), which currently contains 3.4 million people (Puget Sound Regional Council [PSRC] 2005). The central region experienced substantial growth over the last three decades, increasing by over 1.3 million people between 1970 and 2000. In most counties, the population distribution is shifting from large urban areas to outlying areas (OFM 2005). Most of the population growth comes from immigration; people moving to the area accounted for two thirds of the growth in the last decade.

Trends in household size are a good indicator of how population change may affect consumption patterns such as the number of housing units, energy use, vehicle trips, and similar expenditures. According to the 2000 Census, household size in the Puget Sound Basin is declining—dropping to approximately 2.5 people per household from about 3.0 in 1970 (US Census 2000). If such trends persist, the number of households will continue to increase faster than the population in the next decades, with important consequences for the use of land and resources.

Economic and employment growth drive population trends and urbanization patterns. The Puget Sound economy has expanded faster than the national economy. Total employment in Central Puget Sound more than doubled between 1970 and 1990, from 740,927 to 1,445,243 jobs, and to 1.7 million by 2000. The PSRC projects that the region will offer 2.0 million jobs by 2010, 2.2 million by 2020, and more than 2.5 million by 2030. As the number of jobs increases, the economy is shifting away from a manufacturing base to dominance by service and office industries, including software, retail, biotechnology, tourism, internet services, and telecommunications.

## Observed and predicted changes in land cover

Several environmental changes are associated with this human population growth, as land cover changes, people demand more natural resources, and the amount of pollution increases. The presence of more people implies even greater demand for goods and services, increased use of land for housing, and expanded transportation and other infrastructure to support human activities.

Today, the land cover of Puget Sound exhibits a dramatic human presence (Figure 2.5). Although no one has yet written an historical account of change in land cover across the entire Puget Sound basin, the dramatic changes documented for the Puget Sound lowlands in the recent past clearly show the rapid effects of urbanization (Alberti et al. 2003). Between 1991 and 1999 alone, 1% of the total area in the region was newly developed—and the area designated as forest land decreased by a total of 5% (Figure 2.6). Overall, forest cover decreased 8.5 percent between 1991 and 1999. Highly developed land (i.e., land with greater than 75 percent impervious cover) increased by more than 6 percent of the total area in the region, and moderately developed land (i.e., between 15 and 75% impervious cover) increased almost 8 percent. The most intense development has occurred within the Urban Growth Boundary, where forest declined by 11.1%. Almost half of the land conversion to development has occurred in the Seattle Metro Area (Alberti et al. 2004).

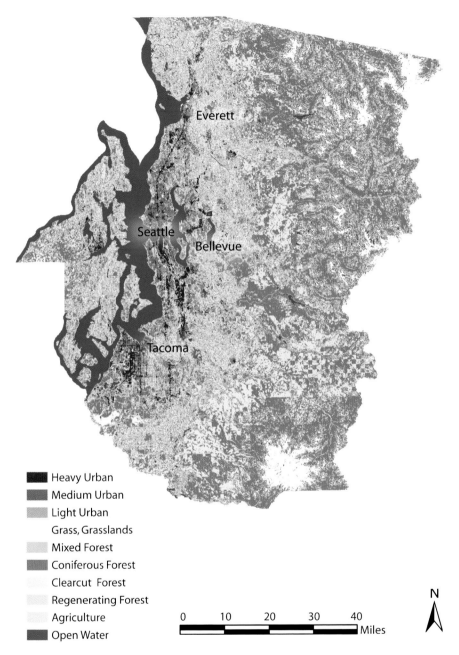

Figure 2.5. Central Puget Sound land cover 2002 (UERL 2005b).

Our simulations of future land cover for the region indicate a conversion from grass, agriculture, and mixed lowland forest to varying intensities of urban developed land cover (Figures 2.6). Most development will occur in areas closer to major roads, inside the urban growth boundary, and closer to larger areas (with larger mean patch size) of contiguous existing urban land cover (Figure 2.7, Alberti and Hepinstall fortcoming). Most landscape change is concentrated in areas surrounding currently developed lands (i.e., expansion growth), with little infill or outlying new patches of development; the latter primarily occur along lower elevations and up river valleys.

The health of Puget Sound is vital to economic, cultural, and recreational activities and to overall human and ecological well-being in the region. While much of Puget Sound is still healthy, the rapid landscape change associated with population growth and urbanization in watersheds and shoreline environments is degrading ecosystem functions; such factors

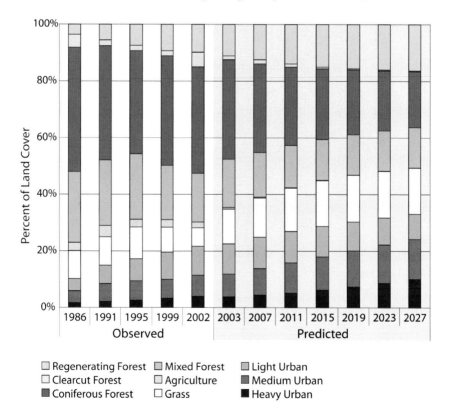

Figure 2.6. Puget Sound land cover observed and predicted (Alberti and Hepinstall forthcoming).

will have important impacts on the ecosystem's future ability to sustain the increasing human population and its activities. These trends are likely to continue over the next several decades as population growth increases and more forested land is converted to suburban development (Vitousek et al. 1997b).

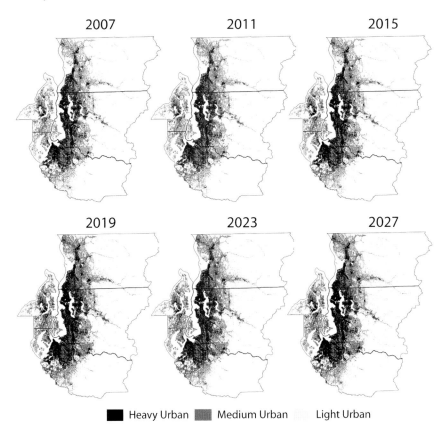

Figure 2.7. Predicted land cover in Central Puget Sound 2007–2027 (Alberti and Hepinstall forthcoming).

# Chapter 3

# URBAN PATTERNS AND ECOSYSTEM FUNCTION

## 3.1 Patterns, Processes, and Functions in Urban Ecosystems

Urbanizing regions present challenges to ecosystem ecology, but the field may gain important insights from studying these areas. Earth's ecosystems are increasingly influenced by urbanization, and their functioning is dependent on the landscape patterns emerging in urbanizing regions. While ecosystem ecology still lacks a theory of ecosystem function that explicitly takes spatial phenomena into account, increasing empirical evidence indicates that the linkages between ecosystem structure and function depend on landscape heterogeneity (Lovett et al. 2005). Landscape heterogeneity is a spatial phenomenon caused by variation in environmental conditions. In turn, it affects the interactions among patches, biodiversity, and ecosystem processes such as energy flows, nutrient cycling, and primary production. Landscape heterogeneity manifests as discrete patterns and gradients across multiple spatial scales, and it matters to populations, communities, and ecosystems.

Urban landscapes exhibit unique spatial patterns, and thus a distinctive heterogeneity. Since the amount and form of urban development affect the mosaic of habitat patches and their ecological properties, we expect alternative urban patterns to have different effects on ecological systems (Alberti 2005). These relationships are not simple, however. Urban development affects simultaneously habitat structure and the processes that control patterns of species diversity and abundance including species interactions, microclimate, and availability of natural resources (Rebele 1994, McDonnell et al. 1997, Pickett et al. 2001, Shochat et al. 2006). Thus, understanding urban landscape patterns is critical to integrating the study of ecosystem processes at multiple scales.

In this chapter, based on a review of the empirical evidence established in the literature, I develop a set of hypotheses about the effects of urban patterns on ecosystem function. First, I present a framework for examining the effects of urban patterns on ecological processes in urban ecosystems. Building on complex system theory I explore interactions and feedback between human and biophysical processes. I draw on the field of landscape ecology to propose that the interactions between humans and biophysical processes in urban landscapes are mediated by patterns of urban development. Second, I hypothesize that emerging landscape patterns (e.g., those caused by floods, fire, or invasive species) influence the dynamics of ecosystem processes in urbanizing landscapes. Through a review of the literature I show that relationships between urban patterns and ecosystem dynamics are still poorly understood. A specific discussion of key ecosystem functions in urban ecosystems is developed in Chapters 5-8. Drawing on an empirical study of the Puget Sound metropolitan region currently being developed at the University of Washington (Alberti et al. 2007, Alberti and Marzluff 2004), I discuss directions for future research that can inform strategies to minimize urban impacts on ecosystems.

## Human impacts on ecosystem function

Urban ecosystems differ from other ecosystems in several ways (Rebele 1994, Trepl 1995, Sukopp 1990, Niemala 1999, Collins et al. 2000). Ecologists have described the city as a heterotrophic ecosystem, highly dependent on external inputs of energy and materials and on a vast capacity to absorb its emissions and waste (Odum 1963, Duvigneaud 1974, Boyden et al. 1981, Collins at al. 2000). Compared to a "natural" ecosystem with a typical energy budget ranging between 1,000 and 10,000 Kcal per square meter per year, cities consume a vastly larger amount of energy. The budget of an urban ecosystem in an industrialized country can range between 100,000 and 300,000 Kcal per square meter per year (Odum 1997). Urban ecosystems differ from natural ecosystems in other key ways: Remnant patches of natural habitat are not integrated, non-native species are present in large numbers, and succession is externally controlled (Trepl 1995). Furthermore, such systems differ in microclimate (warmer and wetter), hydrology (increased runoff), and soils (higher concentrations of heavy metals, organic matter, and earthworms).

In urbanizing regions, humans affect ecosystem function through direct and subtle changes in biophysical and ecological processes. Urbanization affects primary productivity, nutrient cycling, hydrological function, and ecosystem dynamics through change in climatic, hydrologic, geomorphic, and biogeochemical processes and biotic interactions (Figures 3.1 and Table 3.1). Human activities in urbanizing regions, for example, alter the availability of nutrients and water, thus affecting population, community, and ecosystem dynamics. Urbanized areas modify the microclimate and air

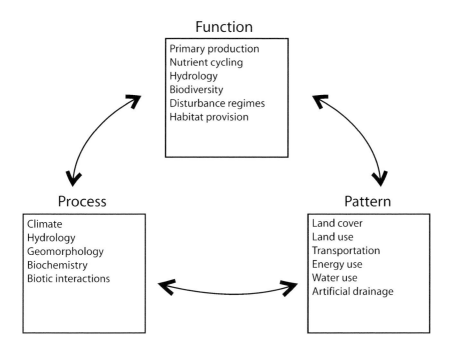

Figure 3.1a. Conceptual model of functions, processes, and patterns.

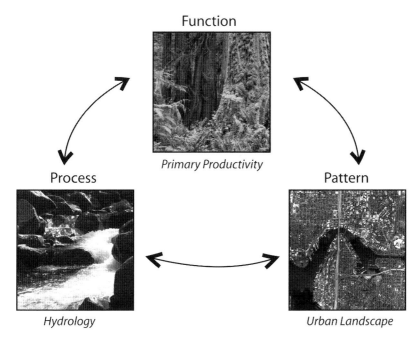

Figure 3.1b. Conceptual model of functions, processes, and patterns.

Table 3.1. Ecosystem patterns, processes, and functions. A synthesis of the established relationships, directions (↑↓), and uncertainties (?) between urban patterns and ecosystem functions though various ecosystem processes.

| Land cover | Land use | Transportation | Energy infrastructure | Water infrastructure | PROCESS / PATTERN |
|---|---|---|---|---|---|
| ↑Impervious surface ↑Grassland ↓Forestcover | ↑Heat ↑CO₂ emissions | ↑Heat ↑CO₂ emissions | ↑Heat ↑CO₂ emissions | | **Climate** |
| ↑Impervious surface ↑Grassland ↓Forestcover ↑Dams | ↑Industrial water use ↑Agricultural water use ↑Residential water use | ↑Impervious surface ↑Grassland ↓Forest cover | ↑Dams | ↑Water extraction ↑Hydrologic connectivity ↑Water budget | **Hydrology** |
| ↓Riparian area ↓Wetlands ↑Gravel extraction | ↑Riparian clearing ↓Pervious surfaces and soil column | ↑Riparian clearing ↓Pervious surfaces and soil column | ↑Dams | ↑Channel modification, ↑Channel incision or "downcutting" | **Geomorphology** |
| ↑Impervious surface ↓Forestcover ↓Vegetation | ↑Fertilizers ↑Pesticides ↑Herbicides ↑Toxic emissions | ↑$CO_2$, $NO_x$, CO, $SO_2$, & VOCs emissions ↑Road salting ↑Oil leakages | ↑$CO_2$, $NO_x$, CO & SO emissions | ↑Septic & wastewater treatment emissions ↑Wastewater overflow emissions | **Biogeochemical** |
| ↓Riparian area ↓Wetlands ↓Forestcover ↑Forest fragmentation ↑Invasives | ↓Riparian area ↓Wetlands ↓Forestcover ↑Forest fragmentation ↑Invasives | ↓Riparian area ↓Wetlands ↓Forestcover ↑Forest fragmentation ↑Invasives | ↓Riparian area ↓Wetlands ↑Forest fragmentation | ↓Riparian area ↓Wetlands ↑Forest fragmentation | **Biotic interactions** |

| PROCESS / FUNCTION | Primary production | Hydro-logic function | Nutrient cycling | Biodiver-sity | Habitat provision | Distur-bance regulation |
|---|---|---|---|---|---|---|
| **Climate** | ? Photosynt. ?Plant growth | ↑Temp. ↓Snow pack ↑Runoff | ↑Nutrient losses | ↓Species div ↑Invasives ↑Geographic shifts in ranges ↑Shifts in the timing of breeding | ↑Temp. | ? Weather events / variability ↑Invasives ↑Fire/drought, insect /path. outbreaks ↑Landslides |
| **Hydrology** | ? Water availability | ↑Runoff ? Base flow ↑Nutrients ↑Toxics ↑Flashiness | ↓N retention | ↓Fish pop. ↓Macroin-vertebrates diversity ↑Tolerant / generalist species | ↓Water qnty. ↓Substrate ↓Habitat div. ↑Temp. ↓Woody deb. | ↑Flooding ↑Drought ↑Erosion ↑Sedimenta-tion |
| **Geomor-phology** | ? Nutrient availability | ↑Runoff ? Base flow ↑Temp. ↑Sediment ↑Nutrients | ↓Organic matter ↓Nutrient uptake | ↓Fish pop. ↓Macroin-vertebrates diversity ↑Tolerant / generalist species | ↑Channel width ↑Pool depth ↓Channel complexity ↓Spawning, rearing & hiding sites | ↑Flooding ↑Erosion ↑Sedimenta-tion ↑landslides |
| **Biogeo-chemical** | ? Nutrient availability | ↑Temp. ↑Sediment ↑Nutrients ↑Toxics, DO, pH, PAHs, PCBs, Chloride, Organics | ↑Soil $NO_3$ & nitrification ↓Soil denit. ↑Eutroph. ↓Organic debris ↑N sources | ↓Species div. ↑Algae ↑Tolerant / generalist species | ↓Habitat diversity ? Food sources | ↑Nutrients |
| **Biotic interac-tions** | ↑Seed production ? Carbon use efficiency | ? Water use efficiency | ↑Organic material ↑Eutroph. ? Trophic interactions ? Ecosystem metabolism | ↓Species div. ↓Native species ↑Synan-thropic species | ↑Competi-tion ? Nest predation | ↑Invasives ↑Insect /pathogen outbreaks |

quality by altering the nature of the land surface and generating large amounts of heat. The urban heat island effect, which in turn serves as a trap for atmospheric pollutants, is perhaps the best-known example of inadvertent climate modification (Horbert et al. 1982, Oke 1987). Furthermore, the increase in impervious land area associated with urbanization affects both geomorphological and hydrological processes, thus causing changes in water and sediment fluxes (Wolman 1967, Leopold 1968, Arnold and Gibbons 1996).

Highly concentrated human populations also modify the magnitude, frequency, and intensity of natural disturbances, and also cause unprecedented human disturbances in ecosystems (Rebele 1994). Disturbances are discrete events that disrupt ecosystem structures and functions (Pickett and White 1985). Human activities can rescale (increase or decrease) the magnitude, frequency, and intensity of natural disturbances, and/or can rescale areas through human-introduced biogeographic barriers creating radical changes in habitat. Urban development introduces novel disturbances, chronic stresses, unnatural shapes, and/or new degrees of connectedness. In addition, homogeneous natural patterns may be introduced through land use, or through the suppression of the natural processes that maintain diversity (Urban et al. 1987).

While scholars of ecology have previously assumed that these impacts change predictably with distance from the urban core (McDonnell and Pickett 1990), some authors have argued that urban ecological gradients are best described by a set of patterns representing complex interactions between human and ecological processes (Alberti et al. 2001). Several studies have addressed the relationships between urbanization and ecosystems (Hemmens 1967, Stone 1973, Keyes and Peterson 1977, Keyes 1982, Newman and Kenworthy 1989). But few have directly asked how alternative urban patterns control the distribution of energy, materials, and organisms in urban ecosystems (Sukopp 1990). Most studies of the impacts of urbanization on ecological systems correlate changes in these systems with simple aggregated measures of urbanization (e.g., human population density, percent impervious surface). However we do not know how alternative urban development patterns influence ecological systems along an urban gradient. We do not know, for example, how clustered versus dispersed and monocentric versus polycentric urban structures differently affect ecological conditions. Nor do we understand the ecological tradeoffs associated with different housing densities and infrastructure.

The question of how patterns of human settlements affect ecosystem functioning is becoming increasingly important in ecology (Collins et al. 2000, Grimm 2000, Pickett 2001). Humans increasingly dominate ecosystems. Without an accurate description of how human spatial patterns affect ecosystem function, we cannot produce reliable predictions of ecosystem change under different future scenarios. Several studies indicate that urban development affects the structure and function of natural systems both directly, through converting the land surface, and indirectly, by modifying energy flows and the availability of nutrients and water. Land conversion rearranges the landscape and reduces its connectivity—a property of landscapes that facilitates or limits the movement of resources and organisms among natural patches (Turner and Gardner 1991, Tischendorf and Fahrigh 2000)—leading to a variety of global and local ecological effects (Godron and Forman 1982). Since ecological processes are tightly interrelated with the landscape, the mosaic of elements resulting from urbanization has important implications for ecosystem dynamics. Land use heavily influences the patchiness of natural systems.

Urban landscapes have been analyzed in terms of patch dynamics (Wu and Loucks 1995, Pickett et al. 1997). This approach focuses on how spatial heterogeneity is created within landscapes, and how that hetero-geneity influences the flow of energy, materials, species, and information across the landscape. Machlis et al. (1997) describe the urban landscape as a complex mosaic of biological and physical patches within a matrix of human infrastructure. Spatial heterogeneity within an urban ecosystem is generated by both biophysical and human processes (Pickett et al. 2000). By altering biophysical processes and disturbance regimes, at one level, human activities reduce the heterogeneity of natural habitat (Pickett and Rogers 1997). At another level, humans increase heterogeneity by introducing exotic species, creating new landforms and drainage networks, controlling or altering natural disturbance agents (e.g., floods, fire, invasive species), and constructing extensive infrastructure (Pickett et al. 1997).

## Mechanisms linking patterns to functions

Social and natural scientists increasingly recognize the complexity of interactions between humans and ecological processes in urbanizing regions and the limitations of the traditional approaches for investigating urbaniza-tion processes and their ecological impacts. Both socioeconomic and eco-logical studies have tended to focus on human impacts on ecosystems, and only recently have started to formalize approaches to study two-way inter-actions in coupled human-natural systems. In addition, these studies have often assumed spatial and temporal stationarity, and only recently have scholars started to pay attention to spatial and temporal heterogeneity. Ecological re-search, on one hand, has tended to simplify the actual spatial pattern of human disturbance in urbanizing landscapes to a few aggregated variables—such as the density of population or built-up areas—that are expected to change predictably with distance from the urban core. In most ecological studies cities are generally described as monocentric agglomerations, whereas most US metropolitan areas over the last few decades have changed from a monocentric to a polycentric structure (Gordon et al. 1986, Giuliano and Small 1991, Cervero and Wu 1995). Furthermore, urban-to-rural gradients are often represented by simple geographical transects from the urban core to exurban rural areas, instead of the complex patterns that emerge from the spatial distributions of land use and land cover.

Social and economic studies, on the other hand, have tended to oversimplify or ignore the diverse biophysical factors that drive, or are affected by, socioeconomic patterns. These studies rarely discriminate among diverse biophysical and ecological processes, or among different species. As a result, strategies devised to minimize the ecological impacts of urban growth often fail, first, to identify the underlying key mechanisms that link urban patterns to ecosystem functions (i.e., interactions between the extent and distribution of impervious surface and stream conditions—roads generating runoff that pollutes streams), and second, to understand the tradeoffs existing among different ecological processes (i.e., tradeoffs among species).

While both ecological and socioeconomic studies have attempted to establish relationships between patterns and processes in urban landscapes, they have done so in their respective domains (Figures 3.2 and 3.3). Only recently, various disciplinary approaches have been combined to study the interactions between complex human behaviors and ecosystem function in urban ecosystems, and the concept of ecosystem function has been expanded to include human and ecological components (Figure 3.4). Urban ecosystems consist of several interlinked subsystems—social, economic, institutional, and ecological—each representing a complex system of its own and affecting all the others at various structural and functional levels. When studying the interactions between humans and ecological processes in urban ecosystems, we need to consider the many socioeconomic and biophysical factors that work simultaneously at various levels and provide important feedback mechanisms. These complex interactions give rise to emergent phenomena whose properties cannot be understood by studying in isolation the properties of socioeconomic or ecological systems.

A new integrated framework is needed to explore interactions between human and ecological patterns and processes in coupled urban systems (Figure 3.5). Scholars of both urban economics and urban ecology have begun to recognize the importance of explicitly representing finer-scale feedback mechanisms in their studies of urban regions (Grimm et al. 2000,

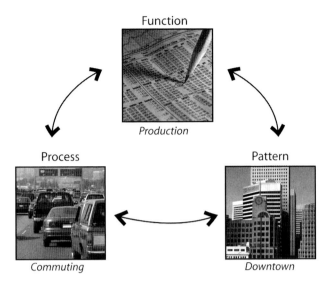

Figure 3.2. Relationship between functions, processes, and patterns in human systems (Bottom right image photographed by Manos, P.).

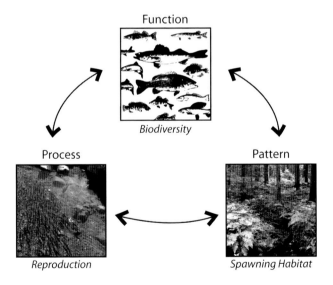

Figure 3.3. Relationship between functions, processes, and patterns in ecological systems.

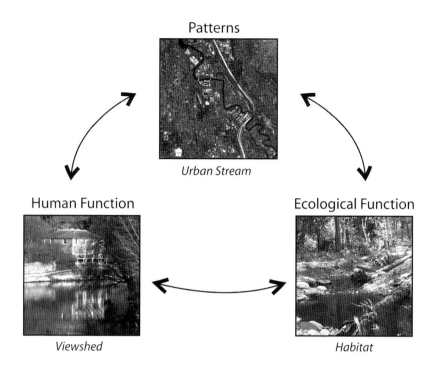

Figure 3.4. Linking patterns to human and ecological functions.

Alberti and Waddell 2000, Pickett et al. 2001). Humans are the dominant driving force in urbanizing regions, and changes in ecological conditions also control human decisions. Furthermore, these interactions are spatially determined. The evolution of land use and its ecological impacts are a function of the spatial *patterns* of human activities and natural habitats, which affect both *socioeconomic* and *ecological processes* at various scales. For example, land-use decisions are highly influenced by patterns of land use (e.g., housing densities), infrastructure (e.g., accessibility), and land cover (e.g., green areas). These local interactions affect the composition and dynamics of entire metropolitan regions.

In ecology, *ecosystem function* is the ability of Earth's processes to sustain life over a long period of time. Biodiversity is essential for the functioning and sustainability of an ecosystem. Different species play specific functional roles, and changes in species composition, species richness, and functional type affect the efficiency with which resources are

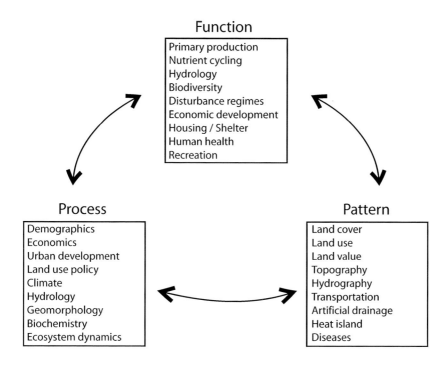

Figure 3.5. Integrated framework of coupled urban ecosystems.

processed within an ecosystem. Thus, the loss of species will impair the biogeochemical functioning of an ecosystem. Furthermore, the distribution, abundance, and dynamic interactions of species can be good indicators of ecosystem condition. Often the disappearance of a species precedes changes in ecosystem function and overall health (Rapport et al. 1985). There are a variety of possible species functional types and measures of ecosystem function (i.e., energy flow, nutrient cycles, productivity, species interactions) to target for assessing system health. Biodiversity is generally considered a good indicator of ecosystem function. Another indicator is net primary production (NPP), which determines the amount of sunlight energy that is fixed by the processes of photosynthesis to support life on Earth.

The concept of ecosystem function has evolved over time to include the interactions between a system's structure and functions and its spatial heterogeneity (Likens 1998, Pickett et al. 2001, Alberti et al. 2003). Ecosystems are no longer considered as closed, self-regulating entities, which at their mature stage reach an equilibrium. Instead they are recognized as open, dynamic, highly unpredictable, and multi-equilibria systems. Disturbance is a frequent, intrinsic characteristic of ecosystems. Successions display multiple pathways and are highly dependent on history and context. Resilience depends on the distribution, abundance, and dynamic interactions of species at several spatial and temporal scales (Holling et al. 2001). In this framework, ecological scholars propose that functional diversity should be the focus of biodiversity conservation, rather than individual species (Folke et al. 1996). Since several species fill similar ecological roles, maintaining the distribution of redundant species across multiple time and space scales can retain key functions despite change (Peterson et al. 1998, Nystrom and Folke 2001). Furthermore, if humans are an integral part of urban ecosystems, resilience depends on maintaining both ecological and human function.

While urban ecologists have focused primarily on ecosystem patterns in urbanizing regions, recent work emphasizes the need to investigate mechanisms that link urban patterns to ecosystem functions by studying the behavioral ecology, species interactions, genetics, and evolution of human dominated ecosystems (Shochat et al. 2006). Some initial mechanistic studies indicate that urban ecosystems are unique settings, in which human activities are coupled with ecological processes in fundamental ways (Pickett et al. 2001, Faeth et al. 2005, Kaye et al. 2006). Investigations of coupled human-natural systems require studying processes at multiple scales of space, time, and organization (Holling 2001, Liu et al. 2007). Urban landscapes exhibit distinctive spatial patterns at different scales, which may be caused by different processes operating at those scales (Wu and Qi 2000). Studying urban landscapes requires that we model hierarchical patch-dynamics, and that we use a scaling approach that deals explicitly with spatial heterogeneity, functional complexity, and the multiplicity of scale landscapes.

The framework proposed here provides new directions for research linking complex patterns, processes, and functions in coupled human-ecological ecosystems. Tables 3.1 and 3.2 provide a summary of the current evidence about the effect of urbanization on ecosystem functions. Other reviews have addressed various aspects of ecology in cities, as well as the ecology of cities (Pickett et al. 2001, Berkowitz et al. 2002). The synthesis proposed here reveals important gaps in the study of urbanization patterns.

Table 3.2. Summary of selected findings on the effects of urban patterns on ecosystem functions (Alberti 2005, p. 176, modified).

| Ecosystem Function | Findings | References |
|---|---|---|
| Primary productivity | Urbanization in the US has reduced the annual net primary productivity (NPP) by 0.04 Pg C or 1.6 percent of its preurban value | Imhoff et al. 1997 |
| | Urbanization is taking place on the best soils | Imhoff et al. 1997 |
| | Urban heat islands extend the growing season in cold regions and increase the winter NPP | Imhoff et al. 2004 |
| Hydrology | Urbanization increases surface runoff | Dunne and Leopold 1978, Arnold and Gibbons 1996 |
| | Bankfull discharge increases with increasing impervious area | Booth and Jackson 1997 |
| | Urban basins are more flashy than non - urban ones | Konrad and Booth 2002 |
| | Stream flow in urban streams is highly variable | Konrad 2000, Booth 2004 |
| | Urban hydrographs vary with impervious area draining directly to streams through pipes | Walsh et al. 2005 |
| Nutrient cycles | Phosphorus concentrations are higher in basins with higher percent of urban land use | Omernik 1976, Meybeck 1998, Wernick et al. 1998 |
| | Concentrations of heavy metals organic matter salts, and soil acidity increase with proximity to the urban core | Pouyat et al. 1995 |
| | Cities have energy budgets 100 to 300 times greater than natural ecosystems | Odum 1997 |

Table 3.2. (Continued)

| Ecosystem Function | Findings | References |
|---|---|---|
| Nutrient cycles (cont.) | Earthworms enhance nitrogen cycling processes by compensating for the effects of air pollution on litter decomposition | Steinberg et al. 1997 |
| | Mass loss and nitrogen release is maximum in urban oak stands and N - mineralization is highest in urban stands | Pouyat et al. 1997 |
| | Positive correlations have been observed between catchment urbanization and concentrations of some streamwater pollutants | Horner et al. 1997 |
| | Urban litter decomposition is slower than in rural areas | Carreiro et al. 1999 |
| | "Hydrologic drought" creates aerobic conditions in urban riparian soils which decreases denitrification | Groffman et al. 2002 |
| | Urban and suburban watersheds have higher N losses than forested watersheds | Groffman et al. 2004 |
| | Urban catchments show high variation in streamwater nutrient concentrations | Grimm et al. 2005 |
| | Urban and suburban watersheds have higher N retention than forested watersheds | Wollheim et al. 2005 |
| Biodiversity | | |
| Vegetation | Exotic species are greater in urban and suburban oak - dominated stands in New York | Rudnicky and McDonnell 1989 |
| | Native flora decreases from the urban fringe to the city core | Kowarik 1990 |

Table 3.2. (Continued)

| Ecosystem Function | Findings | References |
|---|---|---|
| Biodiversity<br>    Vegetation | Native species decrease from the urban fringe to the city core in several Latin American cities | Rapoport 1999 |
| | Plant diversity is greater in larger patches in urban areas | Bastin and Thomas 1999 |
| Birds | Cats and other domestic pets influence bird population in suburban areas | Churcher and Lawton 1987 |
| | Urbanization alters the composition of urban avian communities (decrease native species and increase exploiters) | Beissinger and Osborne 1982, Mills et al. 1989, Blair 1996, Bock et al. 1997, Marzluff 2001, Blair 2001 |
| | Urbanization affects canyon habitat age, total area of chapparel, and predation | Soulé et al. 1988 |
| | Urbanization affects nest predation, brood parasitism, and food availability | Robinson and Wilcove 1994, Newton 1998 |
| | Exotic generalists constitute between 80 to 95 percent of bird communities in cities | Wetterer 1997 |
| | Proximity to urban land use influences bird communities in urban green spaces | Nilon and Pais 1997 |
| | Songbird diversity peaks in landscapes at intermediate urban disturbance levels, because such areas gain more synanthropic and early successional species | Marzluff 2005 |

Table 3.2. (Continued)

| Ecosystem Function | Findings | References |
|---|---|---|
| Biodiversity (cont.) | | |
| Fish and invertebrates | Fish diversity decreases with increase in impervious surface | Klein 1979, Steedman 1988, Schueler and Galli 1992 |
| | Effect of impervious surface on biotic integrity is reduced with intact riparian zones | Horner et al. 1997 |
| | Benthic Index of Biotic Integrity decreases with increase in impervious surfaces | Allan et al. 1997, Yoder et al. 1999 |
| | Effects of impervious surface on fish diversity is minimized in streams with high riparian vegetation | Yoder et al. 1999 |
| | B-IBI declines with increase in urban land cover better explained by sub-basin rather than local scale | Morley and Karr 2002 |
| | Urban landscape patterns (aggregation of impervious area and road density) are associated with stream integrity | Alberti et al. 2007 |
| Habitat provision | Habitat on an urban to rural gradient changes as a result of changes in biophysical processes | Pickett et al. 1997 |
| | The human habitat in cities sets new selective forces affecting behavior and genetic structure of populations | Wandeler et al. 2003, Yeh 2004 |

Table 3.2. (Continued)

| Ecosystem Function | Findings | References |
| --- | --- | --- |
| Habitat provision (cont.) | Heat islands in urban environments buffer extreme climatic conditions affecting biotic interactions | Parris and Hazell 2005 |
| | Absence or reduction of predators and the increased abundance and predictability of resources in urban areas may cause a shift from top - down to bottom - up control | Schocat et al. 2006 |
| | Changes in productivity dampen seasonal and yearly fluctuations in species diversity | Faeth et al. 2006 |
| Disturbance regime | Soil erosion increases catchment sediment yields | Wolman 1967, Leopold 1968 |
| | Channel enlargement increases with increasing impervious surface | Hammer 1972 |
| | Urban areas have a high degree of invasive and immigrant species | McDonnell et al. 1997 |
| | Human sources of disturbance include introduction of exotic species, modification of landforms and drainage networks, control or modification of natural disturbance agents and the extensive infrastructure | Pickett et al. 1997 |
| | Suppressing disturbances alters landscape heterogeneity | Turner et al. 1998 |
| | Vegetation in housing developments is subject to catastrophic disturbances when buildings are demolished and rebuilt | Sukkop and Starfinger 1999 |

## 3.2 Net Primary Productivity

Measures of photosynthetic production, such as net primary productivity (NPP) and net ecosystem productivity (NEP), can be used to quantify the influence of urbanization on ecosystem productivity. NPP is the amount of solar energy converted to chemical energy through the process of photosynthesis (production minus respiration); it directly measures the energy flowing into an ecosystem. NEP represents the net carbon exchange between the ecosystem and the atmosphere, including the costs of respiration by plants, heterotrophs, and decomposers. Changes in photosynthesis affect atmosphere composition (Schimel et al. 2000), fresh water availability (Postel et al. 1996), and biodiversity (Sala et al. 2000). Humans already appropriate a large amount of the products of photosynthesis (Vitousek et al. 1986) by transforming between one third and one half of the planet's land surface (Vitousek et al. 1997). Despite the large uncertainty in these global figures (Rojstaczer et al. 2001), many show that land conversion during the last two centuries has played a major role (Vitousek et al. 1997, DeFries et al. 1999).

Urbanization is a major driver of land conversion. Most importantly, it takes place on the most productive lands (Imhoff et al. 1997). Using the night lights footprint derived from DMSP/OLS satellite images of a digital soils map, Imhoff et al. (1997) estimate that urban areas occupy approximately three percent of the land area of the continental United States. But they also indicate that most of urbanization is taking place on the best soils—with the fewest limiting factors. Urbanization in the US occupies, respectively, 6%, 48%, 35%, and 11% of the land in the categories of high-, moderately high-, moderate-, and low-soil productivity. The change in NPP due to urbanization differs from region to region based on the ecosystem surrounding a city, but overall urban land transformation in the US has reduced the annual NPP by 0.04 Pg C, or 1.6 percent of its pre-urban NPP value (Imhoff et al. 2002).

At local and regional scales, urbanization may increase NPP in resource-limited regions (Imohff et al. 2004). Imhoff et al. (2004) show that through localized warming, "urban heat" extends the growing season in cold regions and increases the winter NPP. They also show that this seasonal carbon flux is dependent on climate and other limiting factors. The localized increase in NPP however is not sufficient to offset the overall negative impact of urbanization on NPP (Imhoff et al. 2004). The localized effect of urbanization pattern on NPP, however, can influence species richness. Several studies found that net primary productivity is highest at the urban fringe, not only because of low-density development in such locations, but also due to resource inputs in highly managed green spaces (Shochat et al. 2006). Most urbanizing regions show a hump-shaped species richness—productivity relationship along urban gradients (Blair 1996, Allen and O'Connor 2000, Pei-Fen Lee et al. 2004). This relationship, however, can be expected to vary across cities with different biophysical settings and development patterns (Shochat et al. 2006).

Urban areas influence NPP not only directly, but also indirectly, by demanding and appropriating natural resources from distant regions. Rees and Wackernagel (1994) have proposed the term "ecological footprint" to quantify such impact. The ecological footprint is the ecologically productive area required to provide the ecological services that support the human population (Folke et al. 1997, 1998, Young et al. 1998, Jansson et al. 1999). The ecological footprints of metropolitan Toronto and Vancouver, for example, are respectively estimated at 181,260 (Onisto et al. 1998) and 36,344 square kilometers (Rees 1995), which correspond to about 300 times their nominal areas.

The ecological footprint is useful in describing the impact of the human population on Earth's ecosystems, but it does not allow us to specify how different patterns of urban development affect primary productivity. A spatially explicit definition of the ecological footprint would be necessary to capture how clustered versus dispersed urban forms affect the "ecological footprint" of alternative urban settings. Furthermore, since natural resources are typically not distributed uniformly across landscapes, Luck et al. (2001) used a spatially explicit approach to show that both the location of urban areas and inter-urban competition may play a crucial role in determining the magnitude of the ecological footprint.

## 3.3 Hydrological Function

Urbanization affects hydrology by altering the water supply through water extraction and contamination, and by modifying the water flow through changes in land cover and the building of infrastructure (See Chapter 5). Cities import large quantities of water and produce large quantities of wastewater. On average, households account for less than 10% of the total water usage in industrialized countries, but most water demand is concentrated in urban areas where it is needed for services, industries, and food production. In arid regions, the amount of water imported for irrigation can exceed the amount of precipitation, and thus can dramatically modify patterns of terrestrial primary productivity and nutrient cycling (Grimm et al. 2005, Kaye et al. 2006). Moreover, in urban areas, the water infrastructure can be old and leaky, increasing water consumption. Municipal sewage systems are potentially subject to discharges from combined sewer-stormwater overflows that can affect groundwater and streamflow levels, and contaminate surface waters with pathogens and increased concentrations of nutrients (Paul and Meyer 2001, Kaye et al. 2006).

Urbanization affects hydrological processes by altering surface runoff. As more surfaces are made impervious, surface runoff increases, lowering the amount of infiltration (Gregory 1977, Dunne and Leopold 1978, Booth and Jackson 1997). In many natural ecosystems, it is estimated that about 90% of water flows from uplands to streams by subsurface flow. In urbanizing basins, an increase in impervious surfaces of 10% to 20% of the total surface area may double surface runoff compared with that in a forested area. It may shorten the lag times between precipitation input and discharge, and may also generate higher flood peak discharges during storms (Paul and Meyer 2001). The impact varies with regional climate and geomorphology. Indeed, while this is typical in a city in a humid region, a city in an arid region has naturally low ratios of infiltration to runoff and short lag times, and hydrographs might not show such dramatic shifts (Kaye et al. 2006).

Hydrological function in urbanizing regions is simultaneously controlled by the landscape processes existing prior to development and the emerging biophysical template created by human settlement. Watershed characteristics such as climate, topography, geology, and land cover pattern affect the resilience of the hydrological function. The frequency of storm events, size and slope of drainage basins, and amount and pattern of vegetation, all influence the amount, route, speed, infiltration, and retention of water, and its final delivery to streams.

In urban areas, stormwater drainage infrastructure is primarily built to convey runoff rapidly to stream channels and mitigate the effects of increased runoff (Hammer 1972, Booth and Jackson 1997). What is built with the objective of minimizing urban impact on hydrological processes, may have the opposite effects downstream. In fact increased hydrological connectivity through stormwater drainage means shorter time for water to reach the stream, causing increased peak flows (Leopold 1968, Bailey et al. 1989, Booth et al. 2004). Less water retention by reduced forest cover implies that floods occur more quickly and more frequently (Konrad 2000, Konrad and Booth 2002). Urban basins are more flashy than non-urban ones (Konrad 2000, Konrad and Booth 2002, Booth et al. 2004), and stream flow is highly variable (Konrad 2000). On the other hand, base flow in urbanizing basins is relatively lower, but research is inconclusive with respect to base flow trends (Dunne and Leopold 1978, Burges et al. 1998, Konrad and Booth 2002).

It is increasingly evident that hydrological processes are important mechanisms that link urban patterns to stream conditions since more aggregated patterns of impervious surfaces may increase runoff, and more fragmented patterns of forest make vegetation less capable of intercepting surface-water runoff. As I show in Section 3.7, differences in 1) the configuration of impervious area and forest patches, 2) the distance between locations, 3) the degrees of connectivity, and 4) the types of hydrologically distinct *non*-impervious land cover can explain the variability in a stream's biotic conditions (when controlling for land-cover composition). Varying urban form along the continuum of "clustered versus dispersed development" is most likely to affect aquatic ecosystems depending on the degree to which the upland watershed surface is pervasively dissected by new, artificial conduits of surface-water drainage (i.e., roads).

## 3.4 Nutrient Cycles

As forests are converted to developed (i.e., urbanized) land, many processes are affected: nutrient cycling, soil erosion, hydrological flow, and climate. Disturbances that change biogeochemical movements and transformations in urban environments have been studied extensively (Chapter 6). Several studies have pointed to new sources and pathways of nutrients across the urban landscape (Newcombe 1977). Humans affect biogeochemical cycles through hydrology, atmospheric chemistry, climate, nutrients, vegetation composition, and land use. But in urban ecosystems these drivers are mediated by three major patterns of human activity including the built infrastructure, urban demographics, and household activities and life styles (Kaye et al. 2006). Recently Baker et al. (2001) tracked the human input of nitrogen in the Phoenix metropolitan area, and have indicated the important role of fertilizers, human food, fuels, and nitrogen oxides produced by fossil fuel combustion. Furthermore, landscape position and spatial patterns affect flows—both horizontal (i.e., nutrients in surface water) and vertical (carbon exchange between the atmosphere and biota). But we know far less about the effect of landscape structure on the redistribution of material and nutrients.

We do know that urban development affects plant–environment interactions and vegetation functions, and that the urban forest influences the microclimate, the atmospheric concentration of pollutants, and the local carbon storage fluxes (Jo and McPherson, 1995). McPherson et al. (1994) estimated that in 1991 the tree cover in Chicago removed 17 tons of CO, 93 tons of $SO_2$, 98 tons of $NO_2$, 210 tons of $O_3$, and 234 tons of (less than 10 micron) PM (particulate matter). These trees also store 942,000 tons of carbon. In addition, a strategically located tree can save up to 100 kWh/year in electricity used for cooling (McPherson and Simpson 2002, 2003). Among other important ecological functions, the urban forest mitigates storm-water runoff and floods. No less important is the role urban vegetation plays in providing critical aesthetic values and community well-being. While the literature substantiates the hypotheses that urban patterns affect plant communities and vegetation functions in urban ecosystems, scholars have not yet determined how different spatial urban structures influence the ecosystem function of the urban forest.

Several studies have indicated that the percent of impervious surface in an urban watershed is a good predictor of its health (Paul and Mayer 2001). Fish and macroinvertebrates have been used to compare the biotic integrity of streams exposed to various degrees of urbanization in watersheds. The two taxonomic groups are used to measure both the biotic diversity and the impacts of pollution on species. Evidence from current studies documents the relationship between land use/land cover and biotic integrity. Because biophysical and biological processes influence fish–stream dynamics, land use activities cause related alterations in fish populations and communities (Schlosser 1991). But while fish reflect conditions on a large scale, macro-invertebrates may better reflect local ecological conditions. In addition, later studies hypothesize that local land use and habitat variables are better predictors of biotic integrity than is regional land use (Richard et al. 1996, Lammert and Allen 1999). In a recent study, summarized at the end of this chapter (section 3.7) and in Chapter 5, my research team in Seattle established that the spatial distribution of the impervious area in urban water sheds and its connectivity to the channel affect the hydrologic response and thus their bio-logical conditions (Alberti et al. 2007). Alternative land-use and land-cover patt-erns associated with urbanization have different effects on aquatic ecosystems.

## 3.5 Biodiversity

The debate on the importance and role of diversity in the functioning and stability of ecosystems becomes even more relevant in the context of increasing human-induced transformation and urbanization. I discuss the relationship between urban development, biodiversity, and ecosystem function in Chapter 8. Ecological studies have provided ample evidence that different species perform diverse ecological functions (i.e., nutrient cycling, trophic regulation, pollination, dispersal, and disturbance). There is also substantial evidence that functional diversity depends on species richness. The question is the extent to which species richness affects stability (Tilman et al. 1996).

*Fragmentation* of natural patches is one of the best-known impacts of human activities on the diversity, structure, and distribution of vegetation

(Leverson 1981, Ranney et al. 1981, Brothers and Spingarn 1992). Ecologists have described its opposite quality—*connectivity*—as a critical property of landscapes, one that facilitates or limits the movement of resources and organisms among natural patches (Turner and Gardner 1991, Tischendorf and Fahrigh 2000). Urban growth affects connectivity directly by modifying the landscape and indirectly by changing biophysical patterns and processes. Ecological studies have established relationships between landscape structure and the distribution, movement, and persistence of species. Although the effects of alternative urbanization patterns on plants are still not fully understood, it is known that converting natural or rural landscape into an urbanized landscape reduces the diversity of native plant species in the urbanized region. The *edge effect* has also been studied, particularly in forests (Ranney et al. 1981, Harris 1984, Brothers and Spingarn 1992, Murcia 1995, Cadenasso et al. 1997). Because forests are primarily vertical in structure, the removal of vegetation and the consequent exposure to natural and human disturbances have important consequences on the structure and composition of plant communities.

Based on the physical changes observed on the urban-to-rural gradient (Pickett et al. 1997), Mckinney (2002) describes a biodiversity gradient with species richness declining from the urban fringe towards the urban core. In the transition from the rural towards the urban core, more and more habitat is lost; it is also replaced by remnant, ruderal, and managed vegetation and built habitat that have varying degrees of inhabitability for most native species. Species composition along this gradient is characterized by urban exploiters dominating the urban core, urban adapters dominating suburban areas, and urban avoiders dominating the urban fringe (Blair 2001, Mckinney 2002).

## Avian communities

Birds are excellent indicators of the effect of urbanization on ecosystems: they respond rapidly to changes in landscape configuration, composition, and function. Urbanization affects birds directly through changes in eco-system processes, habitat, and food supply, and indirectly through changes in predation, interspecific competition, and diseases (Marzluff et al. 1998). The percentage of land cover that is vegetation is in fact a good pre-dictor of the number of bird species. Urbanization increases the number of introduced species and drastically reduces the number of native species (Marzluff 2001), thus altering the composition of urban avian communities. Populations of native species decline because their natural habitats are reduced, and they cannot tolerate human disturbances (Beissinger and Osborne 1982, Blair and Walsberg 1996).

Several scholars have started to explore how patterns of urbanization affect bird survival (Beissinger and Osborne 1982, Bolger et al. 1997, Rolando et al. 1997, Marzluff et al. 1998, Marzluff 2005). They have documented how urbanization modifies the composition of urban avian communities through changes in climate, abundant food and water supply, increased nest sites, and smaller predators. Few studies have looked directly at the effects of urban patterns. Instead they have investigated how habitat fragmentation creates edges, reduces vegetative cover, and introduces changes in food supply, nest placement, and predation.

Beissinger and Osborne (1982) compared the avian community of a mature residential area in Oxford, Ohio, with two control sites in Hueston Woods State Park. The urban community supported nine fewer species than the forest, a difference explained primarily by vegetative cover and habitat patchiness. In a study of breeding bird diversity and abundance in Springfield, Massachusetts, Tilghman (1987) found that woodland size is the most important single variable explaining the number of bird species.

A more direct question related to habitat fragmentation in urban areas was addressed by Soulé et al. (1988) in their study of chaparral-dependent birds in 37 fragments of native chaparral habitat in canyons of coastal, urban San Diego. Focusing on the effect of isolation on species diversity, they found that four variables could explain 90% of the variation in species richness across the fragments: canyon age, total area of chaparral, total area of canyon, and predation. In addition, they found that the absence of coyotes in urban-ized environments allowed greater numbers of other avian predators and grey foxes. By eliminating large predators, urbanization offsets the capacity of large predators to control small predators, and thus increases the impacts of small predators on birds. Cats and other domestic pets also influence the bird population in suburban areas (Churcher and Lawton 1987).

Studies of the effect of fragmentation on birds are extensive and provide evidence that urbanization has various effects on diversity and reproduction in avian communities. However, it is unknown how variations in the con-centration, land-use intensity, heterogeneity, and connectivity of urban development influence the abundance and community diversity of birds and their chances of reproduction and survival.


## Fish and macro-invertebrates

Fish and benthic invertebrates are useful indicators of biological conditions of aquatic ecosystems and of human impacts (Karr 1981). Initially developed for fish, the Index of Biotic Integrity framework (IBI) has been applied to benthic invertebrates and extensively applied to study the impact of human induced impact on streams.

Urbanization causes riparian change. Loss of riparian habitats reduces the ability of watersheds to filter nutrients and sediments (Karr and Schlosser 1978, Peterjohn and Correll 1984). Clearing of riparian vegetation and replacing deep-rooted vegetation result in a reduction in important stabilizing elements, such as woody debris, that provide the stream channel an effect by dissipating flow energy and protecting the channel from erosion (Booth et al. 1997). Furthermore, the loss of overhead canopy eliminates the shade that controls temperature and supplies leaf litter to the aquatic food chain. All these hydrologic, geomorphic, and biological changes in urban streams are associated with an increase in impervious surface (Booth and Jackson 1997, Yoder et al. 1999). Previous studies of the impacts of urbanization on ecological systems have typically correlated changes in ecological conditions with simple aggregated measures of urbanization (e.g., human population density or percent impervious surface). They show that the composition of land cover within a watershed can account for much of the variability in water quality (Hunsaker et al. 1992, Charbonneau and Kondolf 1993, Johnson et al. 1997) and stream ecology (Whiting and Clifford 1983, Hachmoller et al. 1991, Thorne et al. 2000, Morley and Karr 2004). Yet such correlations generally make no allowance for the location of that impervious area within the watershed, its proximity to the stream channel, or its spatial configuration. While several studies have addressed the relationship between watershed urbanization and biotic integrity in streams, few have directly addressed the question of how urban patterns influence ecological conditions and biodiversity.

## 3.6 Disturbance Regimes

Urban landscapes exhibit rich spatial and temporal heterogeneity. The urban landscape is a complex mosaic of biological and physical patches within a matrix of infrastructure, human organizations, and social institutions (Machlis et al. 1997). The spatial heterogeneity of landscape features is typically characterized by sharp boundaries, mostly as the result of human activities. Spatial and temporal heterogeneity within an urban landscape has both natural and human sources. Natural sources include the physical environment, biological agents, disturbance regime, and stresses (Pickett and Rogers 1997). In the urban landscape, human sources of heterogeneity include the introduction of exotic species, the modification of landforms and drainage networks, the control or modification of natural disturbance agents, and the construction of massive and extensive infrastructure (Pickett et al. 1997).

In urban areas, natural disturbance regimes, along with the introduction of invasive species, have altered natural succession. Urban development leads to several changes in disturbance regimes at multiples spatial and temporal scales (Rebele 1994). First, urban development rescales natural disturbances by reducing or increasing their magnitude, frequency, and intensity. It also rescales areas by introducing biogeographic barriers (roads, canals, etc.) and reducing the patch size of natural vegetation. In addition, it introduces new disturbances, chronic stresses, and unnatural shape complexity or degrees of connectivity. Furthermore, changing land use results in changes in patch structure and integration that homogenize natural patterns and modify the natural processes that maintain biodiversity. Rebele (1994) notes that human-induced disturbances depend on economic, political, and social factors, which means that in cities disturbances are very difficult to predict.

Changes in land cover and land use can cause significant alteration of natural disturbance regimes and drive fundamental changes in both biodiversity and ecosystem function. Housing development, road building, urban wastelands, and landfills are only some of the most obvious sources of disturbance in urbanizing regions. In addition, changes in microclimate, hydrological patterns, geomorphology, soil conditions, and habitat indirectly modify natural disturbance regimes. However, as is true for the other categories of effects addressed above, the existing literature does not examine how various patterns of housing and roads influence the extent, distribution, intensity, and frequency of disturbances.

## 3.7 An Empirical Study in Puget Sound

To illustrate the direction of research that links urban patterns to ecosystem function I draw on an empirical study developed at the University of Washington Urban Ecology Research Laboratory: the impact of urban patterns on ecosystem dynamics (Alberti et al. 2007, Alberti and Marzluff 2004, Marzluff 2005). The study empirically explores relationships between urban patterns and ecological conditions in the Puget Sound region. We developed and tested formal hypotheses about how patterns of urban development produce changes in biophysical processes that affect bird communities and aquatic macroinvertebrates, and about what factors determine and maintain an urban ecological gradient.

We investigated four questions:

1. How do variables describing urban landscape patterns vary on an urban gradient?
2. Which pattern metrics best describe the composition and configuration of urban landscapes?
3. What is the relative importance of individual pattern metrics in predicting changes in ecological conditions?
4. At what spatial scales are various ecological processes controlled in urban landscapes?

We had four overarching hypotheses:

1. Urban spatial patterns can be described along distinct dimensions that represent relationships between biophysical (land cover) and socio-economic variables (land use).
2. Urban landscapes are complex patterns of intermixed high- and low-density built-up areas that can best be described using a series of pattern metrics that link urban development patterns to ecological conditions.
3. By including patterns of urbanization we can improve the predictive ability of models that currently relate aggregated measures of urbanization to ecological processes.
4. The predictive ability of a model that relates a pattern variable to an ecological process varies with change in spatial scale (resolution and extent).

We conducted three major research activities: 1) a pattern analysis of land use and land cover, 2) a study of how these patterns affect aquatic macroinvertebrates, and 3) a study of how patterns affect avian communities. We hypothesize that biotic integrity in urban stream is mediated by changes in hydrology and runoff driven by an increased amount and connectivity of the impervious surface (Figure 3.6). We further hypothesize that decline in bird diversity is driven by reduced habitat and change in food sources, nest predation, and species competition (Figure 3.7).

We started by quantifying the landscape signatures of alternative develop-ment patterns in the Puget Sound metropolitan region using a series of land-use and land-cover pattern metrics. Land-use data at the parcel level were obtained from the assessor offices in King County and Snohomish County. Land-cover data were interpreted from Landsat Thematic Mapper (TM)

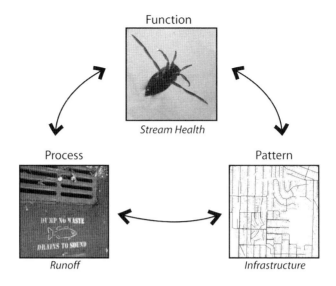

Figure 3.6. Linking macroinvertebrate diversity to stormwater infrastructure through hydrological changes.

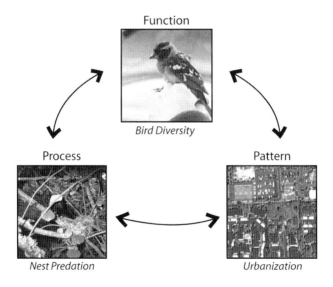

Figure 3.7. Linking bird diversity to urban sprawl through nest predation.

imagery for the Puget Sound region for 1998. The procedure creates an eight-class land-cover system that discriminates among three classes of urban land cover, characterized by varying levels of impervious surface and vegetation coverage. These are: *paved urban* (approximately 100% paved cover), *grass/shrub urban* (characteristic of newer suburban areas with limited tree canopy and relatively large lawn coverage), and *forested urban* (characteristic of mature residential neighborhoods with a high degree of canopy cover). In addition the procedure discriminates three types of non-urban vegetation (*grass/shrubs/crops, deciduous forest,* and *coniferous forest*) and *water* (Hill et al. 2002).

We applied six landscape metrics to measure urban landscape patterns: percent land, mean patch size, contagion, Shannon index, aggregation index (AI), and percent of like adjacencies (PLADJ). Details on the spatial metrics and methodology are in Alberti et al. (2007). Percent Land is the sum of the area of all patches of a given land use/cover type divided by total landscape area. Mean Patch Size (MPS) is the sum of the areas of all patches divided by the number of patches. The Shannon diversity index represents the number of land use classes in the landscape. Contagion, AI, and PLADJ all measure various aspects of aggregation of the land cover.

We used these metrics to discriminate six land-development types: single-family residential (SFR), multi-family residential (MFR), mixed use, commercial, office, and open space. We found that the best discriminants for the six types were percent paved land cover, percent mixed urban land cover contagion, and aggregation of land cover. We then used these landscape patterns to assess the relationships between urban development and ecological conditions, specifically aquatic macroinvertebrates and avian diversity.

To establish the relationships between urban patterns and aquatic invertebrates, we delineated 42 sub-basins with varying degrees of urbanization from 42 points with an associated *Benthic Index of Biological Integrity* (B-IBI) value. B-IBI is an index of biotic integrity developed by James Karr at the University of Washington. We chose basins that were no larger than 5 $km^2$, and developed five scales of analysis for investigation with each spatial metric. From large-scale analysis to small-scale analysis, these scales were: basin-wide, 300 m riparian zone, 200 m riparian zone, 100 m riparian zone, and local riparian zone.

The study of the impact of urban patterns on avian diversity was based on 54 km$^2$ study areas randomly selected in the Puget Sound region. We stratified the area by dominant land cover (forest, urban, urban forest), mean size of urban patches, and pattern of forest-settled area contagion. We restricted our selection to low elevation sites (<500 m) dominated by coniferous forest. (Details of metrics and selection approach are in Alberti 2001, Donnelly and Marzluff 2006). At each study site we measured the relative abundance of birds during one breeding season (2000–2001).

We described some of the complex relationships—between land use and land cover in urban landscapes—revealed by the distributions of land cover across parcels of differing land uses (Alberti and Marzluff 2004, Alberti 2005). Using various percentages of land-cover types, we were able to discriminate between the different land use parcels. Our study showed that single-family residential parcels have significantly lower amounts of impervious surface than do multi-family parcels. We also found a large percentage of impervious surface on mixed-use (i.e., combined residential and commercial activities) and industrial parcels. As we expected, a high percentage of forest cover is found in single-family residential areas, but this percentage drops significantly for other development types. Our results also show that land-cover composition is highly variable within the same land-use types. We found that three factors—the parcel size, the location of the parcel along an urban-to-rural gradient, and the year built—significantly influence the distribution of land cover within land-use types. More importantly, our results show that land development types will generate different level-of-fragmentation signatures, and that different land use scenarios can preserve different amounts of natural land cover.

Then, using multi-regression models we established empirical relationships in the selected sub-basins between the metrics of landscape patterns and a series of aquatic ecosystem stressors (Alberti et al. 2007). Our study clearly indicates that both the amounts of impervious surface and the patterns of urban development and roads are correlated with ecological conditions. The best individual predictors of B-IBI are the number of roads crossing the stream and the road density. We also showed that landscape configuration—measured as mean patch size, aggregation, and percent adjacency of urban and forest patches—explains the variability of B-IBI not explained by percent of impervious area. Metrics of landscape configuration including mean patch size of urban land cover and aggregation index of both forest and urban land cover are good predictors of biotic integrity in streams. Percentage total impervious area (TIA) and percentage forest cover are both highly correlated with B-IBI. The results of this study are described in more detail in Chapter 5.

Findings for the avian study indicate that fewer species occur within forest fragments than in settled areas (Alberti and Marzluff 2004, Marzluff 2005). The amount of forest patch in the developed area is significantly correlated with bird diversity: the number of bird species increases with an increasing amount of forest, while the arrangement of forest in the area is less important than the total amount. Bird diversity remained high in the settled Puget Sound region when the percentage of forest in each 100 ha (hectare) unit remained at approximately 30% or more. In our study sites this happened despite variation in forest connectivity—that is, the degree to which forest and settlement are interspersed, as measured by the forest aggregation index. It is not surprising that connectivity mattered less to birds in a region with a vast forest matrix. I discuss these results further in Chapter 8.

In this chapter I have suggested that explicitly linking urban patterns to ecosystem function is critical in advancing urban ecological research and developing strategies to minimize the impacts of urban growth. Current ecological research already provides increasing evidence of the impact of urbanization on ecosystem function. However, this research simplifies the consideration of urban structures to such an extent that the results are not useful to urban planners and managers. Furthermore, it fails to recognize the complex interactions between the urban patterns and ecological processes that occur across multiple scales. To understand how species populations and community characteristics change in response to urban development, we need to expand our knowledge about the drivers and the effects of ecosystem structure and functions on the urban landscape.

Building on the existing evidence provided by the urban planning and landscape ecology literature, it is possible to articulate testable hypotheses on the mechanisms that link urban patterns to ecological function. In particular, we can start to systematically test hypotheses linking urban development patterns to patch structure in urbanizing landscapes, and the consequences for primary productivity, biodiversity, nutrient and material cycles, and disturbance regimes. We can ask, for example, what degree of contagion or dispersion of urban structure best allows urban landscapes to maintain the integrity of patch structure. We can also investigate how land-use intensity and urban patterns interact to affect ecological conditions. For example, we can investigate how a modification of the landscape structure (i.e., the amount of impervious surface and vegetation) at a subwatershed scale interacts with local effects of land use on the riparian zone. Moreover, we can establish what roles transportation and infrastructures for artificial drainage of surface water play in the overall impact.

It is clear from the current knowledge, however, that the interactions between urban economic, social, and ecological processes are extraordinarily complex. Interactions between urban patterns and ecosystem function are controlled by multiple stressors. In future research, we will need to investigate 1) relationships between urban patterns and human-induced stressors, 2) interactions among multiple stressors associated with these patterns, and 3) whether thresholds exist in these relationships and what they are.

Results from current research also indicate that these human-ecological interactions are process-specific. In order to assess the impacts of alternative patterns of urban development and determine their tradeoffs, we need to consider the different roles that diverse species play in ecosystem processes. We also need to consider that dynamic interactions between urban patterns and ecosystem function occur at multiple spatial and temporal scales. The concepts of target species and functional diversity provide a new framework for studying the impact of urban patterns on ecosystem function, and for designing more effective conservation strategies. In particular, the concept of target species suggests that it is important to establish what degree of redundancy—multiple species per functional group—is necessary if we are to make urban ecosystems sustainable over the long term.

# Chapter 4

# LANDSCAPE SIGNATURES

## 4.1 Hybrid Urban Landscapes

Urban landscapes exhibit some fundamental features of complex systems. They are open, nonlinear, and highly unpredictable (Hartvigsen et al. 1998, Levin 1998, Portugali 2000, Folke et al. 2002, Gunderson and Holling 2002). Furthermore, they are highly heterogeneous, spatially nested, and hierarchically structured (Wu and David 2002). Disturbances are frequent—an intrinsic characteristic of such systems (Cook 2000). Many factors cause these disturbances, which can follow multiple pathways as they develop and depend greatly on historical context; that is, they are path-dependent (Allen and Sanglier 1978, 1979, McDonnell and Pickett 1993). As other complex systems, urban landscapes are made up of many interacting heterogeneous components, and these interactions lead to emergent patterns that cannot be predicted from an understanding of the behavior of the individual parts. The evolution of such landscapes is in itself an emergent property. Furthermore, uncertainty is important, since any change that departs from past trends can affect the trajectories of landscape dynamics.

What makes urban landscapes particularly complex is that they are hybrid phenomena emerging from interactions between human agents and ecological processes (Forman 1995, Spirn 1998, Bell 1999, Wu and David 2002, Alberti et al. 2003, Cadenasso et al. 2006). Consider, for example, the patterns of natural and built elements in an urbanizing watershed. They reflect simultaneously pre-existent biophysical factors, such as land cover, geomorphology, hydrology, climate, and natural disturbance regimes. At the same time, they reflect the decisions regarding land and infrastructure development of multiple human agents (both individuals and organizations) who interact in economic markets and public institutions (e.g., governments). Both the emerging natural and built elements of the landscape, in turn, affect human and ecosystem functions.

An urban stream and a highway in an urban watershed are driven by different processes and perform different functions, but the emerging road and stream networks in urban landscapes share their hybrid nature. An urban stream channel is the result of both its natural geo-morphological processes and the human action of straightening, deepening, and widening the stream to increase drainage and prevent flooding (Klein 1979, Booth and Jackson 1997). The highway corridor is primarily driven by decisions and socio-economic objectives to maximize accessibility while minimizing transportation costs, but accessibility and cost of infrastructure are ultimately influenced by topography and pre-existing land cover. Both the emerging urban stream morphology and highway corridor reflect complex interactions between land cover and land use processes, and ultimately affect the respective ecological, hydrological, and transportation functions that streams and road infrastructure perform in the urban landscape.

Patterns of urban landscapes can be detected and measured using a variety of spatial analysis approaches (Turner et al. 1989, Alberti et al. 2001, 2005, Wu 2004, Cadenasso et al. 2007), but decoding them and understanding the underlying causes is a much more difficult endeavor. In this book I argue that we cannot understand the structure, function, and evolution of urban landscapes if we focus separately either on the built or natural component. The composition and configuration of urban landscapes reflect coupled human and ecological processes and their interaction over space and time. To address the inherent complexity of urban landscapes, many complementary theoretical frameworks need to be integrated. In the following sections I propose that gradient analysis, patch dynamics, networks, and hierarchy theory can be integrated to identify emerging patterns, and to test hypotheses about the underlying processes at multiple scales and hierarchical levels.

Landscape patterns contain information about essential processes and structures occurring at multiple scales and levels. But before we can test hypotheses about mechanisms that govern their dynamics, we need to detect and accurately quantify landscape pattern and its change over time. Two major questions are: What dimensions of pattern should be measured, and at what scale? Landscape studies have indicated distinct spatial patterns occurring at multiple scales, as a function of the dominant underlying processes (Kolasa and Pickett 1991, Wu and Loucks 1995, Cullinan et al. 1997, Werner 1999). This has important implications for the scale of observation, and emphasizes the importance of characterizing spatial pattern and processes at multiple scales and hierarchical levels. Using a strategy called pattern-oriented modeling (Grimm et al. 2005), we can explore alternative models of landscape dynamics by simultaneously testing hypotheses about underlying structures and processes operating at multiple scales and hierarchical levels.

## 4.2 Gradients, Patches, Networks, and Hierarchies

Scholars from various disciplines have made important progress in modeling coupled human-ecological systems, but they have not formally tested hypotheses about the interacting emergent behaviors of human and biophysical agents. Simulating emergent behaviors in ways that reasonably capture patterns observed in urban landscapes remains a significant research challenge. One major problem in modeling urban landscape dynamics is explicitly representing the human and biophysical agents at a level of disaggregation that allows us to explore the mechanisms linking patterns to processes (Portugali 2000). A second challenge arises as we attempt to model the interactions between human and natural systems: many factors operate simultaneously at different levels of organization. Simply linking these models in an additive fashion may not adequately represent the behavior of a given system, because interactions may occur at hierarchical levels that are not represented (Pickett et al. 1994). Additionally, since urban landscapes differ widely, changes in driving forces may be relevant only at certain scales or in certain locations (Turner et al. 1995), but we currently have limited understanding of the interactions between spatial scales.

To simulate the behavior of urban ecological systems, we must explicitly consider the temporal and spatial dynamics of these systems, and also identify the interactions between human and ecological agents across the different temporal and spatial scales at which various processes operate. Gradient analysis, patch dynamics, network theory, and hierarchy theory provide different perspectives for studying urban ecosystems (Figure 4.1). Scholars in ecology have applied *gradient analysis* to understand the relationships between environmental variation in factors such as elevation, soil moisture, and salinity and ecosystem functioning (Whittaker 1967, Pickett and Bazzaz 1976, Austin 1987, Vitousek and Matson 1990). The theory of *patch dynamics* provides a framework to address spatial heterogeneity and to explicitly represent the structure, function and dynamics of patchy systems (Levin and Paine 1974, Wu and Levin 1994, 1997, Pickett and Rogers 1997, Pickett and White 1985). *Network theory* reveals the underling patterns and topology of interactions in ecosystems and the implications for the systems' stability and resilience (Green and Sadedin 2005). *Hierarchy theory* provides a theoretical framework for modeling complex systems conceptualized as nested organizational hierarchies that emerge from the interactions among system components (Simon 1962, 1973, Allen and Starr 1982, O'Neill et al. 1986, Ahl and Allen 1996, Wu 1999).

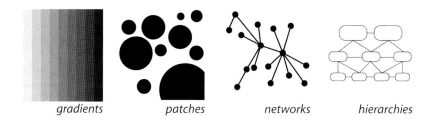

gradients          patches          networks          hierarchies

Figure 4.1. Approaches of urban ecological studies.

## The urban ecological gradient

The gradient paradigm (Whittaker 1967) provides a useful framework for developing and testing hypotheses about the range of interactions between urban development and ecological processes. Gradient analysis can explain the ecological conditions associated with human disturbances in urbanizing environments; it can also reveal thresholds in the biotic and ecological responses to human-induced stresses. The interactions among human factors, and between them and biophysical factors, make urban-to-rural gradients potentially quite complex.

Gradients have been applied to study the human impact on ecosystems. Forman and Godron (1986) hypothesized that predictable landscape patterns can be observed along a gradient of human-induced changes—from urban to suburban, cultivated and managed to natural. They also hypothesized that patch density increases exponentially with the degree of human change. That is, patches exposed to more human impact are also likely to be smaller, more uniform in structure, and less connected.

The gradient paradigm has also been proposed as an effective approach to studying the urbanizing landscape. McDonnell and Pickett (1990) suggest that ecological conditions in urbanizing regions can be systematically analyzed by quantifying changes in ecosystem structure and function in relationship to varying levels of urbanization. They conceptualize the metropolitan landscape as a complex gradient of urban effects on ecosystem function (McDonnell and Pickett 1990, McDonnell et al. 1993, Medley et al. 1995). Using indirect multivariate gradient analysis to characterize the complex urban-to-rural gradient of the New York metropolitan area, McDonnell et al. (1997) reveal a new array of virtually unexplored conditions produced by urbanization. They find that the unique ecosystem structures and functions of urban forests are significantly impacted by urban stresses such as air pollution and elevated levels of heavy metals in the soil, as well as the positive effects of the heat-island phenomenon and the presence of earthworms.

Formal hypotheses about the relationships between urban gradients and ecosystem functions have been developed, and a number of studies have been conducted or are underway to empirically test them in diverse urban regions. A stylized representation of a number of the hypothesized relationships is represented in Figure 4.2. Scholars in ecology have applied gradient analysis to study microclimate (Ziska et al. 2004), nutrient loads (Groffman et al. 2004, Wollheim et al. 2005), plant distribution (Sukopp 1998, Porter et al. 2001 Burton et al. 2005), stream health (Limburg and Schmidt 1990, Walsh et al. 2005), avian diversity (Blair 1996) and richness (Jokimaki and Suhonen 1993, Marzluff 2001, Crooks et al. 2004), and a variety of ecosystem properties (McDonnell and Pickett 1990, Pouyat and McDonnell 1991, Pouyat et al. 1995, McDonnell et al. 1997, Zhu and Carreiro 1999) in relation to urban gradients.

Describing human-induced disturbances associated with urbanization, however, is a much more complex undertaking. We cannot always place landscape states on a simple continuum of population and/or built up densities because humans affect the landscape through multiple stressors operating at multiple scales, causing both habitat loss and habitat modification. To address this issue, McIntyre and Hobbs (1999) propose a conceptual model linking land use and landscape alteration that takes both factors into account. They categorize the degree of alteration as ranging from relictual to fragmented to variegated to intact; they also identify two levels of landscape alteration. The first level defines the landscape's status, based on the level of habitat loss (i.e., fragmentation). The second level of analysis evaluates the patterns of habitat modification imposed on the remaining habitat (i.e., degradation). However, no operational framework or quantifiable metrics have yet been developed and empirically tested to represent such complex gradients.

Ecological researchers have tended to simplify the actual spatial pattern of disturbance in urbanizing landscapes to a few aggregated variables such as population density or built-up density, both of which are expected to change predictably with distance from the urban core. Cities are described as monocentric agglomerations with concentric rings of development surrounding a dense central business district (CBD). The assumption for current applications of gradient analysis is that, overall, urban-induced disturbances decline with distance from the urban core. For most US metropolitan areas, however, this is not accurate: over the past few decades

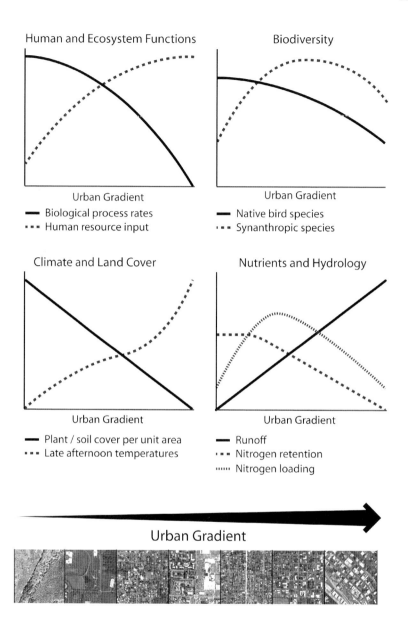

Figure 4.2. Urban gradient hypotheses. An illustration of hypothesized relationships between human and ecosystem functions and the urban-to-rural gradient.

they have changed from a monocentric to a polycentric structure (Gordon et al. 1986, Giuliano and Small 1991, Song 1992, Cervero and Wu 1995). Consequently, exposure to human influences is not necessarily a function of distance from a definable urban core. In addition, patterns of urbanization can cause variations in the interactions between human and biophysical processes.

This simplification makes it difficult to disentangle the factors associated with urbanization that generate ecological heterogeneity. Medley et al. (1995) describe the quadratic relationships between four factors: population density, land use heterogeneity, highway connectivity, and distance from the urban core. They were able to identify distinct characteristics of urban and suburban environments based on the landscape structure of native forest vegetation. They concluded that while urban exposure parallels the transect, the disturbances associated with urbanization show a complex spatial pattern not clearly related to distance from the urban core alone.

Urbanization results in a complex pattern, in which patches of high-density built-up areas are intermixed with patches of lower density. Alternative urban development patterns affect ecological processes both directly—by replacing native habitat with simplified, human-dominated systems, and indirectly—by rearranging the biophysical attributes that cause a variety of interrelated local and global effects (Godron and Forman 1982). Recent ecological studies have demonstrated that the ecological conditions of any patch are related to patch structure (Turner 1989): i.e., the size, composition, persistence, and interconnectivity of the patch. In turn, the structure is important to the survival of species that utilize those patches. Since ecological conditions are linked to patch structure and function, the landscape changes produced by land conversion and the pollution loads associated with patterns of urbanization have differential impacts on ecological conditions.

Gradient analysis may be a useful organizing framework to investigate how variations in ecological conditions are related to variations in exposure to urban disturbance, but to effectively represent the complex interactions associated with urbanization we need to explicitly represent the composition and spatial configuration of disturbances associated with urbanization (Alberti et al. 2001). I propose that individual factors such as population density or distance from the urban core cannot, taken alone, effectively describe the degree of impact of human stressors on ecological processes. Instead, I suggest that the interactions between human and biophysical processes generate complex gradients and discontinuities in ecological conditions along an urban-to-rural gradient.

Urban patch dynamics

Landscapes are mosaics of patches. Patches are discrete areas of relatively homogeneous environmental conditions that are relevant to a given organism or ecological phenomenon (McGarigal and Marks 1995). In urban landscapes, both human action and biophysical processes create discontinuities as a result of colonization processes, disturbance regimes, and succession. Landscape patches are modified and new patches emerge as humans and ecological processes interact. Thus we cannot simply study the effects of one set of processes on the other without considering the interactions and feedback mechanisms. Furthermore, any definition of a patch must consider scale: How closely will variation be measured? At what scales do both human and biophysical processes have an impact on landscape heterogeneity and connectivity, and how can those scales be identified?

The patch dynamics framework provides the theoretical foundation to explore the interactions among patches in urbanizing landscapes (Wu and Loucks 1995, Pickett et al. 1997). Patch dynamics focuses on the spatial structure, function, and change of relatively discrete elements in landscapes (Pickett and Rogers 1997). Pickett et al. (1997, 2001) propose a patch-dynamic approach to study the Baltimore urban ecosystem. They describe the urban landscape as a mosaic of patches that results from physical, ecological, and socio-cultural processes within a matrix of natural flows and built infrastructure (Machlis et al. 1997). A three-dimensional patch dynamic (physical, ecological, and socio-cultural) occurs in human-dominated ecosystems, generating a set of unique patterns and processes within and across scales (Pickett et al. 1997). Such complex relationships cannot be captured or explained by focusing on each individual process separately.

Scholars of both landscape ecology and urban morphology are interested in the causes and consequences of spatial heterogeneity, a measure of the patchiness of a given landscape (Turner 1989, Pickett and Rogers 1997, Gustafson 1998). But while landscape ecology studies look at how the spatial heterogeneity of landscapes affect energy and material fluxes, species distribution, and ecosystem functioning (Turner 1989), scholars of urban morphology study the form and structure of the built fabric, and the people and processes shaping them (Moudon 1997). Urban morphology has been primarily descriptive in its intent, but urban scientists studying the morphology of cities are interested in the relationship between urban pattern and function, in the same way that landscape ecologists are interested in ecological pattern and function. In spite of the similarities, however, landscape ecology and urban morphology have evolved separately, focusing on the opposite ends of an urban-to-rural gradient. The challenge for those studying the ecology of urbanizing landscapes is to fully integrate human and ecological processes in studying the causes of landscape patterns and their relationship to ecological functions. At the same time, those working in urban morphology will need to fully appreciate how biophysical processes cause the underlying structure of the built landscape and its relationship to human function (Alberti 2007). Attempts to understand the urban landscape will need to build on the theories that constitute the foundations of the two fields and integrate their approaches and methods.

## Networks

Landscapes can also be described as complex networks of human and natural agents connected by biogeophysical and socioeconomic processes. Understanding the topology of the interactions between components is an essential step in decoding urban landscapes and the relationship between pattern and functions. Graph theory has proven to be useful to explore the properties of diverse systems of interacting components. While initially networks were thought to have no apparent design principles and were represented by random graphs, scientists studying real networks have learned through empirical studies, models, and analytic approaches that they are far from being random (Barabasi and Albert 1999). Real networks display some organizing principles, which should be at some level encoded in their topology (Albert and Barabasi 2002).

To develop a theory of real networks, Albert and Barabasi (2002) suggest that we need to identify relevant parameters that, together with the network size, give a statistically complete characterization of the network. The emphasis on understanding complex networks also indicates that the description of topology needs to be complemented by a description of network dynamics (Albert 2006). To fully understand a complex landscape, we need to explore both the evolutionary path of the topology and how topology influences the landscape dynamic and functioning.

Network theory is emerging as a potentially very promising approach to uncovering general rules of complex landscapes (Andersson et al. 2006). Real-time data acquisition together with increased computing power are making it possible to investigate the topology and dynamics of networks containing millions of nodes, and to explore questions that could not be addressed before. But what has enormously contributed to the most recent advances of network theory is the crossing of disciplinary boundaries and the increasing access to data in many domains, which has helped scholars in many diverse disciplines to search for generic properties of complex networks (Barabasi 2005).

## Hierarchies

In hierarchy theory, complex systems are conceptualized as nested hierarchies in which individual domains (e.g., scales in landscapes) interact relatively weakly and infrequently with domains at a higher and a lower level, and more strongly and frequently within a level (Allen and Starr 1982). Domains in the hierarchy are separated by thresholds (abrupt changes in system processes) and different characteristic process rates and spatial extents (Meentemeyer 1989, Wiens 1989). Higher levels are characterized by slower rates and larger entities, whereas lower levels have faster rates and smaller entities (Wu 1999).

Urban landscapes can be described as near-decomposable, nested spatial hierarchies, in which hierarchical levels correspond to structural and functional units operating at distinct spatial and temporal scales (Reynolds and Wu 1999, Wu and David 2002). Near-decomposability is a key tenet of hierarchy theory that allows for diversity, flexibility, and higher efficiency in representing complex dynamic systems (Simon 1962, 1973). In applying hierarchy theory to urban landscapes, holons (horizontal structure) can be represented by patches (the ecological unit) and parcels (the economic unit). Patches and parcels interact with other patches and parcels, and between them at the same and at higher and lower levels of organization, through loose horizontal and vertical coupling (Wu 1999). In the urban landscape, the lowest hierarchical level and the smallest landscape spatial unit vary with socioeconomic and biophysical processes, from households and buildings to habitat patches or remnant ecosystems. At a coarser spatial scale, parcels and patches interact with each other to create new functional levels and units, such as neighborhoods or sub-basins. Neighborhoods and sub-basins initiate and are constrained by regional economic and biophysical processes. Since landscapes are nonlinear systems, they can simultaneously exhibit instability at lower levels and complex meta-stability at broader scales (Wu 1999, Burnett and Blaschke 2003).

To model the landscape as a hierarchical mosaic of patches (Allen and Starr, 1982) we can apply the approach proposed by Wu and David (2002) and link it to an agent-based approach (Parker et al. 2001, Waddell 2002). This allows us to break down the complexity of landscape dynamics by explicitly representing the hierarchies of interactions between human and biophysical agents, and multiple-scale patterns and processes, as well as top-down constraints and bottom-up mechanisms.

## Pattern-oriented modeling

The different theories presented provide accurate but partial views of landscape complexity. The quest for general principles calls for a unifying framework to uncover the underlying structure and mechanisms governing functioning of urban landscapes. Pattern oriented modeling (POM) is a strategy to decode complex patterns in systems in which multiple agents operate simultaneously over multiple scales (Wiegand et al. 2003, Grimm et al. 2005). POM provides a unifying framework to develop a theory of landscape dynamics by building on several hypotheses advanced by the four theoretical frameworks described above. By testing alternative hypotheses of agent behaviors that explain observed patterns at multiple scales and reducing parameter uncertainty, POM provides a powerful approach to systematically study the organization of coupled human-natural systems.

## 4.3 Urban Landscape Signatures

To develop a mechanistic urban ecology—one that aims at uncovering the underlying mechanisms that govern urban ecosystems—we need a formal approach to detecting landscape patterns across multiple regions. Empirical tests of emergent properties and self-organization resulting from internal variations in human-natural system interactions imply observing scale-free behavior across many orders of magnitude (Bak et al. 1987, Sole´ et al. 1999). Since landscape pattern is spatially correlated and scale-dependent, understanding landscape structure and functioning requires a multiscale approach (Wu 1999, 2004, Shen et al. 2004). Building on the work of scholars of landscape ecology (Turner and Gardner 1991, Plotnick et al. 1993, 1996, Wu and Levin 1994, Gustafson 1998, Fortin et al. 2003, Bollinger et al. 2003) and the advances in spatial statistics and modeling (Liebhold and Gurevitch 2002, Wagner and Fortin 2005), I propose to systematically apply and compare a series of techniques specifically aimed at discriminating patterns of urban landscapes that are relevant for both human and ecological functions.

In this section I hypothesize that distinctive hybrid landscape patterns can be associated with alternate ecological states of the landscape on an urban-to-rural gradient (Alberti and Marzluff 2004). To test this hypothesis a set of metrics to represent such patterns needs to be developed. Building on previous analyses of urban landscape patterns (Luck and Wu 2002, Alberti et al. 2007), urban design (Hess et al. 2001, Song and Knaap 2003), and urban growth (Herold et al. 2003, Herold et al. 2005). I propose that emerging patterns can be described along multiple dimensions including form, intensity, hetero-geneity, and connectivity that represent relationships among sets of many interrelated variables (Alberti et al. forthcoming). Such distinct pattern relationships may be used as representative signatures of urban ecosystem dynamics across urban regions.

The challenge for urban ecology is to develop metrics that capture pattern dimensions relevant to both socioeconomic and ecological processes, and that allow us to explore the feedback mechanisms between them. In the attempt to generalize the relationship between urban landscape patterns and ecosystem functions, scholars of urban ecology have relied primarily on single metrics, such as population density or distance to the urban core. Only recently have they started to recognize that these metrics cannot effectively represent the complex nature and heterogeneity of urban patterns (Alberti et al. 2001). In response, scholars have started to apply other metrics, including measures of land cover (McDonnell et al. 1997), land use (Savard and Falls 2001), and housing density (Bowman and Woolfenden 2001, McGowan 2001); ratios of impervious area to vegetated area (Mennechez and Clergeau 2001), commercial to residential development (McMahon and Cuffney 2000), or native to non-native species (Noss 1989); and pattern metrics depicting landscape configuration (Wu et al. 2000). Together with Erik Botsford and Alex Cohen, I have proposed integrating gradient analyses with pattern metrics to more effectively link urban development patterns to ecological conditions (Alberti et al. 2001). In the rest of this chapter I describe both how these initial metrics emerged and how the concept of urban landscape signatures evolved.

Building on the characterizations of landscape structures developed in both landscape ecology and urban morphology, I propose four dimensions of landscape patterns that are relevant to ecological or human functions or both. These are: form, density, heterogeneity, and connectivity (Figure 4.3). From an ecological perspective, urban development generates a variety of intense and unprecedented disturbances by causing physical changes in the landscape. For example, urban development changes the composition and configuration of habitat patches and thus affects ecological function and biodiversity (Marzluff 2001). From an urban development perspective, biophysical processes impose constraints on where and how humans can build. For example, topography constrains urban development and shapes the transportation infrastructure, and thus affects human function and mobility.

Researchers in landscape ecology have developed a broad suite of metrics with the potential to quantify such patterns and their effects on ecosystem processes and disturbance regimes (O'Neill et al. 1988, Turner 1989, Gustafson and Parker 1992, Li and Reynolds 1993, McGarigal and Marks 1995, Gustafson 1998). Each metric addresses an observable change in the landscape that can signal a hypothesized link between landscape structure and function (Turner and Gardner 1991). Although fewer researchers have studied how landscape pattern affects human function, several urban scholars have applied similar metrics to explore the relationships between certain aspects of urban patterns and land use and transportation functions.

Metrics of landscape patterns aim to measure two major characteristics of the landscape: its *composition* and its *spatial configuration* (Turner 1989). Landscape composition refers to the presence and amount of different patch types within the landscape, without explicitly describing its spatial features (i.e., percentage of the land with a certain cover). Landscape configuration refers to the spatial distribution of patches within the landscape (i.e., degree of contagion). In landscape ecology these metrics are good predictors of how well the ecosystem can support important ecosystem functions (Turner and Gardner 1991). Ecological studies have shown, for example, that forests with larger patch sizes have higher species diversity. The metrics of edge-to-interior ratio and nearest-neighbor probability reflect the degree of landscape fragmentation. Fractal dimension reflects the extent of human impact on patch shape and edge complexity. Contagion is an important measure of contiguity of specific habitat types (O'Neill et al. 1988, Turner and Gardner 1991). These metrics allow us to measure the patterns of ecological disturbance, in both the natural and the built environments. Later in this chapter I discuss in greater detail a selected set of landscape metrics.

The four dimensions I use here to describe patterns of urban landscapes are not meant to form a comprehensive representation of urban landscape structure. Rather, they present different aspects of landscape structure that have been found to have some effect on ecosystem processes. Scale affects the ability of metrics to detect landscape phenomena, and some metrics are sensitive to scale. Interactions among patterns and process are affected by scale, and landscapes exhibit distinctive spatial patterns at different scales (Wu et al. 2000). Various aspects of spatial structures become important as we move across various scales of human functions (Owens 1993). At the regional scale, the broad patterns of settlement are significant; at the local scale what matters are design, siting, orientation, and layout. I have selected four dimensions that are relevant at the metropolitan scale—between the neighborhood and sub-regional scales—and that can be measured and examined in relation to a broad range of environmental variables. The first element, *form*, refers to the degree of centralization of the urban structure. The second element, *density*, is the ratio of population or jobs to the area. The third element, *heterogeneity*, indicates the diversity of functional land uses such as residential, commercial, industrial, and institutional. The fourth element, *connectivity*, measures the movement of organisms and the flow of resources (energy and materials) across a given location, as well as the way people and goods interrelate and circulate there.

Form

Density

Heterogeneity

Connectivity

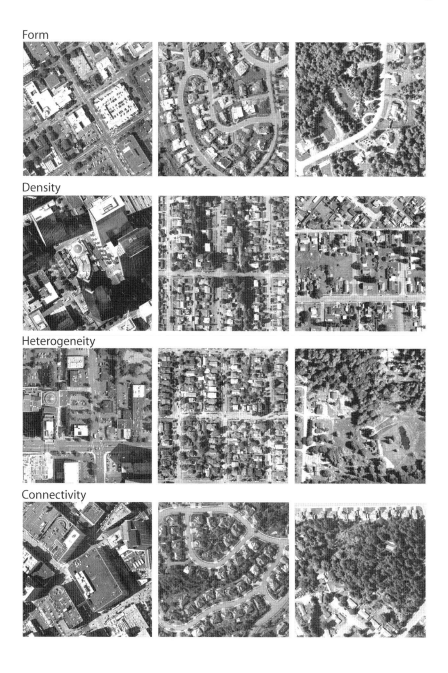

Figure 4.3. Four dimensions of urban landscape pattern.

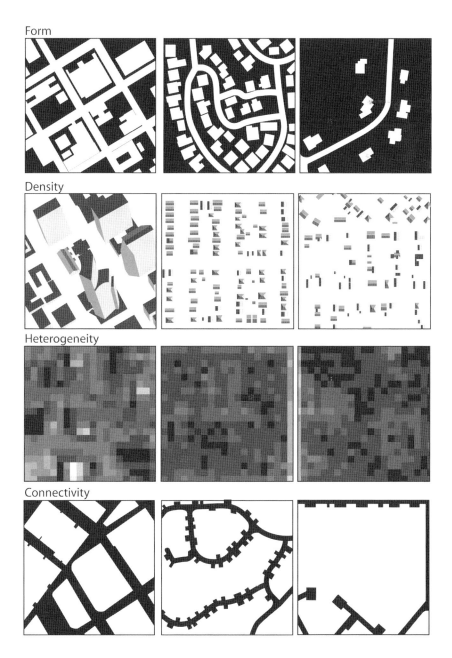

Figure 4.3. (Continued).

## Form

The form of the urban landscape has interested urban scholars for quite some time. A range of approaches has been proposed to describe urban landscape form, its degree of centralization, and its regularity. Different approaches reflect a variety of theories and perspectives, from urban geography and urban economics to landscape ecology and urban design and urban planning (Batty and Longley 1994). Geographers have used point-pattern analysis to measure distance between observed units of development and to compare these observations with theoretical distributions. Observations can be compared to the hierarchically organized development pattern postulated by the central place theory, or to a purely random distribution pattern that can be formulated as a Poisson process (Anas et al. 1998). Batty and Longley (1994) measure fractal dimension—the degree of self-similarity, or repetition of similar geometric pattern elements at smaller and smaller scales—to measure the extent to which urban form departs from regularity. They use fractal dimensionality of multiple cities and its evolution over time to hypothesize that land use control and land use planning may explain greater regularity. But they also see many constraints on the ability of fractal dimensions to measure urban form effectively. One of these is the different scales of processes that influence urban pattern. Still, this use of fractals has been important in pointing out where deterministic models of urban growth cannot account for irregularities in the urban structure.

One key aspect that dominates the characterization of urban form is the extent to which urban structure is monocentric or multicentric. Measures of urban structure have been developed in relation to models of agglomeration economies. Urban scholars are interested in the forces that shape and maintain clusters of economic activities, and they have used theories of agglomeration to develop models that explain urban form. In describing urban structure, Anas et al. (1998) have distinguished between the degree of centralization and the extent to which activities in a city may be clustered in a polycentric pattern or dispersed in a more regular pattern. Ultimately, then, scale is a critical factor in measuring form. As Anas et al. (1998, 1431) suggest, "the distinction between an organized system of (urban) subcenters and apparently unorganized urban sprawl depends very much on the spatial scale of observation."

## Density

Measures of density typically used in several disciplines to characterize urban patterns are population, housing, and employment. Density gradients can effectively measure urban sprawl and detect both change over time and differences among cities. Such density metrics add important elements to test hypotheses about the relationships between urban and ecological functions along an urban-to-rural gradient. With Paul Torrens, I review several models of urban density using a variety of mathematical functions including equilibrium, inverse power, and negative exponents (Torrens and Alberti 2000).

Other approaches to measuring densities include specifying the metrics and unit areas relevant to the urban process being investigated. Song and Knaap (2004) suggest three measures of the density of single-family development units (SFDU) as measures of urban form: SFDU lot size, density, and floor space. They suggest several relationships between these factors: the smaller the median lot size of SFDUs in a given neighborhood, the higher the density; the higher the ratio of SFDUs to the residential area, the higher the density; and, lastly, the smaller the median floor space, the higher the density.

Of course, it is also crucial to consider the scale at which we study density (Torrens and Alberti 2000). We will observe different density patterns if we measure these parameters at the scale of the neighborhood, city, or metropolitan area. We must also decide what boundaries are appropriate for measuring density, and whether to include the overall area of a city, or exclude areas such as water, parks, wetlands, and other non-buildable land, to measure net density (Gordon and Richardson 1997). Areas that could be excluded from calculation vary substantially across cities, and can account for a large share of the metropolitan area depending on both biophysical and socioeconomic factors. Thus, the prevalence of open water and nonbuildable land is important in understanding the relationship between urban pattern and land development processes (Zielinski 1979).

## Heterogeneity

The functions of urban landscapes depend on landscape heterogeneity. In urban landscapes, Pickett et al. (2001) have posited that three factors—hydrological, ecological, and social phenomena—control spatial heterogeneity at different scales. Both natural and social scientists have traditionally defined and applied discipline-specific units to measure the heterogeneity of a landscape, but they recognize the limitations of applying such units in studying the interactions among these factors. To study the Baltimore ecosystems, Pickett et al. (1997, 2001) have proposed linking hydrological, ecological, and social phenomena in an integrated watershed conceptual model, and using the model to explore the relationships between patterns and processes within those three domains. Thus they hope to understand the heterogeneity of an urbanizing watershed, thereby elucidating the relationships between the patterns and processes of hydrology, the ecosystem's structure, and the social strata within such watersheds. This work draws on the ideas of distributed hydrology (Band et al. 1996), ecosystem patch dynamics (Pickett and White 1985), and the political economic theory of place (Logan and Molotch 1987).

Three types of heterogeneity are examined: 1) *Hydrological hetero-geneity* in urban areas is affected by urban patch structure and built infrastructure. Patches of vegetation and pavement within a watershed affect stream flow and nutrient loading, depending on the patches' positions in the watershed, as well as a variety of socioeconomic (i.e., land use) and physical (i.e., topography, imperviousness, and soil moisture) attributes (Black 1991). The Variable Source Area (VSA) approach traditionally used by hydrologists to describe how physical attributes affect the hydrological role of patches in a watershed can be integrated with an assessment of the watershed's biotic (Borman and Likens 1979) and socioeconomic attributes (Grove and Burch 1997). Grove and Burch (1997) provide an example in Baltimore, Maryland of how the concept of VSA can be extended to integrate social, biotic, and abiotic differentiation to study urbanizing watersheds. 2) *Ecological heterogeneity* is driven by the structure, species composition, and spatial organization of vegetation patches, and is in-fluenced by the resulting hydrological heterogeneity. In urbanizing water-sheds, land development and land use change habitat, channel morphology, water quality (with important effects on fish and other biota), as well as the magnitude and frequency of physical disturbances (e.g., floods and droughts). 3) *Socio-cultural heterogeneity* represents the social char-acteristics of the watershed that may explain the extent, distribution, structure, and species composition.

To integrate the hydrological, ecological, and socio-cultural processes in the study of the Baltimore Urban Ecosystem, Pickett et al. (2001) proposed a revised Social Area Analysis (SAA) that takes into account physical, biotic, and social attributes at comparable levels of analysis. Just as a Variable Source Area (VSA) analysis can account for heterogeneity in hydrology, so a revised Social Variable Source Area (SVSA) analysis can be used to measure the impact of social heterogeneity on ecological patterns and processes at different hierarchical levels (Grove 1996, Pickett et al. 2001). As with the other dimensions of urban landscape pattern described above, urban landscape heterogeneity also poses questions of scale. If one of the characteristics of landscape heterogeneity is its scale multiplicity (Kolasa and Pickett 1991, Wu and Loucks 1995, Cullinan et al. 1997, Werner 1999, Wu 1999), how can the multiplicity of scales and the occurrence of scale mismatches among social, physical, and ecological processes (Cumming et al. 2006) be reflected in measuring heterogeneity?

## Connectivity

Landscape connectivity is the degree to which the landscape facilitates or impedes the flow of resources or the movement of organisms between resource patches (Taylor et al. 1993, D'Eon et al. 2002). Such connectivity is important to both a population's survival (Fahrig and Paloheimo 1988) and to metapopulation dynamics (Levins 1970). A population's survival depends on both the rate at which organisms become extinct within individual patches and the rate at which they move among patches. Species are not likely to persist in isolated patches. In landscape ecology, connectivity is seen from the perspective of the organism or ecological function (Tishendorf and Fahrig 2000). Landscape connectivity is a relative concept. In a human-dominated ecosystem, landscape connectivity may either facilitate or impede the flow of resources and organisms, depending on which ecological or human process is the focus. While urban development (e.g., the building of road infrastructure) enhances the connectivity of human activities, it fragments and isolates habitat patches, thus decreasing the chances of species survival and increasing species extinction. Whether or not a landscape is connected depends on the point of view.

Landscape connectivity can be measured directly—by measuring either resource flows or the movement and dispersal of organisms, or indirectly—by measuring landscape structures known to facilitate movement and dispersal. Landscape ecologists have developed mathematical models of animal movements to measure connectivity as the probability of organism movement between two resource patches and their dispersal success (Fahrig and Paloheimo 1988, Henein and Merriam 1990). These are defined as the rate at which organisms migrate to new resource patches, and the time spent in transit between patches, respectively. Any direct measure of landscape connectivity must consider how organisms move through the landscape, but it is often hard to measure these movements, as well as what specific landscape characteristics facilitate or impede movement. Measuring connectivity directly may be difficult and time consuming. Landscape connectivity studies have relied primarily on landscape indices to characterize the spatial pattern of a landscapes (Tishendorf and Farhig 2000).

D'Eon et al. (2002) propose a functional perspective to identify and measure connectivity. The idea is that organisms perceive patches as functionally connected from the standpoint of the organisms (Wiens 1989, Schumaker 1996, With et al. 1997). Furthermore the scale at which the organism interacts with the landscape matters (Davidson 1998, With et al. 1999). The landscape is not inherently fragmented or connected, but can only be assessed in the context of organism vagility and scale. Non-linear patterns in connectivity and discontinuities indicate that there are thresholds (With and Crist 1995, Keitt et al. 1997). Such discontinuities in connectivity depend on organism vagility and scale (D'Eon et al. 2002).

In highly heterogeneous urbanizing landscapes, the landscape matrix can obscure the relationship between interpatch movement and landscape fragmentation (Bender and Fahrig 2005). Goodwin and Fahrig (2002) have developed several experiments and simulations to examine the relationship between landscape structure and landscape connectivity. They measured six landscape metrics: amount of habitat, number of habitats, habitat patch-size distribution, average interpatch distance, variance in interpatch distance, and ratio of habitat edge to area. They found that interpatch distance is consistently the best predictor of landscape connectivity (Goodwin and Fahrig 2002). These findings are consistent with other studies by Doak et al. (1992), Schippers et al. (1996), With and King (1999), and Tischendorf and Fahrig (2000). However, other research has highlighted that the specific characteristics of the interpatch landscape matrix (e.g., the composition and configuration of land cover types) across which organisms must disperse can also impact dispersal success (Ricketts 2001, Walters 2007). Such factors add further complexity to the process of measuring connectivity.

## 4.4 Measuring Urban Landscape Patterns

Several landscape metrics can be used to measure the four dimensions of urban landscapes and quantify *landscape signatures*. These metrics should include measures of pattern composition (diversity, evenness, etc.) and spatial configuration (fragmentation, dispersion, interspersion, fractal dimension, etc.). Measures of landscape composition identify the presence and amount of each land use and cover patch type within the landscape, but without being spatially explicit. Measures of landscape configuration identify the spatial distribution of land use and cover patches within the landscape. Here I discuss several metrics available to quantify the landscape. Later I describe the set of metrics selected for our study in central Puget Sound.

A large number of metrics have been proposed to quantify landscape patterns and monitor landscape change (McGarigal and Marks 1995, Gustafson 1998, McGarigal et al. 2002). While the measurement varies from a simple summation to complex equations, three basic concepts create the language of measurement: patch, class, and landscape. A more detailed look at the metrics identifies a finer typology. Composition metrics aim to quantify the number of elements, diversity, and distribution across the landscape. Configuration metrics focus on landscape aggregation (i.e., aggregation index), landscape shape (i.e., shape index and fractal dimension), and on the distance between patches (i.e., proximity index). Some metrics combine composition and configuration (i.e., the interspersion and juxtaposition index). Specifically, these metrics focus on how the composition of the landscape is influenced by its configuration; a landscape with both a diversity of classes and a fine resolution of adjacency mixing would be reflected by a high index value. Figure 4.4 includes several landscape metrics; the graphical representations give examples of areas with low and high index values for each metric.

*Percent land* of a certain class can be used to measure both land use and the composition of land cover. The percentage of land cover occupied by each patch type (i.e., paved land, forest, or grass) is considered an important indicator of ecological conditions, because the composition of the patches and the abundance of similar patches within the landscape can influence some of the ecological properties of a patch and the characteristics of the landscape as a whole (Gardner et al. 1987). The percentage of land occupied by each land use type (i.e., single-family residential, commercial, industrial) is an important indicator of the land use structure and can indicate that a particular use is dominant. Two other important metrics of landscape composition are the *Shannon Diversity Index* and *Dominance*. The Shannon diversity index measures the degree of diversity in a given landscape. Dominance measures the deviation from the maximum possible landscape diversity (McGarigal and Marks 1995).

*Patch Density* and *Mean Patch Size* are useful measures of landscape configuration. Counting the number of patches of a specific land use and land cover type is a good way to measure the spatial heterogeneity of a given urban landscape. This is interesting to ecologists, as it can affect population dynamics and disturbance regimes. Patch density is also a fundamental aspect of landscape structure; it expresses the number of patches on a per-unit-area basis, which facilitates comparisons among landscapes of varying size. Patch density for a particular patch type could serve as a good index of fragmentation. Similarly, the density of the patches in the entire landscape mosaic could serve as a good index of heterogeneity. Mean patch size is also useful as an index, especially of habitat fragmentation. For example, a landscape with a smaller mean size for a given patch type might be considered more fragmented than another landscape with a larger mean patch size (McGarigal and Marks 1995).

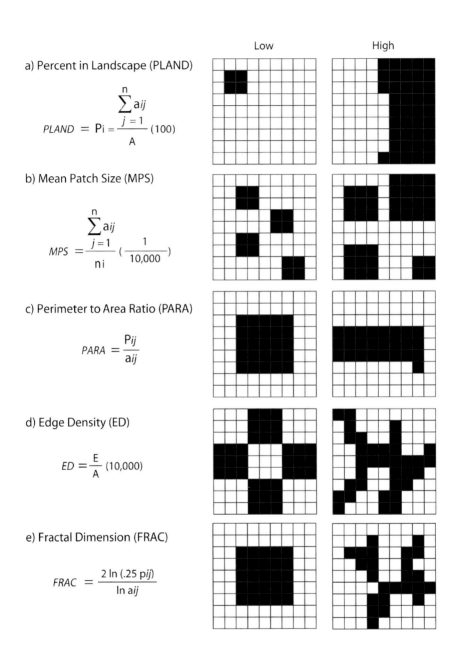

Figure 4.4. Metrics of landscape composition and configuration (Equations are described in McGarigal and Marks 1995).

f) Shannon's Diversity Index (SHDI)    $SHDI = -\sum_{i=1}^{m} ( P_i \cdot \ln P_i )$

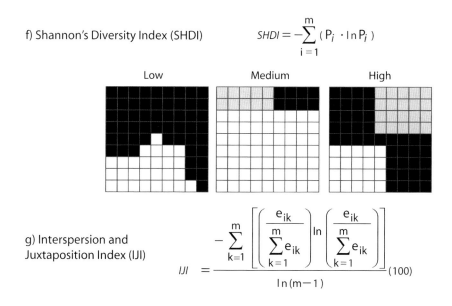

g) Interspersion and
Juxtaposition Index (IJI)

$$IJI = \frac{-\sum_{k=1}^{m} \left[ \left( \dfrac{e_{ik}}{\sum_{k=1}^{m} e_{ik}} \right) \ln \left( \dfrac{e_{ik}}{\sum_{k=1}^{m} e_{ik}} \right) \right]}{\ln (m-1)} (100)$$

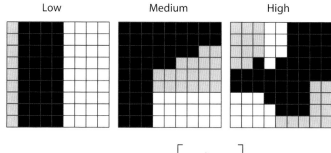

h) Aggregation Index (AI)    $$AI = \left[ \frac{g_{ii}}{\max \to g_{ii}} \right] (100)$$

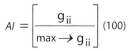

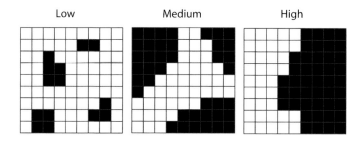

Figure 4.4. (Continued).

The dispersion of the urban structure can be effectively measured using *Contagion* (Turner 1989, Li and Reynolds 1993, 1994) and the *Aggregation Index (AI)* (McGarigal and Marks 1995). In a raster representation of the landscape, contagion is the probability that two randomly chosen adjacent cells or pixels belong to the same class. This probability is calculated by multiplying two other probabilities: the probability that a randomly chosen cell belongs to category type *i*, and the conditional probability that given a cell of category type *i*, one of its neighboring cells will belong to category type *j* (Li and Reynolds 1993, 1994). Contagion is based on cell adjacencies as opposed to patch adjacencies. It measures both patch dispersion and interspersion. Landscapes consisting of large patches of a similar land cover or land use category have a greater number of adjacent cells. Where contagion is low, urban areas can be said to be composed of many small and dispersed patches of various land cover or land use categories. However, because contagion indices do not indicate the degree of connectivity between patches or land use of the same type, they may not effectively indicate much about urban structure. The *Aggregation Index*, the number of like adjacencies of a specified land cover class divided by the maximum possible number of like adjacencies involving that class, may be more effective since it allows to focus on one class at the time.

The *Interspersion and Juxtaposition Index* (IJI) measures adjacency among patch types over the maximum possible interspersion for the given number of patch types. It can be applied at both the landscape and the class level. When applied to land use, it measures the degree of interspersion of various land uses. Unlike contagion, this index measures patch adjacencies, not cell adjacencies, and can be more appropriate than contagion in representing the urban structure. However, the interspersion index measures only interspersion and is not affected by any other aspects of patches, such as size, contiguity, or dispersion. Thus it captures only one aspect of landscape configuration (Torrens and Alberti 2000).

In landscape ecology, *Fractal Dimension* is used to measure landscape complexity. Applied to land cover, it provides a measure of the human impact on the landscape. The assumption is that natural boundaries have complex shapes, and that as human disturbance increases, the fractal dimension decreases. Since edges of one patch are also edges of adjacent patches, edge/area ratios also yield information about the overall shape complexity of a landscape. The fractal dimension represents shape complexity (a departure from Euclidean geometry) that has been effectively used to represent human-induced disturbance.

One of the greatest challenges in using landscape metrics to measure urban patterns is differentiating the spatial patterns of patch dispersion. Using simulated landscape patterns Hargis et al. (1998) found that pattern metrics do provide useful information on a landscape's patch size, shape, and distance relationships, but several selected pattern metrics—including contagion, mean proximity index, and fractal dimension—could not distinguish the overall landscape pattern caused by a unique spatial distribution of patches. Still, each of the measures I have described here quantifies an important component of the landscape, and each can be used to investigate formal hypotheses about the relationships between urban patterns and ecological conditions.

## 4.5 Detecting Landscape Patterns in Puget Sound

The landscape signatures of development patterns can be quantified using selected landscape-pattern metrics that describe the four dimensions described above: form, density, heterogeneity, and connectivity. My research team in the Urban Ecology Lab at the University of Washington conducted a study to identify landscape patterns emerging from the spatial interactions between biophysical and socioeconomic processes in urbanizing regions (Alberti et al. forthcoming). Using landscape metrics, we analyzed urban patterns to characterize their composition and configuration; we were able to identify distinct landscape signatures relevant to various ecosystem processes for different urban patterns. In our study we applied these pattern metrics on an urban-to-rural transect. We aimed to explore the complexity of urbanizing landscapes, and to test hypotheses about the complex patterns emerging from the interactions between human settlement and biophysical factors. Many ecological studies have used a simple distance gradient to represent the degree of urbanization, but we found that such an approach cannot represent the more complex interactions among human and biophysical variables. We also used discriminant function analysis (DFA) to assess whether several pattern metrics can discriminate between different types of development. Finally, using selected landscape metrics, we described patterns of landscape change in the Central Puget Sound region between 1991 and 1999. In the rest of this chapter I describe our findings and what we have learned regarding landscape metrics.

### A complex urban gradient

To characterize the urban-to-rural gradient we developed two levels of analysis. First we developed a gradient analysis combining three variables using principal component analysis (PCA): population density, distance to the closest central business district (CBD), and elevation. Then, we selected a sample of 400,000 pixels of 30 × 30 m resolution in the Central Puget Sound region that lay below 1,000 meters in elevation. By using PCA we sought a linear combination of the vector *x* that would capture the maximum possible variability of the three variables (population density, distance to CBD, and elevation, Figures 4.5 and 4.6). We did this by computing the sample covariance matrix and obtaining the eigenvalue decomposition of the matrix. Using PCA we were able to reduce the three variables to a common index: from 4.0 (urban core) to -4.0 (natural). The estimated index variable captured 93.76% of the total variation in the overall data.

## Selected pattern metrics

To address the complex landscape heterogeneity of the urban landscape pattern along the urban-to-rural gradient, we use both discrete and continuous variables, including percent impervious surface and land cover and land use patches (Figure 4.7-4.9). Using Landsat TM data and land parcel data from the counties, assessors' offices, we calculated selected metrics to measure urban landscape patterns in relation to an urban-to-rural gradient (Figure 4.10) (Alberti et al. forthcoming). We selected our sample using three transects, one in each of three directions—Northeast, East, and Southeast—from the Seattle Central Business District. We used ten metrics: patch size, shape index, building density percent land use, percent land cover, Shannon diversity index, aggregation index, mean interpatch distance, edge contrast, and edge-to-area ratio (Table 4.1). Land use data at the parcel level were obtained from the county assessors' offices of four counties: King, Pierce, Snohomish, and Kitsap. We interpreted the land cover data using Landsat Thematic Mapper (TM) imagery of the Puget Sound region in 1999.

We found that the selected metrics are useful indicators of landscape structure along an urban-to-rural gradient. That is, they effectively capture some of the complex relationships between various aspects of the landscape's structure and functions. The Shannon index shows that the greatest landscape diversity occurs at the urban fringe (Figure 4.10a). Residential building density decreases with distance from the central business districts while patch size increases (Figure 4.10b). Similarly, the aggregation index, used to measure patches of urban land-cover and of forest, provides an opposite trend: urban patches decrease and forest patches increase with distance from the urban core (Figures 4.10c).

## Land use and land cover patterns

Third, we used discriminant function analysis (DFA) to assess how well the selected spatial metrics could discriminate between different types of development. DFA is a multivariate statistical procedure that analyzes the

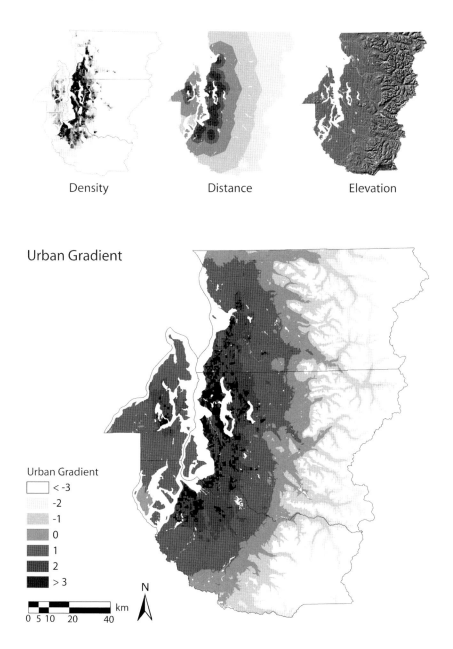

Figure 4.5. Urban gradient for the Central Puget Sound. Map of the urban gradient and the three variables: population density, distance to central business district, and elevation.

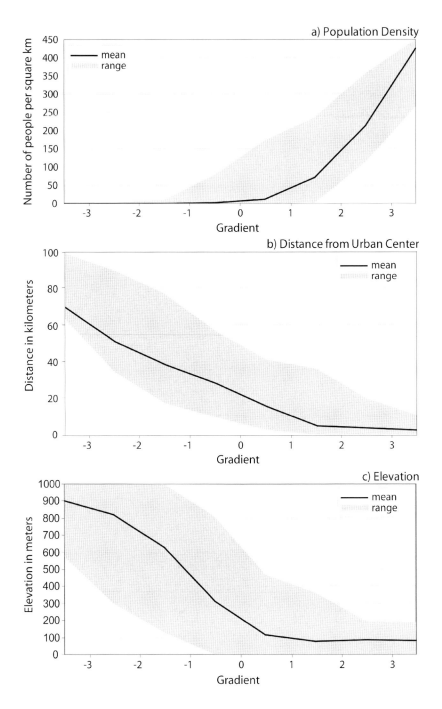

Figure 4.6. Dimensions of the Central Puget Sound urban gradient.

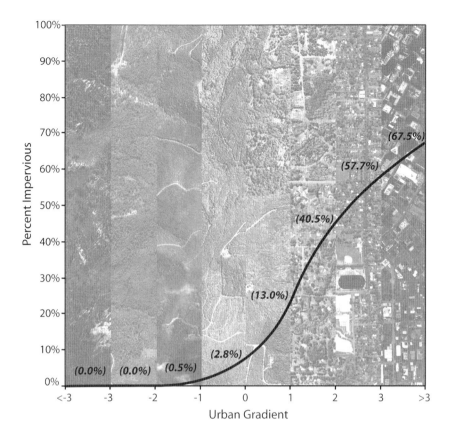

Figure 4.7. Mean percent of impervious surfaces along the urban gradient.

differences between mutually exclusive groups by determining linear relationships between variables to create the largest distance between the groups. The procedure classifies the differences by using a discriminant function to calculate a discriminant "score." The function is a linear combination of variables that has a greater discriminating ability than the original variables. Discriminant analysis also determines the relative contribution that each variable makes to the discriminating ability of the function.

As we used this approach, we were able to observe the land cover distribution across parcels with different land uses, and these distributions revealed complex relationships between land use and land cover. By comparing segments of the landscape with varying amounts of land cover types, we could discriminate different land use parcels (Figure 4.11).

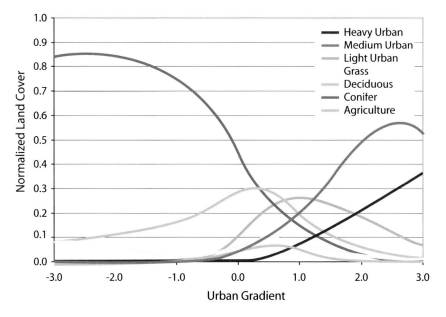

Figure 4.8. Land cover distribution along the urban gradient.

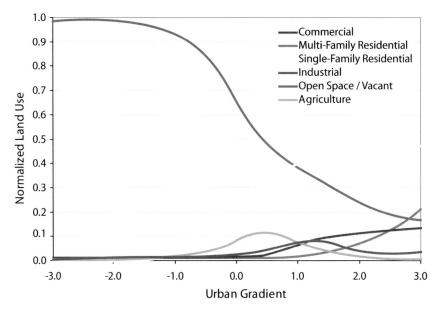

Figure 4.9. Land use distribution along the urban gradient.

Table 4.1. Selected urban landscape pattern metrics.

| Pattern | Attribute | Metric |
|---|---|---|
| Form | Patch size | *Patch area* |
| | Patch shape | *Shape index* |
| Density | Land use density | *Building units / area by land use* |
| | Built up density | *Percent land use* |
| Heterogenity | Land use / cover composition | *Percent land cover* |
| | Land cover / cover diversity | *Shannon Diversity Index* |
| Connectivity | Adjacency | *Aggregation Index* |
| | Distance between patches | *Mean inter-patch distance* |
| | Patch contrast | *Edge contrast, Edge to area ratio* |

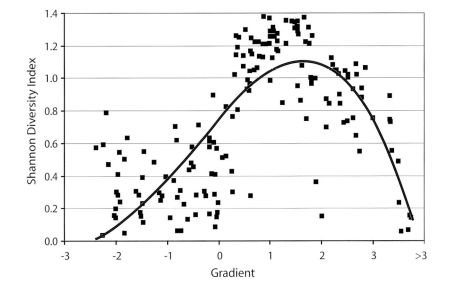

Figure 4.10a. Shannon diversity index along the urban gradient.

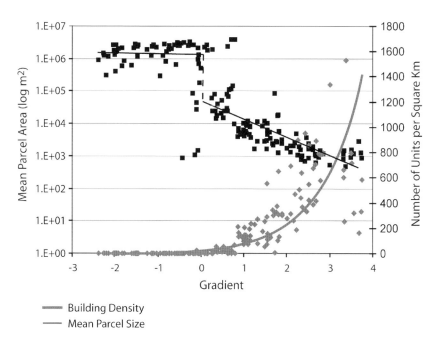

Figure 4.10b. Mean parcel size and building density along the urban gradient.

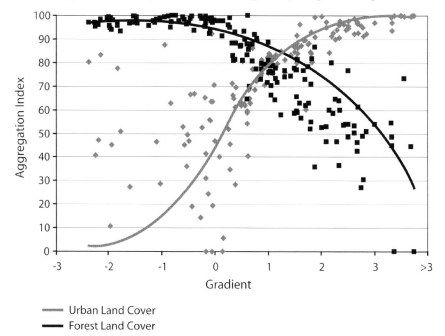

Figure 4.10c. Urban and forest aggregation index along the urban gradient.

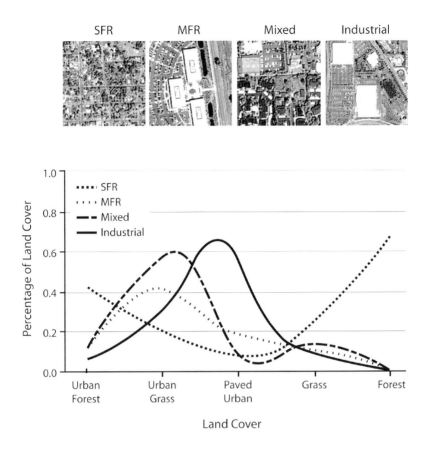

Figure 4.11. Urban landscape signatures (Alberti, 2005, p. 185).

For example, single-family residential (SFR) parcels have a significantly lower percentage of impervious surface than multi-family residential (MFR) parcels, although MFR parcels may accommodate a much larger number of households. Mixed-use (i.e., residential and commercial) and industrial parcels have an even greater percentage of impervious surface.

## 4.6 Monitoring Landscape Change in Puget Sound

Landscape metrics can serve as benchmarks. We can compare and measure urban growth against them to reveal landscape changes associated with urban growth. We can also assess whether the policies aimed at controlling urban sprawl have been effective. For example, at the Urban Ecology Research Lab (2005c, 2005d) we examined landscape change in central Puget Sound over the period 1991-1999 using benchmarks metrics. We selected a set of metrics for monitoring the effectiveness of the Washington State Growth Management Act (GMA). Enacted in 1990, the GMA's overarching goal is to reduce the impacts of uncoordinated and unplanned growth. We also aimed to measure the state's progress towards this goal.

Our study was based on a multi-year land cover classification and analysis of multiple USGS Landsat Thematic Mapper (TM) images at three time-steps (1991, 1995, and 1999) at the regional and county levels within and outside the urban growth boundary. We also conducted a multi-scale analysis using 150-meter, 450-meter, and 750-meter moving windows across the region to evaluate whether changing the spatial scale of analysis had an impact on the results of the analysis. By comparing the results before and after 1995, as well as inside and outside the GMA, we aimed to assess urban growth and the potential impact of the GMA.

The study provides an assessment of how different metrics could assist in monitoring and assessing policy performance by measuring on the ground the effects of the GMA on landscape composition and configuration. We start by specifying metrics that may be considered to measure progress towards 12 of the 14 GMA spatially explicit goals (Table 4.2). While there are several metrics that can target specific GMA goals, a few capture several goals simultaneously.

We started by analyzing land cover change over time (Figure 4.12). We then compared the change in the amount of impervious area and the change in population. We calculated the sum of the impervious area for each county through spectral unmixing of the land cover layers created. We determined population change by intersecting U.S. Census population block-group data for 1990 and 2000 with the counties' boundary layers within and outside the UGB (this information was not available for 1995). We calculated the number of people per sq. km of impervious area by dividing the total population per county by the total impervious area. We found that King County had the largest change in both population and the number of people per impervious sq. km. While King had the highest number of people per sq. km of impervious area, Snohomish increased the most in terms of density, with its population growing faster than its impervious surfaces.

Table 4.2. GMA spatial goals and landscape metrics (UERL 2005c).

| Spatial GMA Goals | Spatial GMA Metrics | Landscape Metrics |
|---|---|---|
| **Urban Growth:** Encourage growth in urban areas where adequate public services are available. | Percent urban land cover outside of UGB. Aggregation of new urban patches. Proximity of new development to urban core. Amount of urban edge encroaching on rural and forested areas. | Percent in Landscape (PLAND), Largest Patch Index (LPI), Interspersion Juxtaposition Index (IJI), Aggregation Index (AI). |
| **Reduce Sprawl:** Reduce low density development. Control urban growth outside of centers. | Perimeter of transition between conflicting uses. Form of urban sprawl. Isolated urban growth or expansion. Location of expansion. Encroachment of urban cover on the non-urban land. Degree of diversity of uses within the urban patches. Change in distance between new growth and existing growth. | LPI, Mean Patch Size (MPS), PLAND, Shannon Diversity Index (SHDI), IJI, AI, Perimeter to Area Ratio (PARA). |
| **Transportation:** Build efficient, multimodal transportation systems. | Distances and isolation between the patches Compactness of patch form. Areas within the urban growth centers that have increased their use diversity | LPI, PLAND, IJI, AI, PARA. |
| **Housing:** Increase the availability of affordable housing and the variety of housing densities and types. | Percent of acreage allotted to single family and multiple family parcels. Availability of jobs next to residential development. Relationship of new developments to the urban core, or central business district. | SHDI, IJI. |

Table 4.2. (Continued).

| Spatial GMA Goals | Spatial GMA Metrics | Landscape Metrics |
|---|---|---|
| **Natural Resources:** Conserve productive forest and agricultural lands and discourage incompatible uses. | Amount of cover attributed to forest and agriculture. Change in acreage or percent over time. Patch size as indication of economic viability of farming and forestry. Encroachment of urban uses onto forest and agricultural resources. | LPI, PLAND, AI, PARA, Edge Density (ED). |
| **Open Space and Recreation:** Protect open space and natural habitat for recreation, fishing, and wildlife purposes. | *Parkland:* Percent attributed to parkland from a land use layer High edge density between parkland and multifamily residential use. *Wildlife Habitat:* Change in large connected patches with compact shapes Change of interior habitat and edge habitat | ED, PLAND, AI. |
| **Environment:** Protect and enhance air and water quality. | Percent of impervious surface cover. Percent of like adjacencies and edge contrast index focusing on impervious cover at the interface between water and the paved surfaces. | PLAND. |

We also measured landscape configuration using the Aggregation Index (AI) and the Percent of Like Adjacencies (PLADJ) on a 150-meter moving window for 1991, 1995, and 1999. We saw a consistent, though very small, increase in the Aggregation Index of urban land cover across the overall region. Forest cover (including both mixed and coniferous) decreased slightly overall, as measured by both indices. Within the urban land-cover classes, the AI of heavy urban land cover actually decreased in the region, showing no significant change either inside or outside of the UGA separately. A multi-scale analysis, testing how sensitive different landscape metrics are to the scale of measurement, shows that scores on the PLADJ and the AI increase slightly with increasing scale. The sensitivity analysis, conducted with a t-test for a random 100-point sample, showed that AI is more sensitive to a change in scale than is PLADJ (Figure 4.13).

Overall, our findings show that the conversion to developed land between 1995 and 1999 occurred primarily in the light-urban class outside the UGB. We also identified an increase in the high-urban land cover of the core urban area. Forest cover is declining. Perhaps the most revealing patterns are the changes we observed in the form of growth: a strong reversal between the first half of the decade (1991 through 1994) and the second half of the decade (1995 through 1999), and a trend from a more disaggregated heavy-urban land cover towards a more aggregated one. This trend is also reflected in a reduced rate of forest fragmentation outside the UGB. This may mean that the GMA has been effective in minimizing the negative impacts of urbanization on the contiguity of natural areas.

This study points to three major factors to consider in selecting metrics as benchmarks for monitoring urban growth. The first factor is the importance of measuring different attributes of the landscape. By measuring landscape composition, intensity, and configuration, we were able to distill some important dimensions of simultaneous trends that occurred in the Central Puget Sound between 1990 and 2000. We observed both intensification of the urban core together with an increase in the amount of low density development, particularly outside the UGB. While we measured both composition and configuration, several other attributes could add important information that would help in interpreting the complex relationships between the metrics and GMA's goals. In general, the four dimensions used to describe urban patterns—form, heterogeneity, density, and connectivity—were consistently and significantly affected by the trajectory of urban growth.

A second factor is the choice of classification resolution and spatial scale. The study points out the importance of testing the data at multiple scales and aggregation levels. For example, if we had begun by looking at all urban land cover aggregated into one class, we never would have seen that some of the urban areas are becoming denser and shifting from medium-urban to heavy-urban land cover. On the other hand, if we had kept the forest land cover classes disaggregated (as mixed and coniferous) we might have missed the important trend toward spatial disaggregation of forest cover. Additionally, it is important to make sure that the spatial resolution is both high enough to capture the details important in the study and low enough to accurately detect a significant change.

A third factor concerns efforts to tie the GMA goals directly to the selected metrics. As we see, slight changes in scale can greatly alter the observed trend. We must be explicit in defining the expected trend both in terms of direction (increase/decrease) and magnitude (percent change). For example, we can interpret the results of the forest.

Aggregation Index (AI) as either a negative indication (the AI of forest cover decreased outside of the UGB, meaning that forestland is being frag-mented), or a positive indication (the AI decreased inside the UGB, reflect-ing the preservation of urban open space), or as no indication (there is not enough information to detect any significant change). Perhaps the most important point of all is to be consistent and transparent in interpreting the data.

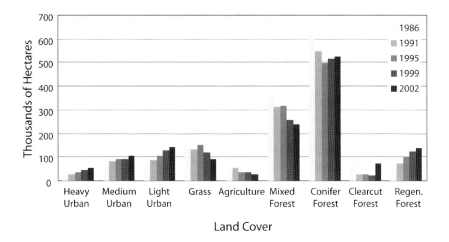

Figure 4.12. Land cover change in Central Puget Sound 1986-2002 (UERL 2005c).

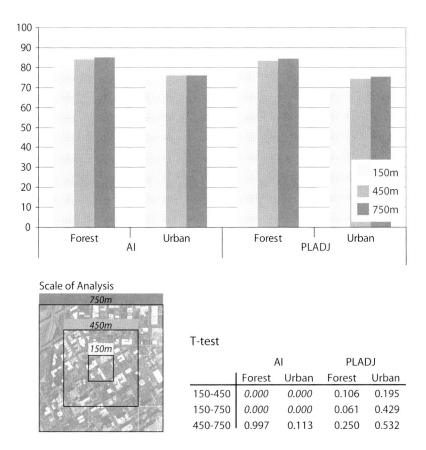

Figure 4.13. Sensitivity analysis of landscape metrics across scales (UERL 2005c).

# Chapter 5

# HYDROLOGICAL PROCESSES

## 5.1 The Urban Hydrological Cycle

The water cycle plays an essential role in the functioning of ecosystems by integrating the complex physical, chemical, and biological processes that sustain life. Water is a key factor in determining the productivity of ecosystems, species composition, and biodiversity. Water is also essential for human well-being: Human activities such as agriculture, fisheries, industries, and hydroelectric power depend on water supplies. An adequate water supply is vital to support households and businesses in urban and metropolitan regions.

Although the overall amount of water on earth will not diminish on shorter than geological time scales, its seasonal variations have important consequences because they determine the amount of water available for human use (Oki and Kanae 2006). Of the Earth's 1386 million $km^3$ of water, only 35 million $km^3$ (2.5 percent) is freshwater and less than one third of that freshwater is available for human use (Korzun et al. 1978). The total amount of water withdrawn for human uses increased by about seven-fold during the twentieth century, from 579 $km^3$/year in 1950 to 3917 $km^3$/year, and is expected to reach 5139 $km^3$/year by 2025 due to future population growth (Shiklomanov and Rodda 2003).

Water circulates naturally and is constantly recharged (Oki and Kanae 2006). It evaporates off the oceans (436,500 $km^3$/year), evapotranspires from land (65,500 $km^3$/year), falls in the form of precipitation back on to oceans (391,000 $km^2$/year) and land (111,000 $km^3$/year), percolates into the ground, and returns to the ocean through rivers and lakes (45,500 $km^3$/year) (Figure 5.1). In this process, water transports large amounts of sediments and nutrients and alters the earth's surface through erosion and deposition (Dunne and Leopold 1978). While the cycling of water is predominantly governed by biophysical elements and processes such as climate, landforms, vegetative cover, and geology, humans have a significant influence on the global water cycle through their global and local impacts.

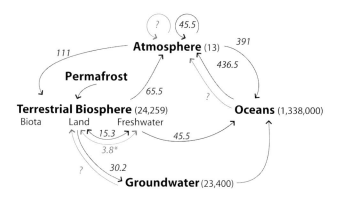

Figure 5.1. Global fluxes of the hydrological cycle. The directional arrows indicate global hydrological fluxes which total 1000 km³/year. Gray arrows represent additional and isolated anthropogenic fluxes. Storages for the major components of the system are included in parentheses in 1000s of km³. Fluxes and storages are synthesized from Oki and Kanae 2006. Question marks represent uncertainty in anthropogenic fluxes. Due to global warming, rates of evaporation from the oceans increase by 7% for every 1°C rise in temperature, while the water vapor in the atmosphere increases by 5% (Solomon et al. 2007). Our understanding of groundwater resources is limited, since scientists are only beginning to synthesize well-log, groundwater discharge/recharge, and aquifer property data for global applications (Foster and Chilton 2003, UNESCOIHP 2004). The *3.8E3 km³ of water exchanged between land and freshwater systems includes water for industrial (0.77), irrigation (2.66), and domestic (0.38) use.

As humans develop cities, they alter the hydrologic cycle by extracting water for urban uses, and by modifying biophysical structures or substituting built infrastructure in order to transfer and control the water flow (Figure 5.2). The extraction of water to meet the needs of urban residents and activities affects flow regimes in urbanizing watersheds. Urban surfaces alter the microclimate and the rate of precipitation, reduce infiltration, and encourage runoff. Compared to water in undeveloped areas, water in cities primarily moves over impermeable surfaces, traveling much faster and collecting a variety of human-made organic and inorganic pollutants before it reaches a treatment plant or body of water. A complex network of culverts, gutters, drains, pipes, sewers and channels extracts, redistributes, and moves the water between the natural hydrological process and the artificial hydro-logical system that supports human settlements and activities. Changes in the hydrological cycle due to urban development have significant physical, chemical, and biological consequences for ecosystems (Karr 1991, Paul and Meyer 2001, Allan 2004).

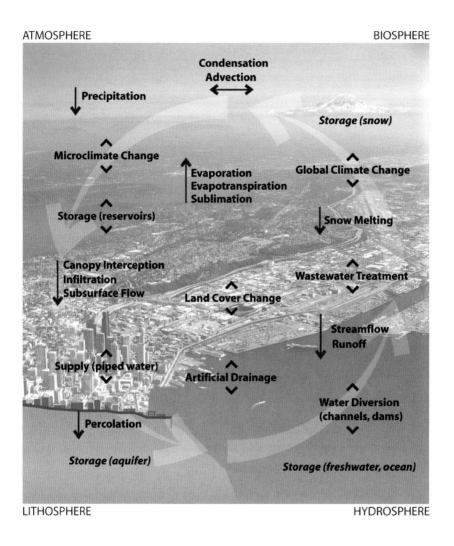

Figure 5.2. Hydrological cycle in the urban landscape (Background photo: © Aerolistphoto.com).

Urbanization influences the cycling of water by changing microclimate and precipitation in and around cities. Evidence is increasing that rainfall patterns and daily precipitation trends have changed in urban areas over the last few decades, but the complex mechanisms that link urbanization to local and regional climate changes are not fully understood. We do know that urban areas tend to be warmer than surrounding areas (Oke 1997). Urban structures are major causes of the observed changes in microclimate known as urban heat islands (See Chapter 7). Asphalt, buildings, and aerosols all help to alter the land-atmosphere exchange of water; downwind we find increased precipitation and decreased evaporation, depending on the basin's degree of urbanization. However, we know little about the ways that patterns of urban development mediate such changes.

Urban development also significantly modifies the amount and delivery of surface runoff (Hall 1984, Lazaro 1990, Konrad and Booth 2005). Impervious surfaces such as paved roads, parking lots, sidewalks, and rooftops stop rainfall and snowmelt from infiltrating soil and ground water, increasing the volume and speed of runoff. In addition, the artificial networks of ditches and culverts quickly convey this water to streams and rivers (Konrad and Booth 2002). Increased runoff associated with urbanization alters the rate, volume, and timing of stream flow in urbanizing watersheds; peak stream flows occur more often and floods are larger (Konrad and Booth 2005). Urbanization is also associated with an increase in dry-season base flows due to artificial drainage, leaky water pipes, and irrigation of lawns (DeWalle et al. 2000, Konrad and Booth 2002).

Changes in stream flow and hydrological disturbance have an impact on the functioning of ecosystems (Resh et al. 1988, Poff and Ward 1990, Poff et al. 1997, Thomson et al. 2002). Stream flow affects the structure, composition, and productivity of freshwater communities as they regulate habitat conditions, food sources, and natural disturbance regimes (Resh et al. 1988, Newbury 1988, Power et al. 1999). Konrad and Booth (2005) identify four hydrologic changes resulting from urban development that are potentially significant to stream ecosystems: increased frequency of high flows, redistribution of water from base flow to storm flow, increased daily variation in stream flow, and reduction in low flow.

Urban runoff also leads to changes in water quality. Runoff transports a variety of contaminants to streams and other water bodies. Urban land use and activities add a large amount of sediments, metals, hydrocarbons, pesticides, nutrients, toxics, and bacteria to runoff. Major sources in urban areas include construction sites, rooftops, cars, backyards, parks, and golf courses (Pitt et al. 1995, Paul and Meyer 2001). The composition of runoff depends on the land use and land cover, but the pollution load reaching the water body also depends on the drainage of stormwater and the systems of wastewater treatment. For example, combined sewer overflows are a major concern for cities when combined sewer systems exceed their capacity since they carry not only storm water but untreated human and industrial waste.

Another significant influence on water flow in urban watersheds is channelization: straightening, deepening, and/or widening stream channels to prevent flooding and facilitate urban drainage (Klein 1979, Arnold et al. 1982, Booth 1990). Channelization modifies flow regimes over time. Urban development affects geomorphology at various stages (Dunne and Leopold 1978, Booth 1990, Booth and Jackson 1997). For example, bridge construction can alter stream channels directly, while also restricting stream flow, reducing flow velocities upstream, and increasing sedimentation. In addition, urban development disturbs soil, increases the movement of sediments, and removes vegetation from the streambank (Jacobson et al. 2001). As such vegetation is lost stream temperatures can rise, in turn altering many in-stream processes, including leaf decomposition (Webster and Benfield 1986, Chadwick et al. 2005), and affecting the distribution of freshwater communities (Lamberti and Resh 1985). Such direct and indirect changes to stream habitat can have dramatic effects on aquatic organisms (Walsh et al. 2005).

Although urban streams share several characteristics in what is generally called the "urban stream syndrome," they vary a great deal across cities and urbanizing regions (Walsh et al. 2005). Emerging hypotheses to explain the variability range from hydrological variability and biophysical interactions to land use legacies and human behaviors. The extent to which development patterns and infrastructure play a role in explaining such variability has not been fully investigated. This knowledge is essential for planning strategies to manage growth. In this chapter I start by examining key hydrological functions that support urban ecosystems. Then I discuss how landscape patterns and processes affect key hydrological functions in urbanizing watersheds and the implications for ecosystem function.

## 5.2 Urban Hydrological Functions

Earth's watersheds support both natural and human functions. They capture, store, use, and clean water, and cycle nutrients, while also regulating water movement, controlling floods, and providing habitat for aquatic and terrestrial organisms (Table 5.1). For urbanizing watersheds to be resilient, hydrological processes must support both human and ecological functions simultaneously. The *watershed* is the basic unit in hydrology, but humans redefine it in urbanizing regions. In fact, the urban infrastructure redefines the boundaries of watersheds by changing the basic elements that govern water drainage across the landscape (Figure 5.3). Here I describe these functions before identifying the key impacts of humans in urbanizing watersheds.

Table 5.1. Landscape patterns, processes and hydrological functions. A synthesis of the established relationships, directions (↑↓), and uncertainties (?) between urban patterns and hydrological functions though various ecosystem processes.

| Land cover | Land use | Transportation | Energy infrastructure | Water infrastructure | PROCESS / PATTERN |
|---|---|---|---|---|---|
| ↑Impervious surface ↑Grassland | ↑Heat Isl. ↑$CO_2$ emissions | ↑Heat Isl. ↑$CO_2$ emissions | ↑Humidity ?Rainfall patterns | | **Climate** |
| ↓Forest cover ↑Impervious surface ↑Grassland | ↑Imperv. surface ↑Grassland ↑Hydrolog. connectivity | ↑Imperv. surface ↑Grassland ↓Forest cover | ?Riverflow patterns | ↑Hydrological connection | **Intercepti on/Runoff** |
| ↓Riparian area ↓Wetlands ↑Gravel extraction | ↑Channelization ↓Riparian area ↓Wetlands ↑Gravel extraction | ↑Channelization ↓Riparian area ↓Wetlands ↑Gravel extraction | ↑Dams ↑Channelization | ↑Channel modificat. ↑Channel incision or "downcutting" | **Geomorphology** |
| ↑Impervious surface ↓Forest cover | ↑Riparian clearing ↓Pervious surfaces and soil column | ↑Riparian clearing ↓Pervious surfaces and soil column | ↑Dams | ↓Riparian area ↓Wetlands ↓Channels & stream banks | **Riparian** |
| ↑Dams | ↑Industrial use ↑Agr. use ↑Energy use ↑Resident. use | ↑Transportation use | ↑Dams ↑Water loss | ↑Water extraction ?Hydrological connec. ↑Water budget | **Water extraction /diversion** |
| ↓Forest cover ↓Vegetation ↑Impervious surface | ↑Fertilizers ↑Pesticides ↑Autos. ↑Industrial ↑Road salting | ↑$CO_2$, $NO_X$, $CO$, $SO_2$, & VOCs emissions ↑Road salt ↑Oil leakages | ↑Saltwater intrusion | ↑Septic & waste $H_2O$ treatment, emissions & overflows | **Biogeochemical** |

Table 5.1 (Continued).

| FUNCTION / PROCESS | Provision and storage of water | Soil protection | Stream flow regulation | Nutrient cycling | Water quality | Riparian and floodplain | Habitat provision |
|---|---|---|---|---|---|---|---|
| **Climate** | ↓ ET & storage<br>↓ Ground-water recharge<br>↓ Snow pack | ↑ Flooding<br>↑ Channel and bank erosion | ↓ Instream flow<br>? Base flow | ↑ Nutrient losses<br>? Nutrient fluxes | ↑ Temp. | ↑ Flooding | ↑ Temp.<br>↑ Plant growth rates<br>? Comm. structure |
| **Interception/Runoff** | ↓ Capacity & storage | ↑ Erosion<br>↑ Sedimen-tation | ↑ Mag. of High flow<br>↓ Time to concentra-tion.<br>? Base flow | ↓ Nitrogen retention | ↑ Temp.<br>↑ Sediment<br>↑ Nutrients<br>↑ Toxics | ↓ Riparian<br>↑ Flooding | ↓ Fish pop.<br>↓ Biotic diversity<br>↓ Substrate<br>↓ Habitat diversity |
| **Geomor-phology** | ↓ Riparian groundwa-ter levels | ↓ Wetland and hydric soils | ↑ Runoff<br>↓ Instream flow<br>? Base flow | ↓ Organic matter<br>↓ Nutrient uptake | ↑ Temp.<br>↑ Sediment<br>↑ Nutrients<br>↑ Toxics<br>↑ Turbidity<br>↑ Heav. Met<br>↑ Minerals | ↑ Redistri-bution of sediments | ↑ Channel width,<br>Pool depth<br>↓ Spawn / rearing sites |
| **Riparian** | ↓ Riparian groundwa-ter levels | ↓ Plant transp.<br>↓ Salamand.<br>↑ Riparian hydrologic drought | | ↑ Soil $NO_3^-$<br>↓ Soil denitrifi-cation<br>↓ Organic debris<br>↑ N sources | ↑ Temp. | ↓ Riparian function | ↑ Temp.<br>↓ Woody debris<br>↓ Habitat diversity |
| **Water extraction/diversion** | ↑ Ground-water extraction<br>↓ Ground-water levels | | ↓ Instream flow<br>? Base flow | ↑ Organic debris<br>↑ Dams<br>↑ Denitrifi-cation | ↑ Sediment<br>↑ Saltwater intrusion | ↓ Riparian function | ↓ Water quantity<br>↓ Habitat diversity |
| **Biogeo-chemical** | ↑ Nutrients<br>↑ Toxic | | | ↑ Organic Material<br>↑ Eutroph.<br>? Ecosyst. metabo-lism | ↑ Temp.<br>Sediments,<br>Nutrients,<br>Toxics,<br>PAHs, PCBs,<br>Organics<br>↓ DO, pH | | ↑ Algae<br>↑ Tolerant species |

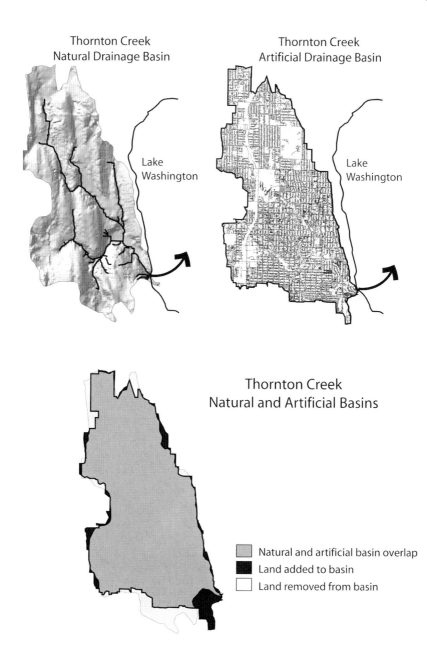

Figure 5.3. Relationship between natural and artificial drainage basins. This figure describes the difference in the extent of the natural and artificial basins for the Thornton Creek, WA.

## Provide and store water

Watersheds serve both humans and wildlife by providing and storing water. Among the hydrological processes that maintain these functions are interception, infiltration, and percolation. As water moves, it either is intercepted and evaporated by vegetation or reaches a water body. Infiltration involves the movement of water from the atmosphere through the soil surface into the soil. The rate of infiltration is governed by the intensity of rainfall and surface slope and by the soil's permeability and saturation. Water percolates through surface soils through gravity. Groundwater is replenished when precipitation is sufficient to infiltrate the soil and seep down into subsurface storage areas or the saturated zone. The process of aquifer recharge is affected by soil, geology, topography, and land uses. Site-specific conditions such as steep slopes, tight clay soils, or unfractured bedrock reduce infiltration. Recharge is drastically impaired when vegetation cover is replaced with impervious cover. Urbanization significantly affects the ability of the water to circulate through the various components of this cycle and replenish the aquifer.

## Prevent soil loss and erosion

Surface runoff is a major cause of erosion. Urban runoff into storm drains and ditches, or directly into stream channels, rapidly increases the stream discharge, which erodes stream banks by cutting and then widening channels. Runoff can transport significant quantities of sediment or other water pollutants. Frequent floods cause channels and banks to erode at a faster rate, creating a positive feedback loop, in which streams flow much more quickly and transport more sediment. This is particularly evident in channels that have been straightened, and where vegetation has been removed from channel banks (Konrad and Booth 2002). Changes in flow regime alters the structure of stream channels (Wolman and Schick 1967, Booth and Jackson 1997); they become wider (Leopold 1972, Booth 1990), deeper (Hammer 1972, Douglas 1974, Booth 1990), and less stable (Dunne and Leopold 1978, Konrad 2000). The result is a shift in habitat condition and overall stream health (Wolman and Schick 1967, Ralph et al. 1994). Streambank erosion represents a serious threat, not only for stream ecology, but also for urban areas—particularly for roads, bridges, and other infrastructure.

## Sequester and recycle nutrients

Hydrologic processes control ecosystem functions through biogeochemical processes. Water movement influences the fluxes of nutrients (i.e., carbon and nitrogen), which in turn induce further changes in the water cycle. As urbanization changes land use, more nutrients are exported from land to water bodies (i.e., phosphorous and nitrogen), changing nutrient regimes and causing changes in ecosystem processes. These changes then feed back into the water cycle through evapotranspiration, soil moisture levels, and a variety of hydrological responses. Although we can draw connections between the land use changes associated with urbanization and water and nutrient cycling, we do not fully understand the interactions and feedback mechanisms among water, carbon, and nitrogen fluxes at multiple temporal and spatial scales. The challenge is compounded because ecological processes are not modeled at the same spatial scales as hydrologic processes.

## Remove toxins and sediments from water

Wetland plants and soils play a key role in purifying water because they can filter toxins and other pollutants from the water. However, urbanization contributes significantly to the loss of such important functions when wetland areas are replaced and drained. Since the early 1800s, the United States has lost more than half of its wetlands (Dahl 1990). Of the remaining 100 million acres of wetlands, according to the US EPA (2005), about 100,000 acres per year are being lost today. Only recently have water managers started to recognize that preserving and restoring wetlands is a much less expensive way to maintain drinking water quality than building water treatment plants. Faced with the choice of technological or natural options for filtering water, New York City has recently chosen a watershed protection approach that preserves and restores nature's services.

## Support natural riparian and floodplain function

Urban basins have only limited capacity to store water. Water runs off paved and other impervious surfaces very quickly, causing urban streams to rise more rapidly during storms and have higher peak discharge rates compared to streams in predominantly rural and undeveloped basins (Booth 1990).

Development along stream channels and floodplains causes a series of geomorphic modifications, altering both a channel's capacity to convey water and its ability to prevent flooding (Konrad 2000). Several positive feedback mechanisms exacerbate this effect. Konrad (2003) describes how built structures (i.e., bridges) encroaching on the floodplain narrow the width of the channel, increasing the flooding upstream. Large amounts of sediments and debris transported by floodwaters can further constrict channels. Especially small stream channels rapidly fill up with sediments and exacerbate the magnitude and frequency of floods during storms (Konrad 2003).

Streamside vegetation provides flood control in several ways. Riparian areas that are maintained intact in the uplands and floodplains reduce the volume of floods by filtering and retaining water. Riparian vegetation also slows the movement of water and thus reduces flooding. Moreover, plant roots, whether of streamside trees or aquatic vegetation, help pull water into the soil. Since plants use large amounts of water in transpiration, they also perform flood control by reducing the amount of water that otherwise would flow downstream. Urbanization reverses all these functions by reducing the amount of riparian vegetation.

## Provide habitat for fish, waterfowl, and wildlife

Hydrological processes also support habitats for fish and wildlife. Coupled with geomorphic processes, hydrological processes play a key role in shaping aquatic habitats (Gregory et al. 1991, Naiman et al. 1999). As water, sediment, and wood debris interact, they create ecological disturbances that fundamentally influence the dynamics of ecosystems (Montgomery 2002). In addition important biotic feedbacks on physical structures are emerging (Naimen et al. 1999). Urbanization affects many geomorphic processes including erosion and fire, hydrologic responses, the transport of sediment, the stocks of woody debris, and the evolution and modification of channels. Changes in the hydromorphic regimes are influenced by urban impacts on local conditions, as well as impacts at the basin scale.

Hydrological processes interact with biological and geomorphic processes in complex ways by creating and maintaining aquatic habitats for a variety of species. These interactions are mediated by soil, vegetation dynamics, and nutrient cycling. Many aquatic organisms have coevolved and adapted to exploit the habitats created by these processes. Urbanization affects these interactions by changing both land cover and nutrient fluxes, and particularly by replacing wetlands that provide essential habitat for both fish and wildlife. While the impact of urban development on freshwater and coastal wetlands has been documented extensively, the impact that alternative urban development patterns and related hydrological modifications have on natural habitat is not explicitly addressed in current research, nor is it addressed in legislation aimed at mitigating the impact of urban growth.

## 5.3 Human-Induced Changes in Urban Watersheds

Rivers, lakes, and underground aquifers are the source of freshwater for irrigation, industrial activities, and households. They provide many essential products and services for humans, including food, transportation, flood control, water purification, energy, and recreation. Urbanization increases the human demand on water supplies and services, simultaneously increasing both the pressures on and threats to such services. Humans are fundamentally altering hydrologic processes through a range of behaviors: we are consuming more water, changing land cover, building artificial infrastructures, and introducing both point and non-point source effluents. As a result, the amount of precipitation varies more, runoff increases, and changes occur in flow regimes, drainage pathways, and even the composition of the water and sediments. In urban watersheds, these modifications of the water cycle threaten hydrological functions and, ultimately, the human and ecological functions that depend on them.

### Urban water use

Cities rely on piped water for their water supply. In the United States, 54,000 community water systems provide about 90% of Americans with their tap water. About 3,000 of these community systems provide more than 75% of the nation's water. Of these major systems, 80% are owned by municipalities and 20% by investors (AWWA 2007). Of the 341 billion gallons of water withdrawn in the US only 1% is used for drinking water. About 41% is used for agriculture, 39% for hydroelectric power, 6% for industrial use and 6% for household purposes. In the United States, residential end use of water is equivalent to more than 1 billion glasses of tap water per day (AWWA 1999).

The average American household uses approximately 107,000 gallons of water each year, 58% of it outdoors (gardening, swimming pools) and 42% indoors. It is ironic that water clean enough to drink is used for many other purposes: to flush toilets (~30%), to water lawns (~15%), and to wash clothes (~22%), dishes (~2%), and cars (~3%). Furthermore, about 14% of the water we pay for is never even used—it leaks down the drain (AWWA 1999).

Average per capita use varies among cities depending on many factors: climate, a city's mix of domestic, commercial, and industrial uses, the sizes of its lots and households, income levels, and public uses. Differences in the allocation of water use among households also depends on the use of water-efficient technologies. For households that utilize water-efficient technologies, faucets are the top water user (10.8 gallons per capita per day), followed by washing machines (10.0 gallons), showers (8.8 gallons), and toilets (8.2 gallons). In households not using efficient technologies, toilets use the most water (18.5 gallons per person per day), followed by washing machines (15 gallons) and showers (11.6 gallons), (Vickers 2001).

Until a few decades ago, many cities could meet their requirements for piped water from natural resources such as rivers, lakes, and aquifers. Today, cities have to collect such water in reservoirs to respond to peaks in consumption. In such basins water can evaporate or leak away; during hot dry periods the water levels in basins can drop dramatically. The need for clean water sources has also led cities to extract deep groundwater. In 2000, public suppliers withdrew about 43 million gallons a day of freshwater. Many cities have to pipe water over great distances, reducing the efficiency of the systems. As distances increase, more transformation stations and pumps are needed, requiring more energy and materials. Longer distances also imply more risks and losses, especially when the infrastructure is not well maintained.

## Land cover change

Changes in land cover associated with urbanization affect the hydrological cycle by altering precipitation, evaporation, evapotranspiration, and infiltration, as well as radiation flux (Pielke et al. 2006). As I discuss in more detail in Chapter 7 (7.3), researchers have long recognized the effects that urban areas have on the microclimate, energy regime, and air pollution caused by urban development. Land cover change and loss of canopy cover also have dramatic effects on the hydrological functions provided by vegetation and soils, particularly its wetlands and riparian areas. Because wetland areas are often inundated or saturated by surface water and ground water, they have unique physical, chemical, and biological conditions that allow them to support important hydrological, biogeochemical, and habitat functions.

Wetlands, which include swamps, marshes, and bogs, serve many important functions. They recharge aquifers and provide surface water for wildlife and human uses, including recreation, irrigation, and industrial processes. They also slow down water flow, attenuate floods, filter water, and cycle elements. As mentioned earlier, they provide habitats for amphibians, aquatic invertebrates, and fish. Similarly, riparian areas act as critical transition zones between upland and aquatic environments, keeping streambeds stable, preventing shoreline erosion, supporting the spawning and rearing ground for salmon, creating shelter for wildlife, and providing nutrients for macro-invertebrates. The riparian forest also provides food and energy to streams in the form of leaves, wood, and nutrients (Naiman 1992).

The functioning of aquatic ecosystems depends on the hydrological functions provided by both wetlands and riparian vegetation. A healthy, functioning ecosystem depends upon a stable substrate, an irregular water depth, a heterogeneous habitat, movement or sinuosity in the channel, and shoreline cover. Development along stream channels and shorelines changes these conditions, by removing vegetation, and by shifting land cover from forest to pavement and grass (Morley and Karr 2004, Booth et al. 2004).

## Impervious surface and runoff

Urban settlement modifies the landscape structure and processes that influence the amount, quality, and delivery of runoff, through a variety of mechanisms operating at local and watershed scales (Leopold 1968). The introduction of impervious surfaces has a significant impact on the hydro-logical regime (Leopold 1968, Dunne and Leopold 1978, Booth and Jackson 1997, Burges et al. 1998). Increasing the amount of impervious surface is the most direct way that urbanization reduces the ability of watersheds to intercept, retain, and filter rainfall. In addition, with more impervious surface, precipitation runs off more quickly. In urban areas with large proportions of impervious surfaces up to 55% of rainwater may drain off, as opposed to 10% to 15% in unbuilt areas (Schueler 1994). The annual volume of stormwater runoff can increase to up to sixteen times its pre-development volume (Schueler 1994). Figure 5.4 illustrates the influence of impervious cover on the hydrologic cycle and the relative amount of infiltration that occurs.

The impact that the impervious surface has on the amount, quality, and timing of flow depends on several factors: climate, morphology, geology, vegetation, and land use (Dunne and Leopold 1978, Ziemer and Lisle 1998, Booth and Jackson 1997). As infiltration is reduced, less water is available to recharge deep-water aquifers, thus lowering base flow means and re-ducing water available to streams during dry periods (Dunne and Leopold 1978, Klein 1979, Burges et al. 1998, Konrad and Booth 2002). Instead, water is delivered to streams quickly, often, and in large quantities. As a result, streams become more "flashy" as their flow changes more quickly and more often (Konrad 2000, Konrad and Booth 2002, Booth et al. 2004).

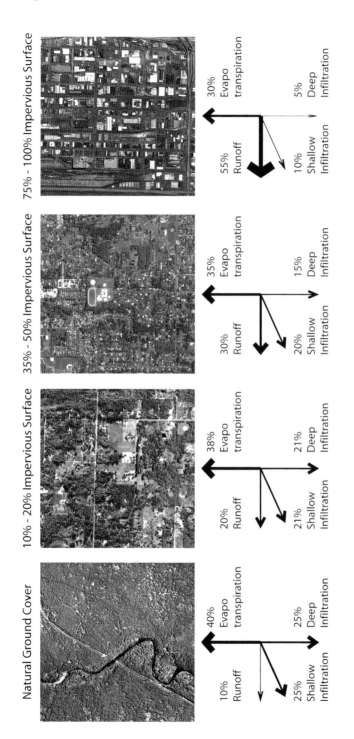

Figure 5.4. Relationship between impervious surface and runoff.

In an urban basin "flashiness" results from a combination of more areas having impervious surfaces and piped stormwater drainage systems that more efficiently transport the runoff from those impervious surfaces (Dunne and Leopold 1978). Post-development hydrographs consistently show changes across urbanizing basins with peak flows during high-flow events increasing sharply in magnitude (Figure 5.5). In addition, channels more often flood over, or nearly over their banks, and are exposed to critical erosive water speeds for longer intervals (Hollis 1975, Booth et al. 1996, MacRae 1996).

Increases in impervious cover levels create other key changes in urban streams: High flows occur more often, water is redistributed from periods of base flow to periods of storm flow, and the daily variation in stream flow increases (Konrad and Booth 2006). Increased impervious surface increases the relative variability of stream flow and, as more water reaches a stream more quickly, the peak flows increase (Konrad 2000). Without the capacity for surface retention that the forested landscape originally provided, much smaller storms result in observable changes in stream flow (Hollis 1975, Konrad and Booth 2002, Konrad and Booth 2006).

In predicting hydrologic changes due to urbanization, a useful indicator is total impervious area (TIA) in the catchment, although its influence varies depending on the permeability of the drainage basin land cover (Booth et al. 2004), and the amount of impervious area that drains directly to streams through pipes rather than through pervious land (Walsh et al. 2005). Konrad and Booth (2002) have proposed a hydrologic metric (TQmean) to measure flashiness, which may be a more accurate descriptor for the hydrologic effects of urban land use (Booth 2005). This metric measures the proportion of time the mean discharge is exceeded. Three factors contribute to flashiness: shorter lag to peak, shorter peak flow duration, and more frequent high flows (Konrad 2000, Konrad and Booth 2002, 2005, Booth et al. 2004).

The extent to which hydrological metrics can predict stream ecological conditions has been investigated extensively, but much remains unknown. When high flows become more common, they exert a strong ecological effect (Roy et al. 2005, Walsh et al. 2005); they are also likely to cut into channel edges and erode banks (MacRae and Rowney 1992). Flashiness can also increase the frequency of small overland floods. Small rain storms are not likely to cause large hydraulic stress in streams, even if their catchments are highly impervious. But if such events happen frequently, they can potentially impact biotic conditions, although the relative importance of these events is not known (Walsh et al. 2005).

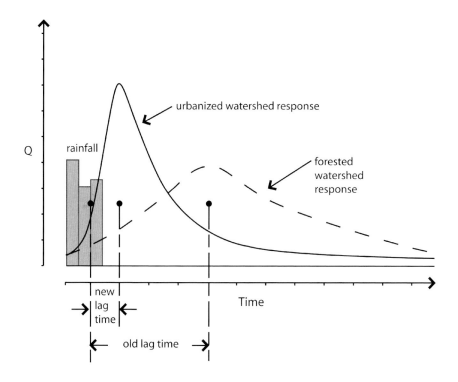

Figure 5.5. Urban hydrograph. The urban hydrograph is a stylized representation of change in discharge over time (line) and precipitation (bar graph). The lag time represents the time difference between when the most precipitation occurs and when the most discharge is recorded. The urbanized watershed discharge rates reflect a shorter lag time between the intensity of precipitation and the intensity of discharge.

## Changes in flow regime and sediment loads

Naturally variable flows create and maintain the spatial and temporal dynamics of in-channel and floodplain conditions and habitats; in turn, these influence the success of aquatic and riparian species (Poff et al. 1997). Five critical components of the flow regime regulate ecological and geomorphic functions in river ecosystems: magnitude, frequency, duration, timing, and rate of change of hydrologic conditions (Poff and Ward 1989, Walker et al. 1995). In urban watersheds these are all affected by urban development and related human activities, which alter both the gross and the fine-scale geomorphic features that constitute habitat for aquatic and riparian species.

Channel morphology is important to aquatic communities because it governs many other factors: flow regime, channel and substrate composition, and riparian and bank stability in streams. In turn, changes in flow regimes and sediment loads affect the shape and composition of the channel, including its substrate, sinuosity, cross-sectional area, and habitat sequence. Several factors determine the channel's morphology and shape, including the flow, quantity, and character of the sediment moving through the channel, and the composition of the channel's streambed and banks, including riparian vegetation (Leopold et al. 1964).

The urban flow regime is a key factor in changing the structure of the channel. The width and depth of stream channels adjust in response to long-term changes in both sediment supply and water flow (Dunne and Leopold 1978). Urban streams typically become wider and deeper to accommodate higher flows. The cross-sectional area of urban stream channels may increase by a factor of two to five, depending on the degree of impervious cover and the age of the development in the upland watershed (Arnold et al. 1982, Gregory et al. 1992, MacRae 1996). Some stormwater management strategies are designed to reduce channel erosion primarily by controlling flow rates from large events, but the small, frequent, high-flow events may cause channel incision and impact the ecology of urbanizing basins (MacRae and Rowney 1992).

In urban streams development can cause drastic changes to the flow regime. Numerous studies describe changes in sediment and erosion processes resulting from urbanization (Wolman and Schick 1967, Leopold 1972, Booth and Jackson 1997, Booth 1990). Observed changes include altered channel density (Dunne and Leopold 1978, Meyer and Wallace 2001), increased channel width and bank height (Leopold 1972, Booth 1990), reduced stability (Dunne and Leopold 1978, Booth and Jackson 1997, Konrad 2000), incision (Booth 1990, Neil and Yu 1999). Coupled with hydrological changes, modification of the stream channel has also drastic consequences for the aquatic habitat (Wolman and Schick 1967, Ralph et al. 1994, Morley and Karr 2004, Booth et al. 2004).

Scholars have documented changes associated with different stages of urbanization (Paul and Meyer 2001). During the construction phase, soil erodes down hillsides, increasing sediment supply, decreasing stream depths and leading to greater flooding and deposit of sediments on stream banks. After construction is completed, the supply of sediments is reduced, but the surface is more impervious, so more water flows over riverbanks. This leads to increased channel erosion, as channels are cut more widely and deeply to accommodate increased flow. The changes occurring after development has taken place and the erosion pattern has readjusted are the subject of most recent investigations (Hammer 1972, Finkenbine et al. 2000, Henshaw and Booth 2000). Some initial evidence indicates that channel morphology continues to change over time in urbanized areas, suggesting that channels may eventually restabilize (Finkenbine et al. 2000). However much uncertainty characterize these results (Hartley et al. 2001)

How stream channels respond to the new flow regime over time, and how much they recover hydrological function, is still not known. To address this question traditional cross-sectional studies need to be complemented with longitudinal studies. Greve (2007) evaluated changes in channel morphology on a temporal scale in several urban streams in the Puget Sound Lowland. She founded some encouraging results. Although not all the elements of channel morphology adjust to urban conditions over time, her findings indicate that channel stability improves. Thus time may be a key factor in assessing the impacts of urbanization on channel morphology.

## Urban effluents and wastewaters

A great proportion of the water withdrawn from surface and groundwater sources for human use (~70%) returns to water bodies. But only a small portion (~17%) is treated from publicly owned treatment works (POTWs), that are regulated through state and federal regulations and satisfy water quality standards. Households not served by public sewers usually rely on septic tanks. Septic systems fail and are often a major contribution to both ground water and surface water contamination. As of 2000, the US had an estimated 16,225 POTWs in operation, treating more than 40 billion gallons of wastewater daily. Almost all are municipally owned; together they provide wastewater collection, treatment, and disposal service to 190 million people (EPA 2000).

Urban catchments show high concentrations of nutrients (e.g., phosphorus and nitrogen) similar to, if not greater than, those seen in agricultural catchments (Omernik 1976). Although nutrients occur naturally in the environment and are essential to plant life, excess amounts of phosphorus and nitrogen cause rapid growth of phytoplankton, creating unsuitable habitat conditions for fish and other species. In addition to wastewater (LaValle 1975), lawns and streets are important sources of nutrients that wind up in urban streams (Waschbusch et al. 1999). Soils are important in retaining phosphorus and nitrogen in areas where septic systems are prevalent (Hoare 1984, Gerritse et al. 1995).

Furthermore, urban areas have many sources of metals (Mason and Sullivan 1998, Horowitz et al. 1999); automobiles are probably the most ubiquitous source of metal contamination in urban streams. Lead and zinc are washed out of roads and parking lots in large quantities (Sartor et al. 1974, Forman and Alexander 1998). Brake linings and tires contain zinc, lead, chromium, copper, and nickel (Muschak 1990, Mielke et al. 2000).

Other metals (i.e., arsenic and iron) found in high concentrations in urban stream sediments result from a variety of construction and production processes (Khamer et al. 2000, Neal and Robson 2000).

Urban runoff also contributes a large amount of pesticides and insecticides. Households and businesses account for one third of US pesticide use, on lawns and golf courses (LeVeen and Willey 1983, USGS 1999); application rates in urban environments are estimated to exceed those in agricultural applications by an order of magnitude (Schueler 1994). Other organic contaminants frequently detected in urban streams include polychlorinated biphenyls (PCBs), polycyclic aromatic hydrocarbons (PAHs), and petroleum-based aliphatic hydrocarbons (Whipple and Hunter 1979, Frick et al. 1998).

## 5.4 Urban Patterns and Stream Biotic Integrity[1]

The effects of landscape change on stream ecosystems have been documented extensively (Omernik 1987, Roth et al. 1996, Paul and Meyer 2001). As I described earlier in this chapter, the composition of land cover within a watershed can account for much of the variability in water quality (Hunsaker et al. 1992, Charbonneau and Kondolf 1993) and stream ecology (Whiting and Clifford 1983, Shutes 1984, Hachmoller et al. 1991, Thorne et al. 2000). Both human and natural processes drive landscape changes and affect the functions of aquatic ecosystems. Changes in vegetation cover due to suburban development and clear-cutting can produce dramatically more runoff than in unaltered watersheds (Franklin 1992). Then, the loss of riparian habitats greatly increases runoff and makes watersheds less able to filter nutrients and sediments (Karr and Schlosser 1978, Peterjohn and Correll 1984). As impervious surfaces increase, they affect the volume of water and sediment that streams can move (Arnold and Gibbons 1996). In total, an increase in impervious surfaces leads to many kinds of hydrologic, geomorphic, and biological degradation in streams (Booth and Jackson 1997, Yoder et al. 1999).

The question of how patterns of human settlements affect aquatic ecosystems is increasingly important to both ecology (Forman 1995, Collins et al. 2000, Grimm et al. 2000, Pickett et al. 2001) and urban planning (Collinge 1996, Alberti 1999b, 1999c, Opdam et al. 2002). Many researchers have addressed the relationship between watershed urbanization and the associated biotic

---

[1] This section summarizes the results of a study published in Alberti, M., D. Booth, K. Hill, B. Coburn, C. Avolio, S. Coe, and D. Spirandelli. 2007. The impact of urban patterns on aquatic ecosystems: An empirical analysis in Puget lowland sub-basins. Landscape and Urban Planning 80(4):345–361. Modified paragraphs extracts are reprinted with permission.

conditions in streams (e.g., Karr and Schlosser 1978, Arnold and Gibbons 1996, Booth et al. 2001, Paul and Meyer 2001), but few have investigated how the patterns of urban development control hydrological, geomorphological, and ecological processes in human-dominated watersheds. For example, we do not know how clustered versus dispersed urban patterns affect runoff and in-stream ecological conditions, even though clustered design has become a mainstay of what is presumed to be low-impact development.

My team at the University of Washington conducted an empirical study of the impact that urban development patterns had on ecological conditions in the Puget Sound lowland region (Alberti et al. 2007). We explored 42 urbanizing sub-basins within the King County and Snohomish County lowlands and used population measures of benthic macroinvertebrates (B-IBI) as indicators of a stream's biological integrity. After examining a broad range of landscape patterns, we proposed that ecological functions are influenced by four dimensions of urban development patterns: landscape composition, landscape configuration, land-use intensity, and land-use connectivity.

Since ecological processes are tightly interrelated with the landscape, the mosaic of elements resulting from human action has important implications for ecosystem dynamics. Our overarching hypothesis was that the pattern of human activities—extent, distribution, intensity, and frequency—affects hydrological and ecosystem functions. We hypothesized that we could explain the variability in stream biotic conditions by looking at differences in the configuration of impervious area and forest patches: between different locations, between areas with differing degrees of connectivity, or between different types of hydrologically distinct land cover—as long as we controlled for land cover composition. Hydrological processes are important mechanisms that link urban patterns to stream conditions since more aggregated patterns of impervious surface may increase runoff, and more fragmented patterns of forest reduce the ability of vegetation to intercept surface-water runoff. The more pervasively the upland watershed surface is dissected, the more likely we are to see aquatic ecosystems affected by the variations in urban form along the continuum of clustered versus dispersed development.

We found significant statistical relationships between the ecological conditions in streams and the urban landscape patterns—for both the amount and the configuration of impervious area and forest patches (Figures 5.6a, 5.6b, 5.6c). The proportion of trees and impervious surfaces within the riparian corridor has a direct effect on a stream's habitat structure. Higher B-IBI scores are associated with a greater amount of intact forest in the basin and lower percentages of impervious area. Generally, an increase in impervious surfaces within the stream buffer causes forest fragmentation and removes—or prevents the addition of—fallen logs and other woody structures. A healthy, functioning ecosystem depends upon several factors that are affected by impervious surfaces: stable substrate, irregular water depth, habitat heterogeneity, channel movement or sinuosity, and shoreline cover. Basins with a generally intact riparian forest and less total impervious area have better biological conditions (reflected by a higher B-IBI value).

We also found a strong relationship between the average size of urban patches in a basin and the basin's biological conditions. Urban patches are predominantly composed of impervious cover, and large urban patches represent contiguous areas of impervious surfaces. Basins that have a higher average of urban patches exhibit lower B-IBI scores (Figure 5.6d). These large surfaces directly impact the hydrologic function of a basin by altering its stream flow pattern, especially by increasing its storm runoff. Far more of the rainwater runs over the surface rather than being taken up by trees and other land cover, dramatically impacting a basin's pattern of flow; therefore these surfaces increase flood frequency, and sometimes more than double its flood peaks (Booth 2000). As a basin's hydrology is altered, it has significant impacts downstream; shoreline banks become vulnerable to erosion and produce an excess of sediment and nutrient inputs. Subsequently, microhabitats diminish, along with the availability of shelter (Karr 1991).

## Multiple stressors

Over the past decade, many studies have linked urbanization with the condition of aquatic ecosystems (Hunsaker et al. 1992, Charbonneau and Kondolf 1993, Booth and Jackson 1997, Wang et al. 1997, Karr 1998, Yoder et al. 1999, Finkenbine et al. 2000, Thorne et al. 2000). Although impervious surface emerges as perhaps the most prominent stressor, it is clear that no single variable can explain the complex relationships between urban development and ecological conditions in watersheds. Aquatic ecosystems can be altered by human actions in several ways (Karr 1995). Our study confirmed the strong correlation between urban land cover and B-IBI found by others (Morley and Karr 2002). TIA also explains a large part of the variance in B-IBI across the 42 basins (Figure 5.6a). However, while TIA is highly correlated with multiple factors in urbanizing landscapes (i.e., population density, housing density, and road density), it does not fully represent the complex relationships between land use and land cover. Different landscape patterns (i.e., mean patch size of urban land) and connectivity (i.e., number of road crossings) contribute to ecological conditions across basins (Figures 5.6, 5.7, and 5.8). A highly aggregated urban landscape also impacts the hydrologic function of a basin. Figure 5.6b illustrates the strong relationship between highly aggregated impervious surfaces and poor biological conditions. Further, impervious surfaces that dominate the basin's landscape are direct sources of pollution, especially when vegetation or other land cover does not interrupt them. Such surfaces connect polluted urban run-off directly to streams and other vulnerable aquatic ecosystems.

## Roads as key urban stressors

Roads are another key stressor in urbanizing landscapes (Jones et al. 2000, Trombulak et al. 2000). This is particularly relevant given that studies of land-use/land-cover indicates that road intensity is correlated with total impervious surface in basins. Since roads increase the total amount of impervious surface, and ditches are built to channel water from roads into streams, the rate of water runoff is higher in basins with a greater number of roads. More specifically, we found that road density—especially the number of road crossings—is a better predictor of B-IBI than raw total impervious area (Figures 5.6d and 5.8). The important effect of road crossings can be related to the cumulative effect of various road-related stresses: Stream banks and channels are altered, petroleum products leak onto road surfaces, and the amount of both pollution and sediment increases.

## Landscape fragmentation

Moreover, our study clearly indicated that at the scale of the watersheds supporting individual tributary streams, patterns of urban development affect ecological conditions on an urban-to-rural gradient. At this scale, previous research has shown that impervious surfaces result in characteristically altered and often extreme hydrologic conditions that provide an endpoint on a disturbance gradient (Meyer et al. 1988, Booth and Jackson 1997, Konrad and Booth 2002). However, the indicators of percent

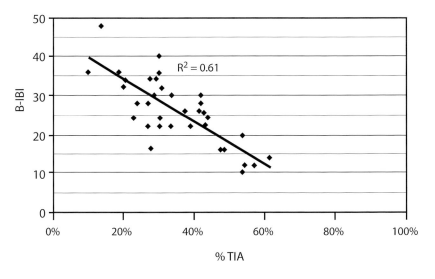

Figure 5.6a.

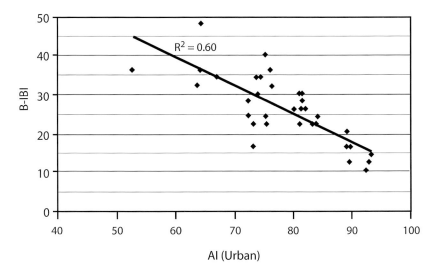

Figure 5.6b.

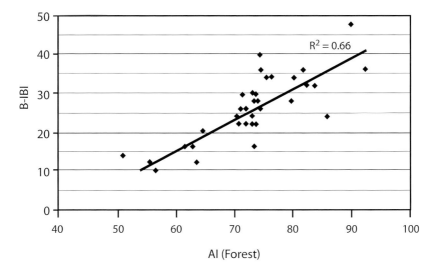

Figure 5.6c.

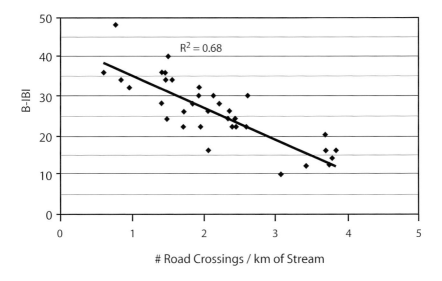

Figure 5.6d.

Figure 5.6. Relationship between Benthic Index of Biotic Integrity and landscape metrics. a) Benthic index of biotic integrity (B-IBI) decreases as the percentage of total impervious area in a watershed increases. b) B-IBI decreases as the aggregation of urbanized land cover increases. c) B-IBI increases as the aggregation of forest land cover increases. d) B-IBI decreases as the number of road crossings over a stream divided by the kilometers of stream length increases. (Alberti et al. 2007, p. 257–8).

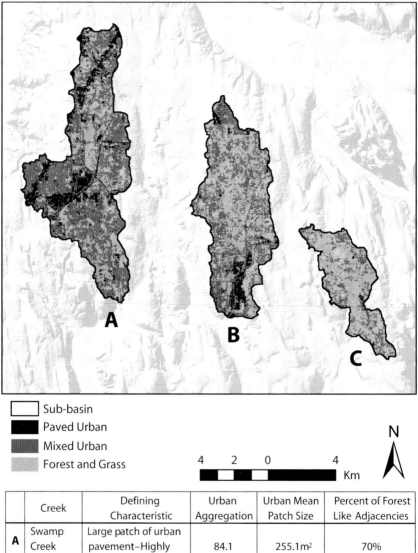

| | Creek | Defining Characteristic | Urban Aggregation | Urban Mean Patch Size | Percent of Forest Like Adjacencies |
|---|---|---|---|---|---|
| A | Swamp Creek | Large patch of urban pavement–Highly aggregated | 84.1 | 255.1m² | 70% |
| B | Little Bear Creek | Medium patch of urban pavement Medium aggregation | 75.4 | 98.5m² | 73% |
| C | Big Bear Creek | Highly compact forest cover | 64.4 | 30.8m² | 81% |

Figure 5.7. Landscape patterns in three sub-basins in Puget Sound Washington.

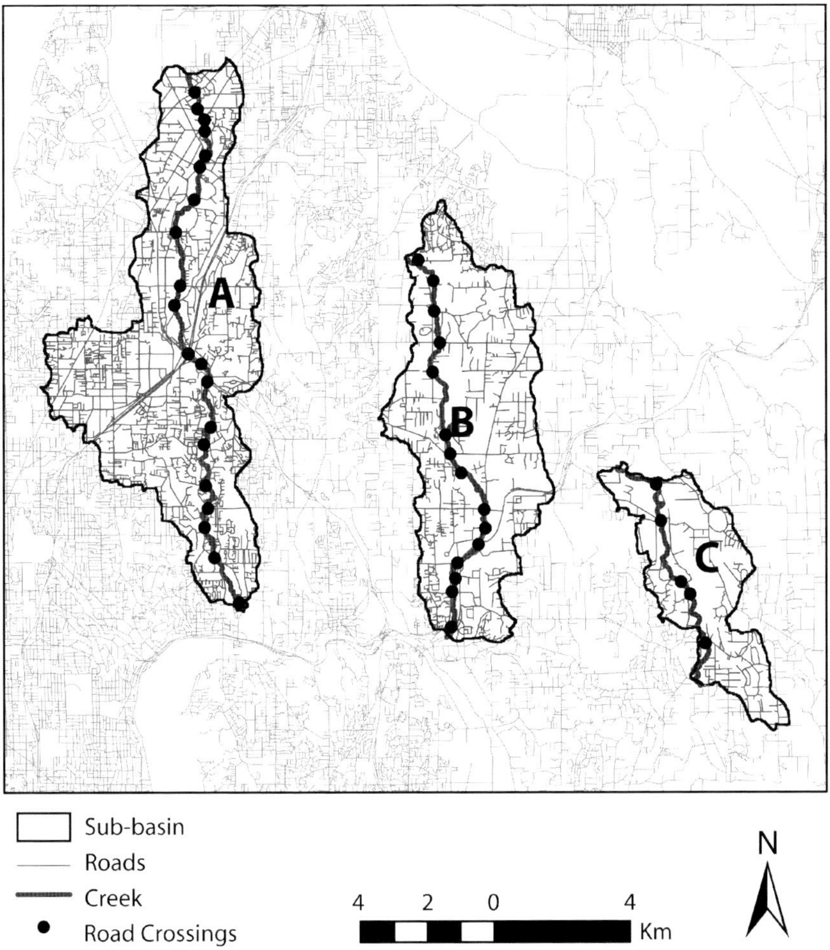

| | Creek | Road Density (km/km²) | Road Crossings (crossing/km of stream) |
|---|---|---|---|
| A | Swamp Creek | 7.85 | 2.44 |
| B | Little Bear Creek | 5.48 | 2.34 |
| C | Big Bear Creek | 4.22 | 1.4 |

Figure 5.8. Road crossings and road density in three sub-basins in Puget Sound Washington.

impervious area and percent forest in the contributing watershed are only coarse predictors of biological conditions in streams, in part because hydrological change is only one of several factors that affect stream biota. The aggregation of urban land and forest land cover may explain some of the variability in B-IBI that TIA cannot explain.

We also found significant statistical relationships between selected landscape patterns and ecological conditions in streams. While the findings clearly suggest that patterns of urban development matter to watershed function, this relationship does not indicate a specific threshold; instead it shows that both the increase in the percentage of impervious surface and its aggregation have a direct impact on the macroinvertebrates in the stream. In particular, as the probability of urban land cover being adjacent rises from 50 to 100 percent, typical B-IBI values decline from 50 (excellent) to 10 (very poor). However, we found a high correlation between the aggregation index and amount of paved urban cover in the basins. Given this finding, we cannot reject the hypothesis that no significant relationship exists between basin level aggregation of paved land and B-IBI values beyond those already explained by the amount of TIA in the basin. We did find, however, that mean patch size and number of road crossings are better predictors than percent TIA alone.

## Basin and riparian effects

Our multiple-scale analysis aimed at discriminating across patterns that operate at different scales—from local riparian zone to basin. Since landscape metrics are scale-dependent, we systematically examined the relationship between each variable and B-IBI at each scale. Except for the local riparian zone, all the variables are highly correlated with B-IBI across the various scales. However, we were not able to determine whether the effects of land cover composition and configuration vary with scale. Particularly since the riparian and sub-basin variables are closely correlated ($R = 0.95$, $P < 0.001$), it is difficult to discriminate between riparian and sub-basin effects because of the nested effect, even though the processes that affect aquatic ecosystems are clearly different.

## Streams as dynamic systems

While the processes of urbanization are certainly associated with ecological degradation in streams, the interactions between urban development and stream ecosystem function are by no means homogenous in terms of their spatial and temporal patterns. Streams are heterogeneous and dynamic systems.

If the aquatic ecosystems in urbanizing watersheds are to remain healthy, we must understand the important role that hydrologic flow regimes and stream flow variability play in maintaining critical physicochemical characteristics of rivers, such as water temperature, channel geomorphology, and habitat diversity. The flow regime plays a critical role in linking urban patterns to ecosystem function. Stream flow quantity and timing are critical components of water supply, water quality, and the ecological integrity of river systems.

A new research agenda is emerging at the intersection of hydrology and ecology: how to maintain ecosystem function while supporting human needs in increasingly human-dominated watersheds (Palmer and Bernhardt 2006). Part of this research will need to address and resolve what reference condition for "ecosystem function" we should aim at when evaluating stream conditions and the effectiveness of restoration strategies; we also need to identify the feedback associated with critical thresholds in variables that may cause stream systems to shift between alternative states. More research is necessary to explore the mechanisms by which patterns of urban development affect the ecological conditions of streams. This knowledge is critical in determining the processes that need to be maintained in order to ensure that ecosystem services can simultaneously support humans and other species.

# Chapter 6

# BIOGEOCHEMICAL PROCESSES

## 6.1 Urban Biogeochemistry

Urbanization affects Earth's ecosystems by changing fundamental processes that control the cycling of elements. Biological, hydrological, atmospheric, and geological processes play essential roles in terrestrial biogeochemical cycles by regulating the synchrony between release and uptake of nutrients by microorganisms and plants (NRC 1986, Melillo et al. 2003, Dahlgren 2006). Across multiple scales from molecular to the entire ecosystem level, biological processes regulate nutrient cycles by providing fuel and materials to Earth's ecosystems. Humans have altered biogeochemical processes in fundamental ways, by burning fossil fuels, changing land uses, extracting metals, and producing and applying synthetic chemicals (Figure 6.1) (Vitousek et al. 1997).

Scientists have long recognized the global implications of the human impact on biogeochemical processes, but we have only a limited understanding of how urban ecosystems contribute to these changes and are affected by them. Today we have more available data and more sophisticated measurement techniques and analytic tools that allow us to explore relationships between human and biogeochemical processes in ways not possible before. Urban ecology scholars have challenged biogeochemical models, suggesting that they are too simplified to apply in urban ecosystems (McDonnell et al. 1997). Scholars in Phoenix and Baltimore suggest that a distinct urban biogeochemistry is emerging due to human-controlled fluxes of energy and elements (Groffman et al. 2004, Grimm et al. 2005, Kaye et al. 2006). Despite the great variability across the bio-physical settings and socioeconomic activities of cities, the coupling of human and natural processes may be creating a unique biogeochemistry. Humans in urbanizing regions affect biogeochemical processes directly by adding nutrients, and indirectly by modifying the mechanisms that control the spatial and temporal variability of nutrient sources and sinks. Kaye et al. (2006) propose that in cities, complex interactions between society and the environment are mediated by distinctive factors such as the degree of impervious surface, the built infrastructure, demographic trends, and the individual choices of

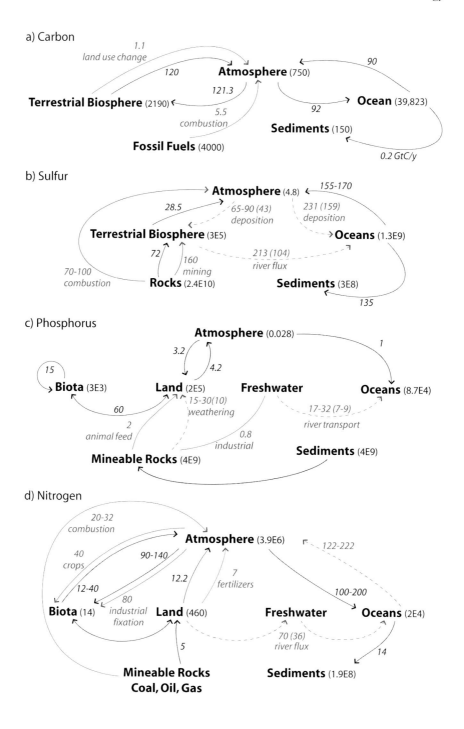

a) Carbon

b) Sulfur

c) Phosphorus

d) Nitrogen

Figure 6.1. Global fluxes of nutrient cycles. Global Fluxes are in Tg per year (Tg = $10^{12}$g) and are associated with directional arrows. Solid gray arrows represent additional and isolated anthropogenic fluxes. Dashed gray arrows represent anthropogenic influences onto existing fluxes. Storages for the major components of the system are included in parenthesis in Tg. Fluxes and storages are synthesized from MEA 2005 unless otherwise noted. a) Global fluxes of the Carbon cycle, synthesized from NASA 2007 b) Global Fluxes of the Sulfur Cycle c) Global Fluxes of the Phosphorus Cycle. d) Global Fluxes of the Nitrogen cycle.

households. If we are to fully understand the role that cities play in the cycling of chemicals, I propose that urban ecologists must also examine how the distinctive spatial heterogeneity caused by human choices, including patterns of urban land use and infrastructure, defines urban biogeochemistry.

Urban ecosystems have a unique metabolism that we can characterize and measure in terms of stocks and flows of materials and energy that move between ecosystems and socioeconomic systems (Fischer-Kowalski 1998). Cities cycle and transform raw materials, fuel, and water into the urban built environment, and into human biomass, consumable products, and waste (Decker et al. 2000). Over the last fifty years, the study of urban metabolism has evolved to more explicitly link biogeochemical and human processes (Wolman 1965, Odum 1975, Newcombe et al. 1978, Hultman 1991, Folke et al. 1997, Fischer-Kowalski 1998, Decker et al. 2000, Kaye et al. 2006, Kennedy et al. 2007).

The urban metabolism is becoming increasingly central to the emerging field of industrial ecology (Ayres and Simmonis 1994, Graedel and Allenby 1995, Fischer-Kowalski 1998). Urban dwellers have distinctive preferences and behavioral patterns associated with the demographics and economic characteristics of urban populations that make them more energy intensive, but perhaps more energy efficient, than their rural counterparts (Liu et al. 2003). Production and consumption processes cannot be fully understood without explicitly accounting for the global shift of the population towards urban regions. Urban metabolism has also emerged as one of the dominant approaches in the study of the ecology of cities, primarily focusing on net primary productivity, biomass accumulation, nutrient cycling, energy efficiency, and system resilience (Kaye et al. 2006).

Studies of the mass balance of cities have provided important insights about the role that urban dwellers play in the cycling of chemicals, but before we can articulate and test hypotheses about a distinctive urban biogeochemistry, we will need to more explicitly represent the complex feedback mechanisms and spatial interactions between human and bio-physical factors that control biogeochemical processes in urbanizing watersheds (Likens 2001). For example, studies of carbon balance indicate clearly that cities import large amounts of carbon and constitute major sources of the $CO_2$ emitted into the atmosphere (primarily because of fuel combustion and changes in land use and land cover). Further, cities have a limited ability to sequester carbon, and urban vegetation represents only a small fraction of the carbon stocks worldwide (Nowak 1993). The form and structure of cities may well influence the amount of $CO_2$ emitted and urban forests can act as sinks, but we do not clearly understand how large a role urban form plays in the urban carbon budget. Furthermore, while we know that cities are sinks for nitrogen (Baker et al. 2001, Groffman et al. 2004), for phosphorus (Faerge et al. 2001, Warren-Rhodes and Koenig 2001), and for metals (Graedel et al. 2004), we do not know the mechanisms by which other factors—demographics, human activities, and wastewater infrastructure—affect nutrient cycling in urban ecosystems.

Studies of urban and suburban ecosystems have been extremely limited compared with the extensive studies of biogeochemical cycles in non human-dominated terrestrial and aquatic ecosystems, (McDonnell and Pickett 1990, McDonnell et al. 1997, Baker et al. 2001, Groffman et al. 2004). However, existing studies provide a set of testable hypotheses. Biogeochemical signatures can be associated with urban land use patterns owing to the distinctive land cover composition and configuration of urban landscapes (Alberti 2005). Nutrient cycling in urban and suburban areas may be altered by atmospheric deposition of nitrogen (Lovett et al. 2000) and the urban heat island effect (Oke 1987). It interacts with ozone and heavy metals (Pouyat and McDonnell 1991), altering leaf and soil quality and simultaneously decreasing microbial growth and decomposition and mineralization rates (Carreiro et al. 1999). Higher incidence of invasive species may alter soil nutrient cycling due to increased populations of earthworms (Blair et al. 1995, McDonnell et al. 1997).

Observations from several biogeochemistry studies along urban gradients suggest that unique interactions occur between the biological, physical, and chemical environments (McDonnell and Pickett 1990, McDonnell et al. 1997, Kaye et al. 2006). For example, less decomposition and nitrogen-cycling occur in urban forests because they have smaller populations of fungi and micro-arthropods and a lower quality of leaf litter, but the heat island effect serves as a counterbalance, along with the introduction and successful colonization of earthworms in the urban forests (McDonnell et al. 1997). In fact, in forests at the urban end of the urban-to-rural transect, litter decomposes more quickly and nitrification rates are higher than in rural forests. Experimental studies in New York suggest that earthworms may play an important role in urban forests by enhancing the nitrogen-cycling processes and by compensating for the effects of air pollution on litter quality and decomposition (Steinberg et al. 1997). Urban $O_3$ exposures compared to rural $O_3$ exposures explain the differential in tree growth in and around New York City (Gregg et al. 2003).

Biogeochemical cycles in urban ecosystems are emergent properties of coupled human-natural systems (Wu et al. 2003). Understanding their functioning and potential changes requires considering not simply expected changes in individual processes but changes in dynamic interactions among multiple factors such as changes in climate, demographics, economics, land use, and biodiversity. These dynamics can be observed only over a long period of time and by conducting large-scale experiments across multiple urbanizing regions. Current coupled long-term urban biogeochemistry studies at both the local (i.e., Groffman et al. 2004, Grimm et al. 2005) and regional scales (i.e., Carpenter et al. 2006, 2007) are starting to lay out the direction of such work while uncovering its complexity.

In this chapter I synthesize some initial observations emerging from the current understanding of urban biogeochemistry, including key drivers of nutrient cycling in urban ecosystems, their interactions, and their local and global impacts. In urban ecosystems, changes in land cover, hydrology, climate, and soil have greatly changed nutrient cycling. I ask what the sources and sinks of potentially limiting nutrients are and what controls their spatial and temporal variability in urbanizing regions. I propose that urban patterns and the built infrastructure may play a key role in understanding urban biogeochemistry.

## 6.2 The Carbon Cycle

Humans are changing the carbon cycle by burning fossil fuels and biomass and by changing land cover. Increases in $CO_2$ in the atmosphere are now clearly linked to climate change, and humans are unequivocally contributing to those $CO_2$ levels (Intergovernmental Panel on Climate Change [IPCC] 2007). The atmospheric concentration of carbon dioxide has increased from a preindustrial value of about 280 parts per million (ppm) to 379 ppm in 2005. Annual fossil fuel carbon dioxide emissions have increased globally from an average of 6.4 gigatons of carbon (GtC) per year in the 1990s to 7.2 GtC in 2000–2005. Land use changes are estimated to have added 1.6 [0.5 to 2.7] GtC (5.9 [1.8 to 9.9] $GtCO_2$) per year through the 1990s, although these estimates have a large uncertainty (IPCC 2007). North America contributes $1.6 \times 10^3$ million tons (Mt) per year to total emissions, of which more than 85% came from the United States (Marland et al. 2005). The current $CO_2$ concentration levels greatly exceed the natural range over the last 650,000 years (180 to 300 ppm) as determined from ice cores. The atmospheric $CO_2$ concentration will likely double over the next 100 years, increasing Earth's temperature and greatly impairing the well-being of both ecosystems and humans (IPCC 2007).

Urban development affects the carbon cycle through both direct and indirect pathways (Figure 6.2), with increasing fossil fuel emissions among the most significant of such impacts (Pataki et al. 2006). About 40% of total fossil fuel emissions in the United States is attributed to the transportation and residential sectors (WRI 2005). Factors that likely affect per capita $CO_2$ emissions include population and housing densities, the rate of population growth, affluence, and technologies (Pataki et al. 2006). Demographic trends interact with urban forms in ways that have an impact on the emissions of $CO_2$. For example, in the United States household size is decreasing (US Census 2000), which implies that the number of households is growing more quickly than the population, requiring more dwelling units and greater consumption of resources per capita. At the same time, the National Association of Home Builders (NAHB 2005) reports an increase in the average size of single-family homes in the United States from 139 m² (1500 ft²) in 1970 to more than 214 m² (2300 ft²) in 2004. These simultaneous trends in household and housing sizes produce an overall increase in per capita $CO_2$ emissions and an increase in the consumption of land associated with urban development.

The pattern of urban development may be key to determining the amount that cities contribute to $CO_2$ emissions and the ability of cities to reduce those emissions. The number and size of households affect the number and size of housing units and associated energy uses. Furthermore, the spatial distribution of residential and commercial housing units affects commuting patterns and transportation choices with important consequences for fossil fuel consumption. Future trajectories of urban form and infrastructure choices in developing cities across the world may very well be decisive factors in future rates of fossil fuel emissions (Alig et al. 2003, Ironmonger et al. 1995, Liu et al. 2003, MacKellar et al. 1995, Ewing et al. 2003). Understanding the relationship between development patterns and greenhouse gas emissions is an important step towards formulating land use strategies that can minimize greenhouse effects in the long term.

Land cover change associated with urbanization also has impacts on the carbon budget, in terms of both sinks and sources. The most obvious impact of changing land cover is its negative effect on net primary productivity (NPP), although differences can be observed at the local and regional scale depending on regional climate and other limiting factors (Imhoff et al. 2004). Studies of arid and semi arid urban ecosystems show enhanced carbon-cycling rates (Imhoff et al. 2004, Kaye et al. 2005). Imhoff et al. (2004) indicate that urbanization can increase NPP in resource-limited

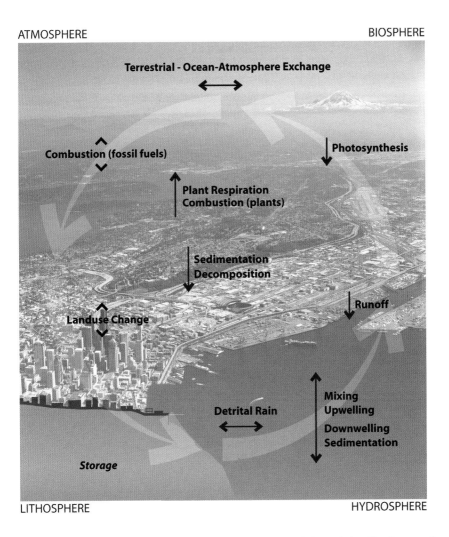

Figure 6.2. The carbon cycle in the urban landscape (Background photo: © Aerolistphoto.com).

regions. In Phoenix, Kaye et al. (2005) found that urban areas had dramatically altered aboveground net primary productivity and soil respiration rates compared with native grasslands and agricultural areas. Variation in C fluxes were observed and explained in terms of higher soil respiration rates and belowground C allocation, with a greater impact on N cycling and soil microbial biomass and a smaller effect on the composition of the soil microbial community.

Urban vegetation and soils in cities may also sequester carbon (Nowak and Crane 2002, Pouyat et al. 2006, Woodbury et al. 2007). Nowak and Crane (2002) estimate that urban trees in the coterminous United States make a significant contribution by storing carbon on the order of 700 Mt (335–980 Mt); the gross sequestration rate is 22.8 MtC per year (13.7–25.9 MtC per year). In a recent review of the carbon cycle in urban ecosystems in North America, Pataki et al. (2006) conclude that urban vegetation and soil are unlikely to be able to offset $CO_2$ emissions but could help reduce them. The emerging understanding of the influence of land use legacies on ecosystem structure and function for decades or centuries (Foster et al. 2003) gives land use history a central role in the assessment of urban forests in regional and global carbon fluxes.

## 6.3 The Sulfur Cycle

Humans affect the sulfur cycle by emitting sulfur dioxide ($SO_2$) from industrial processes and energy production (Figure 6.3). Sulfur dioxide emitted into the atmosphere travels long distances; it is reduced to sulfide, and oxidized to sulfate in the atmosphere as sulfuric acid. Finally, it precipitates on land and water in the form of dry depositions and acid rain. Humans have greatly altered the sulfur cycle, but the magnitude of the human impact and its consequences are not fully known (Stewart and Howarth 1992). We know that anthropogenic sulfur emissions affect human and ecosystem health locally (e.g., urban air pollution and smog), regionally (e.g., acid rain and dry deposits), and globally (e.g., climate change). Aerosol particles of sulfate are one of the most important human-induced factors driving climate change, as they reflect sunlight into space and make clouds more reflective. Although uncertainty is high, the radiative forcing of sulfate aerosols is the second largest ($CO_2$ is largest) and acts to cool the atmosphere (IPCC 2007).

Urban dwellers contribute to the emissions of $SO_2$ primarily as they consume energy and industrial products. More than 90% of the $SO_2$ emitted in the United States is from burning coal which accounts for one fourth of the nation's total energy consumption and about 50% of its total electricity production. Most electricity consumed by the average US household is used for appliances (including refrigerators and lights), which consume approximately two thirds of all the electricity used in the residential sector (EIA 2006). Thus buildings and lifestyles contribute substantially to the emissions of sulfur oxides and the overall impact on the sulfur cycle. In cities, human exposure to both sulfur oxides and nitrogen oxides contributes to respiratory disease and deaths in humans and significant damage to plants.

ATMOSPHERE                                                    BIOSPHERE

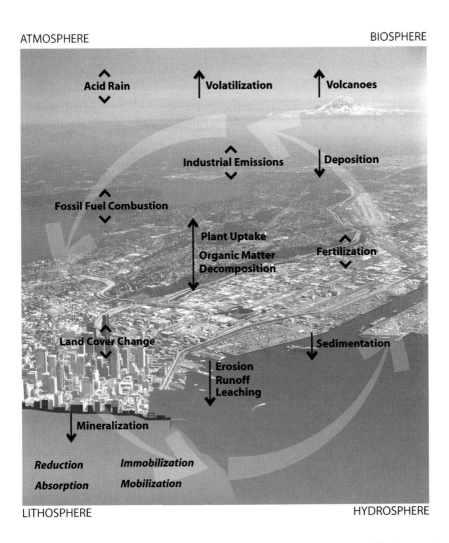

LITHOSPHERE                                                   HYDROSPHERE

Figure 6.3. The sulfur cycle in the urban landscape (Background photo: © Aerolistphoto.com).

While global sulfur emissions increased continuously during the last century until the late 1980s (Lefohn et al. 1999), they have now started to decline as a result of emissions controls in most industrialized countries. Regional agreements in North America and Europe have been developed in response to concerns about acidification. But while global emissions have varied little over 1980–2000, regionally there have been major shifts with a substantial decline in emissions from the United States, Canada, and Europe, and an increase from Asia. Emissions and concentrations of sulfur dioxide have declined in North American and western European cities but remain high in most developing countries, particularly in the largest urban agglomerations such as Beijing, Mexico City, and Seoul.

## 6.4 The Phosphorus Cycle

Phosphorus is essential to the biochemistry of plants and animals. For many organisms it is a key factor limiting growth (Smith 1992), but we do not clearly understand the mechanisms behind this constraint (Vitousek et al. 2002). Although phosphorus cycles much more slowly than carbon and nitrogen, it does move from terrestrial to aquatic systems and then through water, soil, and sediments to living organisms (Figure 6.4). In aquatic ecosystems an excess of phosphorus can cause eutrophication and algal blooms. Worldwide, humans mobilize three times the amount of phosphorus than would naturally flow through systems (Smil 2000). Human-induced erosion greatly increases the amount of phosphorus that moves from soils into air and water; as of 2000 this figure was 30 megatons (Mt) a year, compared to the baseline, where air and water lose about 10 Mt of phosphorus a year. Several mechanisms increase erosion and runoff, including land conversion, recycling of crop residues, untreated human wastes and urban sewage, and applications of inorganic fertilizers (Smil 2000).

The application of inorganic fertilizers (15 million tons of phosphorus a year) is a major reason that phosphorus becomes concentrated in sewage effluents and streams; animal and industrial wastes, including detergents containing phosphorus, make a relatively small global contribution (Bennett et al. 2001). On a local scale, however, urban sources of phosphorus can be comparable to rural ones, although the sources can be very different. The USGS collected data between 1992 and 1998 in 36 major river basins and aquifers distributed across the United States; it found that in about two-thirds of stream sites in urban areas, the concentrations of phosphorus were at least 0.1 part per million (ppm), and about 10% of urban streams sites had concentrations of at least 0.5 ppm.

The U.S. Environmental Protection Agency (EPA) has recommended a limit of 0.1 ppm in order to prevent excess algae growth in streams. In 2000, EPA took steps to facilitate the development of regional criteria, which have not yet been adopted. There is no federal drinking water standard for phosphorus. More than 70 percent of water samples from urban streams exceed the EPA desired goal for preventing nuisance plant growth (EPA 2000).

Figure 6.4. The phosphorus cycle in the urban landscape (Background photo: © Aerolist-photo.com).

Urban development is likely to increase runoff and erosion, exacerbating eutrophication in lakes. In Seattle, levels of lake eutrophication are highest at the urban fringe, most likely because of the dominant septic systems (Moore et al. 2003). Eutrophication of lakes at the urban fringe also suggests an increasing role of non point source vs. point source pollution in urbanizing regions. While most US cities have addressed point sources of nutrient pollution, non point source pollution in urban areas is not explicitly regulated, nor is it effectively managed.

## 6.5 The Nitrogen Cyclc

Humans have also significantly altered the global biogeochemical cycle for nitrogen (Figure 6.5). In the last two centuries, nitrogen inputs to the global cycle have approximately doubled, largely due to humans burning fossil fuels, using nitrate fertilizer, and burning biomass (Vitousek et al. 1997). Humans add an additional flux of about 210 teragrams a year to the annual global flux of nitrogen from the atmosphere to the land and aquatic eco-systems, which was an estimated 90 to 140 teragrams of nitrogen per year in preindustrial times (Vitousek et al. 1997). As a consequence of the ex-ponential growth in nitrogen fertilizer use in the second half of the twentieth century and the shift toward large-scale agricultural production, excess amounts of nitrogen leach into water bodies or return to the atmosphere, some of it in the form of the long-lived greenhouse gas nitrogen oxide ($N_2O$), which is also involved in stratospheric ozone depletion (MEA 2005).

Lavelle et al. (2005) report in the Millennium Ecosystem Assessment an increase of 0.8 parts per trillion (ppt) per year (0.25%) in atmospheric concentration of $N_2O$ during the industrial era due primarily to the extensive use of nitrogen fertilizer to promote large-scale growth of high-yielding crops. The atmospheric concentration of $N_2O$ averaged 314 ppt in 1998 compared to a preindustrial level of 270 ppt (Prather et al. 2001). Emissions of nitrogen oxides ($NO_x$) into the atmosphere also lead to the production of ozone in the troposphere, a major urban air pollutant problem, along with nitric acid ($HNO_3$), which is dissolved in and deposited through precipitation or as dry aerosols on land or sea. When nitrogen is deposited in the environment, it stimulates net primary productivity, since nitrogen is a limiting nutrient in many terrestrial ecosystems.

The leaching of excess nitrogen and phosphorous into adjacent rivers, lakes, and coastal zones causes eutrophication (Carpenter et al. 1998). Nitrogen is the major driver of eutrophication of most estuaries and coastal ecosystems (Howarth et al. 1996, Nixon et al. 1996). An overabundance of algae can reduce the oxygen levels in water to nearly zero, causing "dead zones" where oxygen levels are so low that fish and shellfish cannot survive. Excess nitrogen in the water is also toxic to human beings and other living organisms. Nitrogen pollution in groundwater has been associated with a health condition in infants known as *methemoglobinemia*, an acute nitrate poisoning caused by consuming water with nitrate concentrations of 10 mg/l or greater.

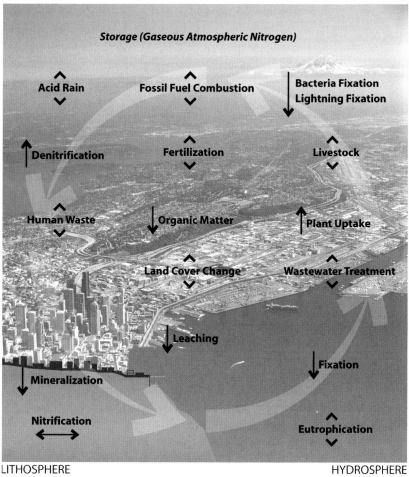

Figure 6.5. The nitrogen cycle in the urban landscape (Background photo: © Aerolistphoto.com).

Urban areas are a major source of nitrogen inputs. In urbanizing regions humans alter the nitrogen cycle by importing fertilizer and food, changing nitrogen fluxes through the atmosphere, soils, and water (Kaye et al. 2006). Sources of nitrogen include runoff from fertilized lawns, failing septic systems, wastewater treatment plants, and industrial discharges. Burning fossil fuels also emits nitrogen to the atmosphere. Nitrogen budgets have been documented in several urban ecosystems in the US (Baker et al. 2001, Groffman et al. 2004), Europe (Nilson 1995, Bjorklund et al. 1999, Folke et al. 1997), and Asia (Faerge et al. 2001, Warren-Rhodes and Koenig 2001).

Baker et al. (2001) estimated that in the Phoenix ecosystem humans mediate over 90% of nitrogen inputs. Using a nitrogen mass balance, they estimated that humans import about half of the total N input (~20 Gigagrams [Gg]) per year, mostly as food and fertilizer, and generate 36 Gg of atmospheric emissions by combustion (Baker et al. 2001). The urban infrastructure, especially sewer and solid waste facilities, mediates the way that nitrogen cycles through waste products. Fertilizer is used much more in agricultural ecosystems, but is an important source of nitrogen when used on urban gardens and lawns, even though the amounts used can vary widely, based on the preferences of multiple land owners (Law et al. 2004). Combustion processes contribute about a third of the total nitrogen input in the Phoenix mass balance and represent a major source of nitrogen input to the atmosphere in the form of oxidized compounds ($NO_2$, $HNO_3$, and aerosol $NO_3K$). About 21% of the total input of nitrogen from these diverse sources accumulates within the Phoenix ecosystem, and about 10% of it is deliberately removed through wastewater treatment (Baker et al. 2001).

## 6.6 Urban Patterns and Nutrient Cycling

In natural ecosystems, nutrient regulation synchronizes both the uptake of nutrients by microorganisms and plants and the release of nutrients through microbial activities and thus limits the overall loss of nutrients (Lavelle et al. 1993). This "synchrony" between release and uptake is determined by complex interactions among physical, chemical, and biological processes. In urban ecosystems, changes in land use, land cover, hydrology, atmospheric processes, climate, and soils have altered the interactions, and consequently the mechanisms, that have regulated nutrient cycling and export patterns.

Urbanization introduces new nutrient sources, and leads to new spatial and temporal variability in nutrient releases. The primary mechanisms are changes in land use and land cover, which affect both emissions and transport. Both point and non-point sources in urban areas contribute to N and P loading in urban rivers (Newman 1995). The emerging landscape pattern and the associated urban infrastructure also modify the ways that nutrients are transported across the landscape. Roads, parking lots, ditches, gutters, stormwater drains, detention basins, and lawns accumulate large amounts of nutrients. When that happens they can become hot spots for denitrification, an anaerobic microbial process that converts reactive nitrogen into nitrogen gases and removes it from the terrestrial system.

Urban land cover dramatically impacts biogeochemistry through vegetation clearing and increases in the amount of impervious land surface (Walsh et al. 2005, Groffman et al. 2002). Two major phenomena increase nutrient concentrations in urban surface waters: the concentration of pollutants generated by human activities and accumulated in runoff, and the increased volume and velocity of surface runoff associated with impervious surface. In cities, the built infrastructure and artificial drainage systems also affect nutrient cycles when nutrients are released from municipal wastewater and from combined sewer-stormwater overflow systems. Moreover, urban stormwater reduces nutrient retention and increases concentration in water bodies, as it makes streams flow more quickly and transport materials downstream (Paul and Meyer 2001, Grimm et al. 2005).

Climatic factors play an important role in mediating the impacts of urbanization on biogeochemical processes. The urban heat island affects the growth and decomposition rates of plants and microbes; this process has important consequences for the biogeochemistry across the urban-to-rural gradients. For example, in New York higher average temperatures at the urban end of the urban gradient partially explain the increased decomposition rates in the urban forest (McDonnell et al. 1997). Kaye et al. (2006) provide another example. During the summer Baltimore's mean maximum temperature is higher than in surrounding rural areas. In contrast, the city of Phoenix is cooler than the surrounding desert. In Baltimore, annual nutrient and carbon cycling increases during the longer growing season, while warming in arid Phoenix can suppress photosynthesis during the summer.

Several studies provide evidence of the ways that urban patterns are affecting nutrient export and retention (McDonnell et al. 1997, Pouyat et al. 1997, Baker et al. 2001, Groffman et al. 2002, Grimm et al. 2005, Inwood et al. 2005, Meyer et al. 2005, Wollheim et al. 2005, Groffman et al. 2006). Wollheim et al. (2005) compared the export and retention of nitrogen in two headwater catchments with different degrees of urbanization in the Plum Island Ecosystem watershed in Massachusetts; they also measured the dissolved inorganic nitrogen (DIN) concentrations in 16 additional headwater catchments with variable amounts of urban land (6%–90%). They found three factors that simultaneously affect nitrogen retention in urbanizing basins: increases in nitrogen loading, increases in runoff, and reductions in biological processes that retain nitrogen. Compared with forested catchments, they found that changes in nitrogen loading were 45% higher in urban basins, nitrogen flux 6.5 times higher, and water runoff 25% to 40% higher. Finally, nitrogen retention was 65%–85% in urban areas vs. 93%–97% in forested catchments (Wollheim et al. 2005).

In Baltimore, Groffman et al. (2004) compared nitrogen input-output budgets for watersheds with different dominant land use types: urban, suburban, forested, and agricultural. They found that urban and suburban watersheds had much higher nitrogen losses (2.9 to 7.9 kg of N per ha per year) than did the forested watershed (<1 kg N ha$^{-1}$ y$^{-1}$), but they were lower than those from an agricultural watershed (13–19.8 kg N ha$^{-1}$ y$^{-1}$). In Baltimore, as in the Plum Island Ecosystem watershed study (Wollheim et al. 2005), the driving factor was the increasing amount of impervious surface, but a surprisingly high amount of nitrogen was retained in the suburban watershed. This accounted for 75% of inputs, of which the largest were home lawn fertilizer (14.4 kg N ha$^{-1}$ y$^{-1}$) and atmospheric deposition (11.2 kg N ha$^{-1}$ y$^{-1}$). They also found low leaching of NO$_3$ and surprisingly high levels of soil respiration and organic matter in suburban watersheds, suggesting that an active carbon cycle facilitates nitrogen retention in grass-dominated ecosystems.

Although nitrogen retention declines with urbanization, it does remain relatively high for reasons that are poorly understood. This apparent anomaly may very well point to the limitation of studies of nutrient cycles in urbanizing regions representing the urban gradient on a simple continuum of population and/or built up densities, because humans affect biogeochemical cycles through multiple mechanisms and over multiple scales. By examining multiple gradients and patterns of urbanization, it is possible to better understand the complex interactions among biophysical processes affecting nutrient cycling in urban ecosystems. Nitrogen budgets vary across urban, suburban, and exurban sites in discontinuous ways due to the spatial arrangement of urban land use and infrastructure (Alberti et al. 2001). The spatial and temporal patterns of many ecosystem processes, such as uptake by vegetation, litterfall, decomposition, mineralization, nitrification, and denitrification, vary in response to atmospheric and related hydrological and biological processes (Band et al. 2001). Further, these processes change with urbanization.

The distribution of nitrogen sinks and sources across the landscape changes and particularly affects the dynamics and function of riparian areas. Such changes in sources and sinks become apparent in analyses of nitrogen budgets in urban ecosystems. In Baltimore, Groffman et al. (2002) found that instead of being sinks for nitrogen, riparian areas have the potential to be sources of it. Hydrologic changes in urban watersheds lead to lower water tables in riparian zones; two key changes are the tendency of streams to cut deeper channels and the reductions of infiltration in uplands as more stormwater infrastructure is built. Moreover, hydrological changes associated with urbanization create aerobic conditions in urban riparian soils, converting soil nitrate to nitrogenous gases and decreasing denitrification (Groffman et al. 2002, 2003). In urban areas stormwater detention basins are hotspots for denitrification (Groffman and Crawford 2003, Zhu et al. 2004). Because of high denitrification potential of urban soils, these areas could function as sinks in urban watersheds. Urban backyards are also important human-created sinks of nutrients, accumulating nutrients in biomass and soil organic matter (Kaye et al. 2006).

Many counterintuitive observations in recent studies of nutrient cycling in urbanizing regions indicate that urban ecosystems have more complex coupled carbon and nitrogen dynamics than previously thought. Surprisingly high rates of nitrogen retention observed in suburban watersheds still puzzle many urban ecologists. Suburban lawns, thought to be major sources of nitrogen in their watersheds, are in fact likely to retain considerable nitrogen. Riparian zones, thought to be an important sink for nitrogen in many watersheds, have turned out be nitrogen sources in urban watersheds because hydrologic changes disconnect streams from their surrounding landscape. In-stream retention of nitrogen, thought to be an important sink for it in forested watersheds, is reduced when urban runoff degrades riparian function.

Some contradictory observations were also documented in studies of forest ecosystem processes along urban-to-rural gradients (McDonnell et al. 1997). As humans add more nitrogen in urbanizing regions, nonlinear effects on plant growth, microbial activity, and soil chemistry are expected because of the complex interactions that take place among biological, chemical, and physical processes (Aber et al. 1989, McDonnell et al. 1997). At the ecosystem level, Aber et al. (1989) found that plants and microbes played a key role in nitrification by taking up nitrogen increases in the $NO_3$ leaching below the rooting zone. Increased nitrification and leaching of $NO_3$ has been documented across a transect from southern Maine to northern New York (McNulty et al. 1990), in forests downwind of San Francisco (McColl and Bush 1978), in the high-elevation spruce forests of North Carolina (Johnson et al. 1991), and in the 140-km transect running from highly urbanized Bronx County, New York, to rural Litchfield County, Connecticut (McDonnell et al. 1997). But the more recent refinements of hypotheses and study methods have started to provide mechanistic explanations.

Based on these studies, the expected influence of urbanization on decomposition and nitrogen-mineralization rates on an urban-to-rural gradient is the reduced ecosystem functioning of urban forests. High levels of heavy metals in urban forest soils, and reduced levels of soil fungi and microinvertebrates measured at the urban end of the transect, are consistent with the hypothesized higher nitrogen-mineralization rates (Pouyat et al. 1994). However, further studies of mineralization and nitrification rates in the forest soils along an urban-to-rural transect revealed results contradictory to those mentioned above (McDonnell et al. 1997, Zhu and Carreiro 2004). The net potential rates of mineralization were higher in the rural forest stands than in the urban stands, but the nitrification rates in the urban forest soils were still higher than those in the rural forest soils (Goldman et al. 1995, Pouyat et al. 2006, Groffman et al. 2005). Thus, high nitrification rates occur in urban soils even when the rates of nitrogen-mineralization are low; they appear to be the result of a history of earthworm activity in urbanizing regions (Bohlen et al. 1996, Pouyat et al. 1996).

More recently, Zhu and Carreiro (2004) observed significantly higher net N mineralization and nitrification rates at the urban end of an urban-to-rural transect in the New York metropolitan area, but the effect of land use on nitrogen-mineralization was not monotonic with respect to distance from the urban center. Furthermore, their findings seem to support the hypothesis that urban sites along this gradient may be approaching a condition of nitrogen saturation due to the combined effects of higher soil nitrogen-mineralization rates, higher atmospheric deposition, and possible constraints on photosynthesis and plant uptake due to air pollutants such as ozone.

To fully integrate human and ecosystem dynamics in urban biogeo-chemistry, urban ecological scholars should integrate mass-balance studies with empirical studies of nutrient cycles on urban-rural gradients coupled with experiments (McDonnell et al. 1997, Kaye et al. 2006). Complex interactions between human and biophysical processes take place on an urban-to-rural gradient, and both communities and ecosystems respond to human-induced changes in biogeochemical cycles in surprising ways. We have only begun to study the structure and function of populations, communities, and ecosystems in urbanizing regions. We do not fully understand the effect of spatial heterogeneity on mechanisms that regulate nutrient export and retention much less in urbanizing landscapes. It is critical that we explore the multiple interactions between simultaneous stresses and disturbances across an urban-to-rural gradient so we can develop predictions of future urban biogeochemistry.

One unresolved question concerns how the urban infrastructure mediates the biogeochemistry of an urbanizing region. A key consideration is the spatial and temporal scales of analysis. Urbanization generally follows a typical progression with various stages of residential development starting with a few outlying homes served by limited transportation and wastewater infrastructure. Homes are likely to be primarily served by minor roads and be on septic systems. As more subdivision increases population density, a substantial increase in roads and potential switches to sewer systems and treatment plants is likely to occur. These simultaneous trends add spatial and temporal complexity to the human-natural biogeochemical coupling puzzle.

Urban biogeochemistry will play an increasingly important role in helping us understand how humans affect long-term biogeochemical processes. We need to study these interactions across multiple spatial and temporal scales to determine human and natural interactions and effects on the dynamics of the earth system; we must also identify the process-level details and feedback mechanisms that regulate nutrient cycling. It is critical to assess the spatially explicit patterns of these interactions, driven as they are by urban development and the urban infrastructure, in order to assess the impact of urbanization on nutrient fluxes locally and globally. The results of such interactions are uncertain, and the impacts on ecosystem function cannot be predicted by extrapolating from non-urban systems. The pattern of urban development and its associated infrastructure may very well be crucial in determining the extent to which nutrient processes will support ecosystem and human functions in urbanizing regions.

# Chapter 7

# ATMOSPHERIC PROCESSES

In cities, complex interactions take place between urban morphology, climate, atmospheric emissions, and human health. Scientists in various disciplines have started to uncover those interactions by examining the relationships among the major chemical, physical, biological, and human processes that drive urban air quality (Figure 7.1). But we will need to develop coupled human-natural models of urban regions before we can fully understand their interactions and feedback mechanisms. For example, we have long known that urban areas affect microclimate; in fact, the urban heat island was first described nearly two centuries ago. We also know that human activities generate emissions that are precursors of ozone, a primary constituent of photochemical smog in cities, and we have increasing evidence that such smog affects the health of both humans and ecosystems. At the same time, as the heat island effect continues to increase in many cities, people demand more energy for cooling, which increases the amount of both fuel combustion and atmospheric emissions.

A vast scientific literature describes the dynamic interactions between the atmosphere and urban processes. In this chapter I review some of the complex interactions between these processes and urban landscape patterns, by focusing on the relationships between photochemical smog, climate change, heat islands, and human health.

## 7.1 Tropospheric Ozone

The most common form of atmospheric pollution in cities—photochemical smog—is a great example of the complex interactions that occur between human activities and the biophysical and chemical processes of urban ecosystems. Cities like Los Angeles have experienced smog for more than a century, but only in the early 1950s was its cause established. Arie Haagen-Smit, a California Institute of Technology chemist, studied plants and determined that ozone ($O_3$) is the primary ingredient of smog (Haagen-Smit et al. 1953). He also found that smog was created by a sunlight-driven

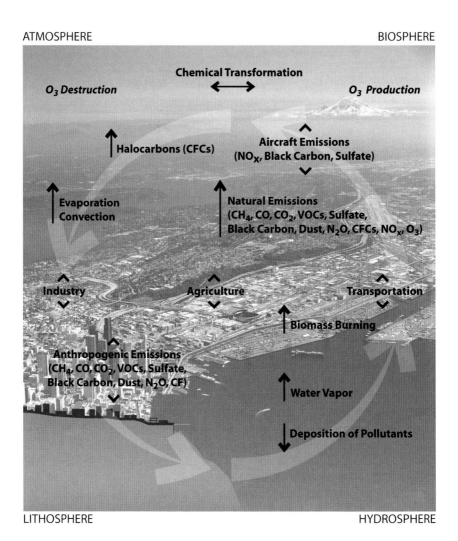

Figure 7.1. The atmospheric cycle in the urban landscape (Background photo: © Aerolist-photo.com).

photochemical reaction involving hydrocarbons from oil refineries, gasoline, and solvents, and nitrogen oxides from various combustion sources. In 1947, it was photochemical smog that led Los Angeles to create the nation's first air pollution control agency: the Los Angeles County Air Pollution Control District.

The increasing concentration of ozone, an atmospheric trace gas, is largely due to the increased emissions of precursor compounds: nitrogen oxides, volatile organic compounds (VOCs), methane, and carbon monoxide. Most cities commonly exceed the air quality guidelines of the World Health Organization (WHO): 75 to 100 parts per billion (ppb) is the one-hour short-term limit concentration, intended to safeguard human health, and 30 ppb is the long-term limit over the growing season, intended to protect vegetation. Compared with preindustrial levels, surface ozone has more than doubled over the past century (Bojkov 1986). Although ozone levels have generally declined in the United States since the Clean Air Act was passed in 1970, high concentrations of ozone persist in several urban areas, exceeding the National Ambient Air Quality Standards (NAAQS).

In the troposphere ozone is a strong oxidant. In humans, ozone reduces lung function, causes several respiratory symptoms, and increases episodes of asthma (Brunekreef and Holgate 2002). Several studies have also linked short-term $O_3$ exposure to premature mortality (Anderson et al. 2004, Bell et al. 2004, 2005). This is a serious urban health concern: in the United States, more than 100 million people live in areas that exceed the current NAAQS for $O_3$: 80 ppb for the daily 8-hr maximum (U.S. EPA 2004).

In urban ecosystems, the precursor emission sources of photochemical oxidants are virtually ubiquitous, highly spatially distributed, and have multiple sources. Two primary active compounds combine to create these oxidants: nitrogen oxides ($NO_x$) are primarily released as fuel is burned, and volatile organic compounds (VOCs) come from motor vehicle exhaust, solvents, plants, and fugitive emissions from the chemical and petroleum industries. In the presence of sun, these gases are converted to secondary pollutants, primarily ozone. Several other products are organic nitrates, oxidized hydrocarbons, and photochemical aerosols. Tropospheric ozone forms primarily in and downwind of large urban areas where, depending on weather conditions, emissions of $NO_x$ and VOCs can produce very high ozone concentrations (over 200 ppb) (NRC 1991). It is extremely difficult to predict and control the formation of ozone in urban areas because of this complex interaction between atmospheric chemistry and meteorological processes.

Concentrations of ozone in the troposphere are likely to increase as humans emit more pollution worldwide, and as global climate change (IPCC 2007) affects climatic conditions in cities (Mickley et al. 2004, Leung and

Gustafson 2005). As future climate changes interact with increased ozone formation, the effects on human health in cities are a great concern. Bell et al. (2007) report that in 50 US cities, the daily summertime 1-hour maximum increased by 4.8 ppb. They also report a 68% increase in the average number of days per summer that exceed the 8-hour regulatory standard (Bell et al. 2007). Using endpoints that were identified by concentration-response functions in epidemiological studies, they predict a significant increase in adverse health effects, including premature death and decreased lung function. These results point to the complex mechanisms that we must consider in monitoring and modeling the urban air quality.

## 7.2 Urban Air Quality and Climate Change

Climate variability and change, including temperature rise, sea-level rise, precipitation change, and changes in extreme events, are expected to have a significant impact on essential human and ecological functions in urban areas. These changes will increase human vulnerability and potentially harm human health in many ways. Here I focus on the interaction between climate change and tropospheric ozone to highlight the linkages between global climate change and urban air quality and their potential consequences for human health in urbanizing regions.

Tropospheric ozone levels in urban areas are expected to increase with climate change because the photochemical reactions that form ozone are affected by temperature. Several studies have shown a positive correlation between temperature and ozone concentrations (Jones et al. 1989, Wakim 1989, Cardelino and Chameides 1990, Holzer and Boer 2001, Seinfeld and Pandis 2006, Dawson et al. 2007). The EPA estimates that an increase in air temperature of 4 C (39 F) in the New York region could result in an increase of 4% in ozone concentrations (Smith and Tirpak 1989). The same increase in temperature in California could increase ozone concentrations by 20% during high-ozone days in August (Morris et al. 1989).

Annual mean levels of tropospheric ozone over the United States are projected to increase by 2 to 5 ppb by 2030 even under a scenario of declining US emissions (Unger et al. 2006). Although the effect of climate change on ozone concentrations has a high uncertainty (Zeng and Pyle 2003, Hauglustaine et al. 2005), several studies in the United States indicate significant changes in frequency and intensity of ozone episodes (>80 ppb) over large areas of the eastern regions and in southern California (Wu et al. 2007). Climate change seems to affect episodes of ozone pollution far more than the mean values of ozone, with effects exceeding 10 ppb in the Midwest and Northeast. Other studies report potential effects, ranging from increased concentrations by 4.2 ppb by 2050 solely from regional climate change, an increase of max-8hr-avg ozone in the summer time in the eastern United States (Hogrefe et al. 2004), and a possible increase in the severity and duration of regional pollution episodes in the Northeast and Midwest (Mickley et al. 2004). Dawson et al. (2007) predict increases of 1 to 3 ppb in the max-8h-avg ozone in the East.

Along with the effects that climate change will have on photochemical smog, several scientists indicate that it will have significant impacts on human health. By the 2050s, Knowlton et al. (2004) predict a 4.5% increase in ozone-related acute summer mortality across the 31-county New York metropolitan region, compared to the 1990s, because the frequency and intensity of ozone episodes will increase, driven by climate change. Bell et al. (2007) studied 50 US cities and found that potential increases in tropospheric ozone due to climate change (Hogrefe et al. 2004) would increase the total amount of cardiovascular and respiratory diseases (e.g., asthma) and mortality.

## 7.3 Urban Heat Islands

Cities are typically warmer than surrounding areas (Landsberg 1981, Atkinson 1985, Oke 1987, 1995). The phenomenon of the urban heat island (UHI) was first described in 1833 by Luke Howard, who noted that at night London was 3.7 F (2.1 C) warmer, and during the day 0.34 F (0.19 C) cooler, than its surroundings (Landsberg 1981). Since then observers have conducted studies in many European (Oke 1973, Horbert et al. 1982, Watkins et al. 2002) and North American cities (Bornstein 1968, Nkemdirim 1976, Oke and East, 1971, Oke 1976, Cayan and Douglas 1984, Balling and Cerveny 1987, Gedzelman et al. 2003, Souch and Grimmond 2006). Urban areas have also been found to be 2 percent drier in the winter, and 8 to 10 per cent drier in summer (Table 7.1). Clouds form above cities between 5 and 10 percent more frequently than in the countryside, with seasonally variable increases in fog. Air pollution also affects the radiation balance and supplies extra cloud condensation nuclei around which cloud droplets may form (Oke 1973). Cities get 5 to 10 percent more rainfall than the areas around them, and 5 percent less snow. But in some situations, these urban aerosols may also suppress precipitation. Aerosols may increase the amount of long-wave radiation.

Urban morphology also affects the flows of energy and air, influencing phenomena like the UHI and rainfall. The UHI is probably the best-documented example of human-induced microclimate change, but urbanization also changes the nature of the land surface and its properties. Changes in urban heat storage and reflectance, moisture storage, and wind patterns all have substantial effects on water and energy budgets. For

Table 7.1. Climate parameters in urban areas. Typical a) surface and atmospheric properties, and b) urban climate effects for a mid-latitude city about 1 million inhabitants. Values for summer unless otherwise noted (Oke 1997).

| Variable | Change | Magnitude/Comments |
|---|---|---|
| Turbulence intensity | Greater | 10-50% |
| Wind speed | Decreased | 5-30% at 10m in strong flow |
| | Increased | In weak flow with heat island |
| Wind direction | Altered | 1-10 degrees |
| UV radiation | Much less | 25-90% |
| Solar radiation | Less | 1-25% |
| Infrared input | Greater | 5-40% |
| Visibility | Reduced | |
| Evaporation | Less | About 50% |
| Convective heat flux | Greater | About 50% |
| Heat storage | Greater | About 200% |
| Air temperature | Warmer | 1-3 degree C per 100 years; 1 - 3 degrees C annual mean up to 12 degrees C hourly mean |
| Humidity | Drier | Summer daytime |
| | More moist | Summer night, all day winter |
| Cloud | More haze | In and downwind of city |
| | More cloud | Especially in lee of city |
| Fog | More or less | Depends on aerosol and surroundings |
| Precipitation | | |
| Snow | Less | Some turns to rain |
| Total | More? | To the lee of rather than in city |
| Thunderstorms | More | |
| Tornadoes | Less | |

example, the materials used in buildings store heat and effectively waterproof the earth's surface. The geometry of streets and buildings can trap radiation and increase wind speeds.

Oke (1987) hypothesized the mechanisms by which urban characteristics change the energy balance, in turn creating the UHI. The geometry of buildings and streets can lead to urban canyons and increased absorption of solar radiation. Because of air pollution, the air in cities absorbs more radiation, and also emits more, increasing the amount of long-wave radiation absorbed from the sky. Meanwhile, the canyon geometry reduces the loss of such radiation. In addition, heat from buildings and traffic further raises temperatures. Construction materials store more heat and decrease the amount of moisture that evaporates. Since the canyon geometry reduces the wind speeds, it also reduces the amount of heat that the air can transport. Dow and DeWalle (2000) examined 51 urbanizing watersheds and found that as cities develop, less water evaporates from watersheds, but the amount of sensible heat in the watersheds increases. As humans produce and consume energy, they release more heat.

Several factors control the formation and intensity of an urban heat island, including climate, topography, land cover, and urban structure. The temperature difference between city and countryside is greater in the evening, especially during clear, calm conditions, and is inversely related to wind speed and cloud cover (Kidder and Essenwanger 1995). There is a threshold wind speed above which the heat island is minimal. Wind speeds greater than about ~4 meters per second (m/s) and clouds that are lower in the atmosphere dramatically decrease intensity, because strong winds efficiently mix heat throughout the atmosphere. The wind speed that limits heat island development increases with population size (Oke 1987). Topography and orography also influence the heat island and variation in temperature within the urban area. The amount and distribution of land cover (i.e., vegetation, bare soil, paved area) and land use (i.e., residential, commercial, industrial) account for variation in temperature across cities and within cities.

In general, scientists find a clear relationship between urban warming and population density. The intensity of the urban heat island, measured as the maximum difference between the background rural temperature and the highest urban temperature, is proportional to the logarithm of the urban population (Oke 1973). Even better established is the relationship between the intensity of the UHI and the density of the city, expressed as the height-to-width ratio of the street canyons in the city center. Large cities of 100,000 to 1,000,000 people can be 8 to 12 C warmer than rural areas. Large cities in North America are a few degrees warmer than comparable European cities (Bonan 2002). The urban-rural temperature difference for a city increases as the amount of impervious surface area increases and as the amount of vegetated area drops (Landsberg 1979, Carlson and Arthur 2000). The explanation is the warming effect of impervious surfaces, as opposed to the cooling effect of vegetation, shown in many cities in North America and Europe (Upmanis et al. 1998). This impact is important at both the local and regional scales.

The surface energy balance differs significantly between urban and rural areas (Figure 7.2) (Grimmond and Oke 1995, 2000, Oke et al. 1998). Both rural and urban areas receive energy from radiative processes: the earth's surface warms during the day and cools at night. The mean amount of incident solar radiation is 7.6 kilowatt hours per square meter (kWh/m$^2$) per day for both urban and rural areas. However, many urban buildings and roadway surfaces do not reflect much sunlight, so urban and rural area differ in their albedo. Compared with the albedo of a rural forest (~0.25) urban area may reflect only a much smaller amount of the incoming solar radiation (~0.05), which results in a difference in net daily energy of 1.5 kWh/m$^2$, with a potential increase in temperature up to 10-13 F.

The urban structure and its surface materials also affect temperature variance (Bonan 2002). Important factors include both the geometry of the built fabric (e.g., size, shape, and orientation of buildings and streets) and the actual urban surfaces (e.g., asphalt, concrete, gravel, grass, etc.). Urban materials have different characteristics, including albedo and capacity to hold and conduct heat and moisture. All these alter the radiation balance at the surface, the storage of heat in the urban fabric, and the partitioning of energy into latent and sensible heat (Landsberg 1981, Oke 1982, 1987, 1995).

The effect that urbanization has on the energy balance varies at different levels in the urban canopy layer (Oke 1976, 1995). At the rooftop level—that is the microclimates created by buildings, roads, and vegetation—shade from buildings in the urban canopy layer can create cooler local temperatures than in open areas. Vegetated areas and parks within a city can have large latent heat fluxes above the rooftop level that integrate the microclimates of the urban canopy layer over a large area. At the level of the urban canopy, street canyons explain the effect that urban form has on microclimates. The height of buildings and orientation of streets create complex shading patterns that affect temperature (Arnfield 1990).

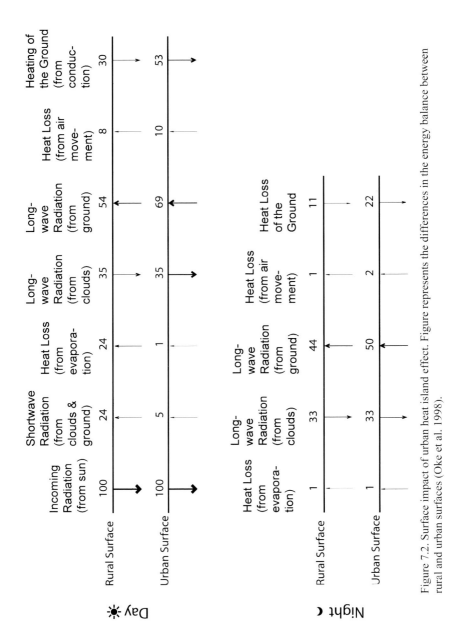

Figure 7.2. Surface impact of urban heat island effect. Figure represents the differences in the energy balance between rural and urban surfaces (Oke et al. 1998).

Depending on its geometry, and the ratio of building height to street width, street canyons can create greater opportunities to trap radiative energy. As a result, the city absorbs more radiation (Oke 1987, Arnfield and Grimmond 1998). Meanwhile parking lots, with their low albedo and low capacity to conduct heat, create their own microclimate.

Providing more detail on these phenomena, studies of the Phoenix metropolitan area, AZ, found that the temperature increased between 1990 and and 2004, depending both on an area's location within urban development zones and the pace of housing construction in a 10 km buffer around fixed-point temperature stations (Brazel et al. 2007). The authors explain the significant temperature variation by the land surface effects in the type of urban development zone, which ranged from urban core and infill sites, to desertland agricultural fringe locations, to exurban. Three studies find an increase in rainfall events downwind of the metropolitan area in respectively, New York, NY (Bornstein and LeRoy 1990), Atlanta, GA (Shepherd et al. 2002), and St. Louis, MO (Rozoff et al. 2003).

Various aspects of urban regions act as controls on urban climate at various scales (Oke 2004, 2006). At the largest scale, these aspects include geographic location and major physiographic divisions; at the intermediate scale, orography, and position on an-urban-to-rural gradient; and at the local scale, built form and structure. Oke (2004) describes four key aspects of the local scale: urban structure (dimensions of buildings and spaces between them, street widths, and street spacing), urban cover (proportions of built-up, paved, or vegetated surfaces, bare soil, water), urban fabric (construction and natural materials), and urban metabolism (heat, water, and pollutants due to human activity).

Ellefsen (1990) provided the first empirically based approach for mapping the urban morphological characteristics relevant to urban climate. Based on a survey of ten US cities, he distinguished morphological zones based on building construction type, venting characteristics, age, density, and street patterns. Expanding on Ellefsen's work, Oke (2004) developed a simple scheme of Urban Climate Zones (UCZ) (Figure 7.3). The scheme constitutes a set of hypotheses of the impact of urban morphology on climate, and adds two elements to Ellefsen's typology: (1) a simple measure of the structure aspect ratio—average height of the roughness elements (buildings, trees), divided by the average street width, which has been shown to be closely related to flow, solar shading, and the nocturnal heat island, and (2) a measure of the surface cover (% built) that is related to the degree of surface permeability. The UCZ categories can be ranked approximately in order of their ability to modify the climate in terms of wind, temperature and moisture.

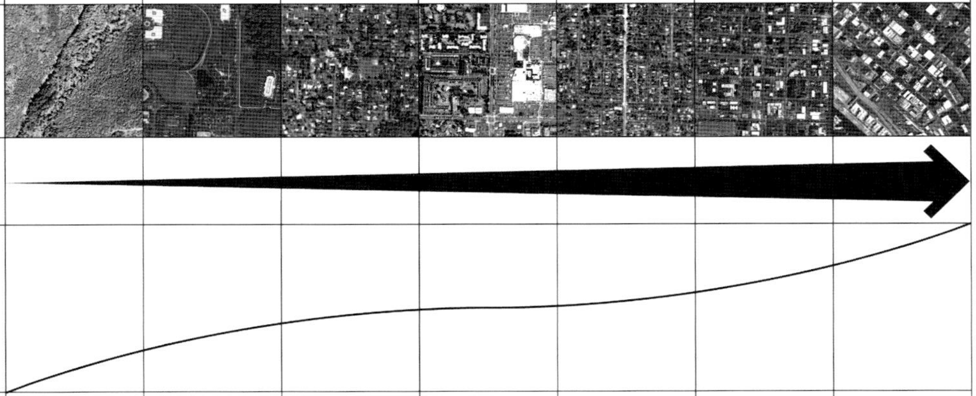

| Urban Gradient | Rural Natural | Mixed Use Institutional | Suburban Residential | Commercial Industrial | Medium Density Urban Residential | High Density Urban Residential | Downtown High Rise |
|---|---|---|---|---|---|---|---|
| Degree of impact | | | | | | | |
| Late Afternoon Temperatures | | | | | | | |
| Roughness | 4 | 5 | 6 | 5 | 7 | 7 | 8 |
| Aspect ratio | >0.05* | 0.1 - 0.5* | 0.2 - 0.6* | 0.05 - 0.2 | 0.5 - 1.5 | 1.0 - 2.5 | >2 |
| % Impervious | <10 | <40 | 35-65 | 70-95 | 70-85 | >85 | >90 |

Figure 7.3. Relationship between the urban gradient and urban heat island. Based on Urban Terrain Zones (Ellefsen 1991) and Urban Climate Zones (Oke 2004) Late Afternoon temperatures are synthesized from Oke 1973. The temperature increases along the urban gradient. Roughness refers to the effective terrain roughness according to the Davenport classification (Davenport et al. 2000). The roughness Aspect ratio = ZH/W is average height of the main roughness elements (buildings and trees) divided by their average spacing. In the city center this is the street canyon height/width. The aspect ratio is related to flow regimes (Oke 1987) and thermal controls (shading and screening) (Oke 1982). Percent impervious reflect the proportion of ground plane that is covered with buildings, roads and other impervious areas. The amount of pervious ground cover affects soil moisture and evaporation.

Over the last 20 years, many studies of the urban heat island have expanded our understanding of it. In an extensive review of more than 20 years of empirical research, Arnfield (2001) found that conceptual advances in boundary-layer climatology have greatly benefited the study of urban microclimate, but they confirm the key generalizations that Oke provided in the early 1980s. These are that UHI is greatest at night, develops primarily in the summer (Schmidlin 1989, Philandras et al. 1999, Morris et al. 2001), and tends to increase along with the city's size and/or population (Park 1986, Yamashita et al. 1986, Hogan and Ferrick 1998).

## 7.4 Urban Patterns and Air Quality

For some time, urban scholars have studied the mechanisms by which the emerging urban landscape patterns affect air quality in cities. The spatial structure and land use patterns influence urban energy flows directly by redistributing solar radiation (McPherson 1994, Kalma et al. 1978), and indirectly by influencing the energy requirements of human activities (Odum 1963, Newcombe et al. 1978, Douglas 1983). On the other hand, spatial structure is an important determinant of future energy supply, distribution systems, and the exploitation of ambient energy sources (Owens 1986). Since energy use is associated with many threats to ecosystems, future relationships between energy and urban land use are expected to have important ecological consequences, both locally and globally.

Although many scholars have addressed the relationships among urban activities, energy use, and environmental impact (Hemmens 1967, Stone 1973, Edwards and Schofer 1975, Keyes and Peterson 1977, Keyes 1982, Newman and Kenworthy 1989a, 1989b), few have looked directly at how urban patterns affect energy demand and emissions patterns (Pisarski 1991). Most empirical studies have focused on transportation-related energy use and have used indirect measures of how humans consume energy and generate emissions. These studies use vehicle-miles-traveled (VMT), modal split, travel distance, and trip frequency as indicators of gasoline consumption and related emissions into the atmosphere (Newman and Kenworthy 1989, Pucher and Lefevre 1996).

In the area of transport-related energy consumption, the work of Newman and Kenworthy (1989a) is perhaps the most controversial. In their first empirical cross-sectional study of 32 cities across the world, they found strong negative relationships between gasoline consumption and both population and job density. In a second study of 63 large metropolitan regions across the world, they described a negative relationship between population density and per capita fuel use (Newman and Kenworthy 1989b).

Based on a cross-section of US and European cities, they suggested that the intensity of land use is correlated with gasoline consumption and that price, income, and vehicle efficiency explain only 40 percent of the variation observed across cities. However, that study contained some methodological problems, primarily because density alone is a very crude measure of urban patterns, and cannot reflect the variety of spatial structures across the diversity of cities included in the sample. In addition, the different definitions of city boundaries make it harder to interpret overall density. Moreover, their statistical model failed to address the link between density and income, and the possible effect of income on gasoline demand.

Several scholars have employed simulation models or travel survey data from a single metropolitan area to establish a relationship between urban form and the amounts of emissions associated with travel behavior (Cervero and Gorham, 1995). Using household and travel survey data, Frank et al. (2000) found significant inverse relationships between household density, employment density, street connectivity, and emissions of CO, $NO_x$, and VOCs in Seattle, Washington. Scholars investigating relationships between patterns of land use and travel have also found that population density and job density are significant factors in explaining the numbers of trips, distance traveled, and modal split (Gordon et al. 1989). Researchers frequently use measures of the local concentrations of selected atmospheric pollutants for which data are available and for which scientific research has provided clear links to effects on human health. They know that concentrations of sulfur dioxide ($SO_2$), particulate matter (PM), carbon monoxide (CO), and nitrogen oxides ($NO_x$) in the urban atmosphere are influenced by topography, location of polluting sources, patterns of artificial heat generation, and climate. These factors also influence the intensity, size, and shape of the urban heat island and thus the processes that trap pollutants in the local environment. In addition, land use patterns directly affect the local sink with water bodies and green areas that mitigate the heat island effect and absorb pollution. McPherson et al. (1994) estimated that in 1991 the tree cover in Chicago removed 17 tons of CO, 93 tons of $SO_2$, 98 tons of $NO_2$, 210 tons of $O_3$, and 234 tons of PM of less than 10 microns. These trees also stored 942,000 tons of carbon. While this evidence substantiates the hypothesis that urban patterns affect how well the urban environment can absorb these pollutants, no systematic study has yet described how alternative spatial configurations impact local sinks.

The spatial structure of metropolitan regions also controls the microclimate and thus the urban air quality, both directly and indirectly. Many aspects of urban form have been associated with the urban heat island (Oke 2004). Differentials in urban land use patterns explain the variability in microclimates and thus the formation of ozone. Heat island effects and warmer temperatures during the summer affect urban air pollution directly since it has been suggested that some species of trees emit VOCs (Cardelino and Chameides 1990). In addition, regional power plants emit more ozone precursors because of increased demands for air conditioning (Rosenfeld et al. 1998). Rising temperatures will make people use more natural gas and electricity in their homes and workplaces. The influence of urban temperatures on tropospheric ozone formation is well documented (Rao et al. 1995, Kelly et al. 1986, Stone 2005). Several studies in Chicago and Los Angeles show the benefits of strategies like increasing canopy cover and making urban surfaces more reflective to reduce ozone formation; these strategies significantly reduced the heat island and reduced the number of days that ozone exceeded recommended limits (Rosenfeld et al. 1998).

Thus considerable evidence is emerging on the important relationships between urban patterns, urban climate, and air quality. These relationships are mediated by decisions of households about housing location, mobility, and energy consumption and their impact on urban climate and air quality. It is clear that land use planning strategies can help prevent the potential increase in photochemical smog and its related health effects. To minimize the heat island effect, multiple strategies can be useful at multiple levels, including expanding green vegetation, minimizing impervious surface, and reducing emissions by designing patterns. However our understanding of these complex relationships is still rudimentary. It is crucial to develop an explicit and comprehensive conceptual model of the interactions among urban landscape patterns, climate, and atmospheric processes. Most studies have focused on individual causes and effects. I propose that such a framework should expand the consideration of environmental interactions to include feedbacks among urban patterns, human behaviors, atmospheric pollution sources, sinks, and fate at the relevant scale for each of these diverse patterns and processes.

# Chapter 8

# POPULATION AND COMMUNITY DYNAMICS

## 8.1 Biodiversity, Ecosystem Function, and Resilience

The study of urban ecology must address the scientific debate on how biodiversity relates to ecosystem function, stability, and resilience (Peterson et al. 1998). Though the term biodiversity has multiple definitions and interpretations, the definition provided by Wilson (1992, 393) may best capture its essence. Wilson emphasizes "the variety of organisms considered at all levels, from genetic variants belonging to the same species through arrays of species to arrays of genera, families and still higher taxonomic level," as well as "the variety of ecosystems, which comprise both the communities of organisms within particular habitats and the physical conditions under which they live."

Ecological scholars disagree on the role that biodiversity plays in the functioning of ecosystems: The issue is not simply what species are involved in what specific ecosystem functions, but the importance of diversity for the functioning of ecosystems and the role that it plays in their resilience (Loreau et al. 2001). Scientists also disagree on what drives the patterns of species diversity and the nature of ecological communities. Different theories can be distinguished based on whether they see community assembly as based on niches (MacArthur 1970, Levin 1970), or dispersal (MacArthur and Wilson 1963), and whether they are neutral (treating individual species as essentially identical) (Hubbell 2001) or non-neutral (assuming that different species behave in different ways from one another) (Ehrlich and Ehrlich 1981, Walker 1992, Levin 1999). The lack of resolution on these different perspectives may simply indicate that they are all true at some level (Hubbell 2001). At the same time, they all share a biased disciplinary perspective: They fail to appreciate the mutual interactions and feedback between ecosystem function and biodiversity. These different perspectives have important implications for conservation, and for efforts to integrate humans into ecological thinking.

Despite these differences, however, a consensus on important aspects on the relationship between biodiversity and ecosystem function is emerging

(Hooper et al. 2005). During the past decade, the debate has shifted its focus from number of species to functional groups and underlying mechanisms (Grime 1997, Loreau et al. 2001, Srivastava and Vellend 2005). Scholars have pointed out the need to formulate new hypotheses, and to formally test theories and models that integrate community ecology and ecosystem science in a unified framework (Loreau et al. 2001). Yet the debate has not fully integrated humans or effectively explored how their inclusion might change our understanding of the relationship between biodiversity and ecosystem function. Building on current advances, in this chapter I focus on how urbanization patterns may affect the relationship between biodiversity, ecosystem processes, and resilience, and the implications for robust generalizations in human-dominated ecosystems. I ask how humans affect both the relationship between biodiversity and ecosystem processes, and the role of biodiversity in the stability of ecosystems. I conclude the chapter with an empirical exploration of bird diversity and human disturbance on an urban-to-rural gradient.

## Species richness and ecosystem function

Ecological studies have provided ample evidence that different species perform diverse ecological functions within the systems they inhabit; for example, they cycle nutrients, regulate trophic mechanisms, pollinate plants, disperse seeds, and control natural disturbance (Hooper et al. 2005). Thus a change in species composition may imply predictable functional shifts when sets of species with certain traits are replaced by sets with different traits (Grime et al. 2000, Loreau 2001). Substantial evidence shows that functional diversity depends on species richness—but to what extent does species richness affect stability (Tilman et al. 1996)? Scientists disagree about the relative influence of functional substitutions and species diversity on ecosystem functioning (Loreau 2001). Furthermore, we do not know whether and to what extent their relative importance changes under changing conditions. Understanding the relationship between species diversity and ecosystem function becomes even more relevant in the context of increasing human-induced impact due to land cover change and urbanization (Hooper et al. 2005).

   If changes in species composition affect the efficiency with which resources are processed within an ecosystem, we would expect that species richness would affect ecosystem function. But much is still unknown about the ways ecosystems respond to changes in species richness. Several competing models were initially proposed to describe this dynamic (Figure 8.1, Naeem et al. 2002). The null hypothesis states that species richness has no effect on ecosystem functions (Vitousek and Hooper 1993). Considerable

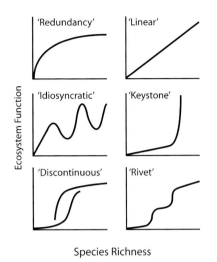

Figure 8.1. Alternative models of species richness and ecosystem function. The graphs represent competing hypotheses regarding the relationship between species richness and ecosystem function—that is, the ecological role that species diversity plays (Naeem et al. 2002, p. 5).

evidence contradicts this hypothesis, however, leading to several alternative hypotheses. The *rivet* and *redundancy* hypotheses suggest that certain species may drive the functioning of an ecosystem, while others have various impacts on the way those functions occur. The rivet hypothesis (Ehrlich and Ehrlich, 1981) suggests that ecosystems are like plane wings: Ecosystem functioning (the plane) may or may not be compromised depending upon which species (rivets) are lost. A plane can lose several rivets before a wing falls off. The ecological functions of different species overlap; therefore, even if a species dies out, the system's ecological functions may persist because other species that perform similar functions can compensate for the lost one.

The redundancy hypothesis is based on a concept similar to the "rivet hypothesis." (Ehrlich and Ehrlich 1981) In addition, it advances that conservation efforts should focus on the species that uniquely represent a given functional type, because of their role in maintaining ecosystem integrity. This hypothesis assumes that above a critical level most species are functionally redundant. The *idiosyncratic* hypothesis emphasizes that the change in ecosystem functioning associated with changes in species diversity is unpredictable because individual species have such complex and diverse roles (Lawton 1994). Under the *keystone* hypothesis, ecosystem function declines rapidly with the loss of species that are crucial to mediating such functions, and diversity is consequently reduced below its natural levels (Walker 1992).

However, these early hypotheses only partially address the complex relationship between biodiversity and stability. During the past decade, important advances have occurred in the debate as observational, experimental, and theoretical studies have helped researchers to better articulate the scientific questions. Recent studies suggest that biodiversity may provide "insurance" or a buffer to maintain ecosystem function in the presence of environmental variability since different species respond differently to environmental fluctuations. Furthermore, scholars recognize that, given the lack of explicit definitions, the debate has not always been productive, which leads to the current scientific controversy (Pimm 1991). And that debate is still influenced by an old paradigm of stability (Loreau et al. 2001). Loreau et al. (2001) point out that the concept of "stability" refers to several properties whose relationships with diversity may change across levels of organization; more importantly, that stability has been approached mainly within a deterministic, equilibrium theoretical framework. Different concepts of stability produce a different diversity-stability relationship (Ives and Carpenter 2007).

The evidence that diversity matters to the functioning of ecosystems is growing. Cardinale et al. (2006) conducted a meta-analysis of 111 studies on the effects that species diversity has on the functioning of numerous trophic groups in multiple types of ecosystems. They found consistent patterns across different trophic groups (producers, herbivores, detritivores and predators) and ecosystem types (aquatic and terrestrial). They conclude that species loss does indeed affect ecosystem functioning (i.e., abundance or biomass of the focal trophic group), but the magnitude of these effects is determined by species identity.

More recently, researchers have started to shift their focus from the number of species to the mechanisms by which biodiversity affects ecosystem function (Loreau et al. 2001). They have started to focus on functional groups, exploring the extent to which functional substitutions alter a variety of ecosystem properties such as productivity, decomposition rates, nutrient cycling, as well as their stability and resilience. Findings indicate that diversity is essential to sustain the functioning of ecosystems undergoing change (Schläpfer and Schmid 1999, Loreau 2000). However, current studies cannot determine whether the effect is due to a few key species or the diversity of species. Current explorations of the mechanisms that link biodiversity and ecosystem function are focusing on two major aspects: (1) deterministic processes, such as niche differentiation and facilitation, which lead to "complementarity" and (2) stochastic processes involved in community assembly, where random sampling coupled with local dominance of highly productive species, can also lead to increased primary production and diversity (Loreau et al. 2001).

If the effects of humans are brought into the picture, a mechanistic approach could help expand our understanding of ecosystem dynamics in urbanizing regions and guide ecosystem management strategies. But to fully appreciate the implications of including humans in such a framework, we need to consider several levels of human interactions with biodiversity and ecosystem function. The first level involves the influence of humans in community assembly. The second is the influence that human settlements have on the species-area relationship. The third is the way that stability domains change in the presence of humans.

Humans can affect species composition and their functional roles in ecosystems both directly by reducing the overall number of species or by selectively determining phenotypic trait diversity. Since individual species may control community- and ecosystem-level processes (Paine 1984, Lawton 1994, Power et al. 1996), ecosystem processes may be highly affected by diversity, since changing diversity affects the probability of occurrence of these species among potential colonists (Tilman 1999, Cardinale et al. 2000). Humans can influence ecosystem processes by both altering the dominance of species with particular traits, and facilitating or impeding complementarity among species with different traits. Furthermore, human activities can influence the sampling effect by selectively reducing or increasing the pool of species representing particular functional traits.

Humans can affect species composition and their functional roles in ecosystems both directly, by reducing the overall number of species, or indirectly, by selectively determining phenotypic trait diversity. Individual species may control processes at both the community and ecosystem levels (Paine 1984, Lawton 1994, Power et al. 1996), so diversity may have a strong effect on those processes, because changes in diversity affect the probability that these species will occur among potential colonists (Tilman 1999, Cardinale et al. 2000). In addition to altering the dominance of species with particular traits, humans affect diversity by facilitating or impeding complementarity among species with different traits. Furthermore, human activities can influence the sampling effect by selectively reducing or increasing the pool of species representing particular functional traits.

Loreau et al. (2001) point out that complementarity and sampling effects can occur simultaneously, since communities with more species have a higher chance of containing a greater diversity of phenotypic traits. Rather than alternative mechanisms, there is increasing evidence that they can represent end points on a continuum from dominance of species with certain traits to complementarity of species with different traits, with intermediate scenarios, where bias in community assembly may lead to correlations between diversity and community composition that involve both dominance and complementarity (Loreau et al. 2001).

As we explore the relationship between biodiversity and ecosystem function in human-dominated ecosystems, it becomes evident that humans influence both ecosystem function and the pattern of biodiversity, as well as the relationship between them. Mutual interactions and feedback occur among biodiversity changes, ecosystem function, and abiotic factors. Humans influence all of these factors by strengthening or loosening some of these interactions and feedback loops and by creating unprecedented interactions between biodiversity and ecosystem function. Thus diversity-stability relationships cannot be understood outside the context of these complex interactions (Ives and Carpenter 2007).

When studying the impact of humans on biodiversity and ecosystem function, it is critical to acknowledge that processes influencing diversity operate at different spatial and temporal scales (MacArthur 1969, Tilman and Pacala 1993). At the local scale, dominant constraints are resource abundance, competition, predation, and disturbance. At larger scales, processes such as emigration, large-scale disturbances, and evolution operate (MacArthur 1969). Humans influence these dynamic interactions across a wide range of spatial and temporal scales through urbanization and land cover change. Recent studies have pointed out that there may be important feedback mechanisms that link ecosystem function and biodiversity across scales. Diversity is correlated with productivity directly and through several factors that influence productivity at a large scale (i.e., climate and disturbance regime). But species diversity and composition also have local effect on productivity. Potential interactions between human settlement and community assembly can be mediated by human effects on environmental processes at the local and regional scales.

Integrating humans into the study of biodiversity could also reconcile key theoretical concepts. Theories of both niche and dispersal-based community assembly can benefit by including humans. Hutchinson (1957) transformed and solidified the niche concept, changing it from a mere description of an organism's functional place in nature (Elton 1927) to a mathematically rigorous $n$-dimensional hypervolume that could be treated analytically. Hutchinson's "realized niche" included only those places where an organism's physiological tolerances were not exceeded (its "fundamental" niche) and where its occurrence was not preempted by competitors. In addition to competition, other potentially important community organizing forces, such as predation, resource variability, and human domination, are altered as a function of human-mediated dynamics. Through all these mechanisms, humans force population-level ecological functions that structure communities.

By integrating humans into the study of processes controlling biological diversity, ecological scholars may be able to resolve important puzzles in island biogeography and explain empirical results regarding the balance between colonization and extinction in human-dominated ecosystems (Marzluff 2005). The balance between extinction and colonization still regulates diversity in a human-dominated world, but as Marzluff (2005) points out: Both colonization and extinction are affected by direct and indirect human actions, including land cover change and the introduction of non-native species. As humans urbanize, they cause the emergence of new selective forces including new habitats, disturbance regimes, predators, competitors, and diseases that may drive native species to extinction (Sax and Gaines 2003, Kuhn et al. 2004, Olden and Poff 2004).

## Environmental variability

An emerging paradigm in ecosystem ecology holds that there may be no single, generalizable relationship between species diversity and ecosystem function; instead, it may depend highly on context (Chapin et al. 1998, Cardinale et al. 2000). Cardinale et al. (2000) show that environmental variability (both spatial and temporal) can change both the form and cause of the relationship between diversity and ecosystem productivity. That context is important was recognized earlier (Risser 1995, Chapin et al. 1998), but only recently scholars have started to study it systematically (Cardinale et al. 2000). In addition, ecological scholars now realize that several aspects of community structure associated with species richness may control the biodiversity-ecosystem function relationship (Naeem et al. 2000, Wilsey and Potvin 2000, Cardinale and Palmer 2002). Several scholars have proposed that community structure may in fact mediate the effects of species richness on ecological processes and thus play an important role in the relationship between species richness and ecosystem function—because the same drivers of community structure might simultaneously affect that relationship (Naeem et al. 2000, Wilsey and Potvin 2000, Cardinale and Palmer 2002). Furthermore, Mulder et al. (1999) and Cardinale et al. (2002) have shown that interspecific interactions between different functional groups of organisms control the effects of species richness on ecosystem function. They hypothesize that diversity affects the efficiency and productivity of ecosystems through facilitation between species.

Environmental context and variability (in space and time) are essential elements defining the current debate on diversity and ecosystem function since variability can greatly influence the partitioning of resources among species in a system. If community structure has the potential to mediate the relationship between species richness and ecosystem function, ecological factors regulating interspecific interactions may also impact such a relationship (Cardinale et al. 2002). Disturbance can introduce variability through mechanisms such as preventing competitive dominance or introducing new potential niches. Cardinale et al. (2000) show how disturbance regimes (e.g., fires, floods, predation, etc.) might affect such relationships by regulating community structure, for example by controlling changes in the relative abundance of species and promoting species coexistence (Paine 1966, Poff et al. 1997). The *intermediate disturbance* hypothesis suggests that biodiversity is highest at intermediate levels of disturbance by precluding competitive dominance, while too much disturbance results in local extinctions (Connell 1978). Spatial heterogeneity may also increase niche diversity and enhance coexistence at the intermediate level of disturbance (Kolasa and Pickett 1991).

## Species-area relationships

Within the debate on biodiversity, another area of contention focuses on species-area relationships: the idea that the number of species increases with the size of the sampling area such that larger areas will contain more species (Arrhenius 1921). This relationship is important for both ecosystem science and management. It is the basis for adequately sampling the species in a particular community, characterizing the community structure, and estimating species richness (Connor and McCoy 1979). For conservation biology, it also provides guiding principles to define the optimal size of reserves (He and Legendre 1996). Urbanizing landscapes provide unique opportunities to expand our understanding of the species-area relationship and to apply the knowledge to better design and manage urban regions.

Several hypotheses have been developed to explain species-area relationships (McGuinness 1984). The simplest is the *random placement* hypothesis (Arrhenius 1921, Coleman 1981): If individuals are randomly distributed, larger samples will contain more species. The *equilibrium* hypothesis (Preston 1960, MacArthur and Wilson 1963, 1967) explains the species-area relationship as a result of a dynamic equilibrium between colonization and extinction, which are determined by the size and isolation of islands. Island biogeography sees remnants as target areas for colonizing organisms. Larger islands support larger populations and large populations are less likely to become extinct than smaller populations. The hypothesis is that extinction rates are negatively correlated with population size due to demographic, genetic, and environmental stochasticity (Harrison 1991).

The *habitat heterogeneity* hypothesis (Williams 1964, Connor and McCoy 1979) maintains that larger areas have greater species diversity because they are more likely to encompass more diverse habitats. Islands of the same size are expected to vary in species diversity because they show different degrees of heterogeneity (McGuinness 1984). More heterogeneous areas are likely to support more species because of variations in climate, soil, topography, and other environmental factors (Williams 1943).

The *intermediate disturbance* hypothesis explains the species-area relationship as a function of variability in disturbance. The hypothesis holds that island size is related to frequency of disturbance. Small areas favor species that can tolerate more frequent disturbance, while the less frequently disturbed large areas are dominated by a few species that most efficiently exploit the resources and out-compete other species. The hypothesis thus contends that intermediate-sized areas support more species because they can support both types of species. This is described as a "humped" distribution (McGuinness 1984).

## Diversity and resilience

The idea that biodiversity provides a buffer or "insurance" against major change in ecosystem function given changing conditions and environmental fluctuations is based on the assumption that separate species utilize separate niches, responding differently to future events. More diverse ecosystems offer more options than simpler ones when placed under stress. Tilman (1996) shows less extreme year-to-year fluctuations in above-ground biomass in more diverse grassland communities, and faster recovery after drought. Under the insurance hypothesis, redundancy of species is a relative concept, depending on time and circumstances.

In theory, redundancy may very well allow for substitution when species belonging to a functional group are lost. This is the foundation of the biodiversity "insurance" hypothesis. Different theories about the apparent redundancy of species, the role of keystone species, and their relationships to ecosystem function have important implications for the strategies available for preventing loss of ecosystem function. But studies have not provided sufficient evidence to resolve the scientific controversy and provide clear policy guidelines (Johnson et al. 1996, Pimm 1991).

The scale at which species perform different ecosystem function may be a key to understanding the relationship between ecosystem function and diversity. Peterson et al. (1998) points out that while most models assume that ecological functions of various species remain the same at various scales, empirical evidence tells us that different species perform these functions at specific spatial and temporal scales (Holling 1992, Peterson et al.

1998). A few key processes regulate ecosystem structures and dynamics and they can be differentiated according to temporal and spatial scales (Levin 1992, Holling 1992). Holling (1992) describes the landscape as a hier-archical structure (Holling 1992). At the finer and fastest scales—centimeters to tens of meters and days to decades—are the biophysical processes that control plant growth and form. At the largest and slower scale —hundreds to thousands of kilometers and centuries to millennia—are geomorphological processes that control topography and soils. Disturbances such as fires, storms, and insect outbreaks operate at the mesoscale.

Peterson et al. (1998) hypothesize that if the species in a functional group operate at different scales, they mutually reinforce the resilience of a function—and minimize competition among species within the group. The presence of different functional groups within a scale and the replication of function across scales provide robust ecological functioning (Peterson et al. 1998). Ecological function is supported by scale-specific processes and structures that different species utilize differently, depending upon the time and spatial scales at which they operate (Morse et al. 1985, O'Neill et al. 1991, Peterson et al. 1998). Species may share the same area but, since they operate at different times and spatial scales, their interactions occur over different scales. For example, within a particular functional group, species that operate at larger scales require resources to be more aggregated in space than do species that operate at smaller scales.

Ecological resilience, as previously defined in Chapter 1, is a measure of the amount of change or disruption that is required to change a system from being organized around one set of mutually reinforcing processes and structures to operating around a different set (Holling 1973). This implies that ecosystems can have alternative self-organized states. In the cross-scale perspective proposed by Peterson et al. (1998), ecological resilience derives from overlapping functions within scales and the reinforcement of functions across scales. Cross-scale resilience complements within-scale resilience, which occurs when ecological functions overlap among the species of different functional groups that are operating at the same scales. Within a multi-taxa functional group, members that use similar resources may exploit different ecological scales. This leads to another form of ecological resi-lience as function is reinforced across scales (Peterson et al. 1998).

Perhaps the greatest challenge in the debate on biodiversity and eco-system function is to reconcile community and ecosystem ecology in a framework that more explicitly includes humans. Traditionally, community ecologists have focused on explaining species diversity as a function of abiotic factors and interspecies interactions. On the other hand ecosystem ecology has focused on the role of biotic interactions in governing eco-system processes and function. Humans influence both the biotic and abiotics forces that govern ecosystems. Understanding the mutual interactions and feedback among biodiversity changes, ecosystem functioning, and abiotic factors is a major challenge to achieving a true synthesis of community and ecosystem ecology. This is an essential step in better understanding the role of humans and articulating hypotheses about the interaction between human and ecological functions and biodiversity.

## 8.2 Urban Patch Dynamics

Over the last three decades, patch dynamics has emerged and evolved as a framework for studying the ways that pattern and process become coupled at different scales. It provides a promising approach to bridge theoretical and methodological gaps and to more effectively integrate community and ecosystem ecology. Patch dynamics explicitly recognizes that ecological systems are hierarchical, non-equilibrial, and vary both in time and space (Pickett and White 1985). The concept is essential to understanding the nature and dynamics of urban landscape ecology and the mechanisms of patch creation and evolution in urbanizing regions. While ecologists have recognized that heterogeneity results from environmental gradients (elevation, climate, etc.) at coarser scales, it is only in the last few decades that researchers in ecology have started to fully appreciate the implications of patch-level spatial homogeneity (Pickett and Rogers 1997).

Urban ecosystems differ from non-urban ones in their structure, processes, and functions (McDonnell and Pickett 1990, Rebele 1994, Trepl 1995). Based on the physical changes observed on the urban-to-rural gradient (Pickett et al. 1997), McKinney (2002) describes a biodiversity gradient in which species richness declines from the urban fringe towards the urban core. In the transition, as more and more habitat is lost, it is replaced by remnant, ruderal, and managed vegetation and built habitat, which vary in how habitable they are for most native species. As I suggested earlier, the key characteristic of urban ecosystems is their hybrid nature. To explain patterns of species diversity in urbanizing regions, we must look at the complex interactions between human processes and ecosystem processes that generate unique spatial and temporal heterogeneity. In turn, emerging patterns and processes of urban landscapes affect both biodiversity and ecosystem function.

What properties distinguish urban landscape patterns and processes from those of pristine ones? Trepl (1995) identifies three key sets of hypotheses related to integration, succession, and invasion. First, he says, in urban ecosystems the urban habitat patches and communities are not highly integrated, i.e. not well organized or connected; the systems are not in equilibrium, and stochastic processes predominate over deterministic ones (Trepl 1994). Second, succession in urban landscapes is hard to predict as the history and legacy of disturbances govern its dynamics. Third, urban ecosystems are open to invasions by unknown numbers of alien species.

Trepl (1995) observes that habitat patches and their species communities are less integrated in cities since patches are often isolated from each other by a matrix of built environment. These new barriers make dispersal difficult and potentially penalize organisms that are less able to move (Gilbert 1989, Rebele 1994). The impact of roads on wildlife dispersal in human-dominated environments has been extensively investigated (Forman 2000, Forman and Alexander 1998), and several strategies have been proposed to limit the negative effects of roads, but the multiple barriers created by urban development are still not fully understood. Offering a specific example, Davis (1978) noted that the best predictor of species richness of ground arthropods in London gardens was the proportion of green areas within a 1 km radius of the sampling site. Building on island biogeography and metapopulation dynamics, Klausnitzer (1993) and Weigmann (1982) have examined the relationship between species richness and patch area and found a consistent positive relationship. For birds, the built environment between green patches is not necessarily a barrier to dispersal, but fragmentation and change of habitat do affect their survival and success. Both dispersal and habitat requirements are modified in urban environments, favoring species that have both the ability to disperse and greater flexibility in habitat requirements (Gilbert 1989).

Human-induced disturbances in urban environments maintain urban habitats at an early successional stage (Trepl 1995, McDonnell et al. 1997, Niemala 1999a, 1999b). Some disturbances, such as fire and flooding, are suppressed in urban areas. At the same time, human-induced disturbances are more prevalent and persistent. Often, one part of the environment will be at an early successional stage (e.g., mown lawn) while another part is at climax stage (e.g., old trees). Furthermore, the patchy distribution of urban habitats, combined with the varying degree of human-induced disturbance and chance, results in a number of succession paths across habitat patches (Niemala 1999a). Even adjacent patches may exhibit very different successional paths depending on the colonization history of plants, which is largely determined by chance events (Gilbert 1989). This historical uniqueness and the overwhelmingly external control of succession are important features that distinguish urban habitats from more natural ones (Trepl 1995).

Urban ecosystems are simultaneously influenced by the environmental changes driven by humans and the ability of plants, animals, and micro-organisms to adapt to and exploit these changes. Environmental conditions that differ between urban and rural areas favor certain species over others. Temperature is a good example: Many species requiring high temperatures thrive in cities where the temperatures are higher than in surrounding areas (Gilbert 1989). Bradshaw (2003) describes the phases of succession in urban areas by identifying the key ecosystem attributes and processes involved (Table 8.1). Urban structures provide unique opportunities for organisms—from abundant food sources to shelter (Bradshaw 2003).

Urban development creates new opportunities and challenges for species competition and predation, both as exotic species are introduced and as invasive species migrate in. Invasive or non-native species take advantage of poorly integrated communities and patches in the urban setting. This can be seen as a colonization process, as more frequent introductions of exotic species translate into invasions (Rebele 1994). Examples of this phenomenon abound. The proportion of alien plant species in Berlin increased from 28% in the outer suburbs to 50% in the built-up center of

Table 8.1. Succession in urban ecosystems (Bradshaw, 2003, p. 81).

| Ecosystem Attribute | Processes Involved |
| --- | --- |
| Colonization by species | Immigration of plants species |
| | Establishment of those plant species adapted to local condition |
| Growth and accumulation of resources | Surface stabilization and accumulation of fine mineral materials |
| | Accumulation of nutrients particularly nitrogen |
| Development of the physical environment | Accumulation of organic matter |
| | Immigration of soil flora and fauna causing changes in soil structure and function |
| Development of recycling process | Development of soil microflora and fauna |
| | Possible difficulties in urban areas |
| Occurrence of replacement process | Negative interaction between species by competition |
| | Positive interaction by facilitation |
| Full development of the ecosystem | Further growth |
| | New immigration, including aliens |
| Arrested succession | Effect of external factors |
| | Reduction of development |
| Final diversification | The city as a mosaic of environments |
| | High biodiversity as a result |

the city (Sukopp et al. 1979). Along a 140-km urban-to-rural environmental gradient originating in New York City, McDonnell et al. (1997) found lower levels of both earthworm biomass and abundance in the urban forests, compared to the rural forests. They attribute the difference to the incidence of introduced species. Insects are also successful invaders because of their ability to survive well around humans (Spence and Spence 1988). For instance, in western Canada, the 20 ground beetle (Carabidae) species of European origin account for the majority of carabids in cities (Niemela and Spence 1991).

## 8.3 Urban Ecosystem Processes and Biodiversity

Urban development affects biodiversity both directly, by altering the land cover and introducing non-native species, and indirectly, by changing ecosystem and biogeochemical processes. Urban impacts on biodiversity occur both locally and globally. At the global scale, the leading drivers of biodiversity loss are changes in landscapes and climate, and the introduction of non-native species. Humans' location choices are a major factor, since people prefer to settle on the most productive soil and in highly diverse areas. Recent findings have indicated that areas with the highest human population density and high biodiversity levels coincide when observed at the regional scale (Luck 2007, Pautasso 2007). This is true for plants, amphibians, reptiles, birds, and mammals across most regions of the world (Balmford et al. 2001, Araújo 2003, Luck et al. 2004, Evans and Gaston 2005, Real et al. 2003). But while this pattern is generally observed at the regional scale, the reverse is true at the local scale where people compete for resources with other species, making for a negative correlation between human presence and biodiversity (Beissinger and Osborne 1982, Clergeau et al. 2001, Fudali 2001, Moore and Palmer 2005). The correlation between human presence and species richness apparently depends on scale (Manne 2003, Váquez and Gaston 2006) which, according to Pautasso (2007), may provide a plausible explanation for findings that species richness peaks at intermediate levels of urbanization.

Although many researchers have tried to estimate the current loss and potential threats to biodiversity globally, we do not know how much species extinction or endangerment can be attributed to urbanization. Ewing et al. (2005) estimate that, of rarest and most imperiled species in the United States, 60% are threatened by rapid growth within the 35 fastest-growing metropolitan areas that are home to 29% of these species. They suggest that sprawl may lead to a great potential loss of species; whether or not these predictions turn out to be true, we know that many effects of sprawl threaten species survival. Though studies of human density and biodiversity have conflicting findings, the predominant pattern is a negative correlation between increasing human population density and species richness at a local scale. But the interactions are complex, and it is imperative that we better understand the mechanisms that govern biodiversity in urbanizing regions if we intend to reduce the impacts of urbanization.

Ecological studies have focused primarily on patterns of species diversity, aiming to establish relationships between the emerging urban landscape structure and the distribution, movement, and persistence of species on an urban-to-rural gradient (McDonnell et al. 1997). However, more recent studies have attempted to articulate the mechanisms governing biodiversity in urbanizing regions (Hansen et al. 2005, Faeth et al. 2005). By focusing on behavioral ecology, biotic interactions, genetics, and evolution, a few studies have revealed that urbanizing environments are a unique setting, in which human actions mediate fundamental patterns and processes in complex ways (Shochat et al. 2006). Despite this complexity, scholars of urban ecology agree that the challenge in the next decades is to formalize a theory of the mechanisms governing biological diversity where humans are present, and to resolve some long-standing unexplained contradictions between observations and ecological theories.

Findings from current urban ecological studies at the two Urban Long Term Ecological Research (LTER) sites in Baltimore and Phoenix, as well as in Seattle, provide important evidence that allows scholars to develop hypotheses on the mechanisms that drive urban biodiversity. Hypotheses about how humans impact species diversity can be articulated around key mechanisms that influence biodiversity: habitat productivity, species interactions, trophic dynamics, heterogeneity, disturbance, and evolution.

## Habitat productivity

Productivity—"the rate at which energy flows through an ecosystem" (Rosenzweig and Abramsky 1993)—is one of the ecosystem properties in urbanizing regions that may explain patterns of species diversity along the urban-to-rural gradient. But the mechanisms that govern the relationship between productivity and species richness are far from being understood (Waide et al. 1999). Although most scholars agree that productivity affects species richness at large scales (Waide et al. 1999), studies show contradictory results. Two types of relationship between productivity and diversity have been proposed: (1) monotonic, where diversity increases as productivity increases at the regional and global scales; and (2) unimodal, where diversity increases along with increasing productivity but declines at the highest productivity levels, primarily at the local scale (Abrams 1995).

This latter hump-shaped relationship has been proposed to describe the ways productivity, urban population density, and species diversity are related to each other (Blair 1996, Marzluff 2001). Since highly developed areas have a net primary production (NPP) close to zero, we should expect to find lower species richness in them. In Seattle, for example, densely populated urban areas arc associated with lower bird diversity (Marzluff 2001). NPP mediates the relationship between anthropogenic land cover change and both faunal and plant species richness (Mittelbach et al. 2001), although the relationship is dependent on scale (Waide et al. 1999). At the urban fringe, in highly managed landscapes, high species richness is often associated with a higher local rate of productivity relative to the surrounding areas. Shochat et al. (2006) suggest that habitat productivity in managed urban green spaces is generally higher compared with the surrounding areas because of the high human resource input. Although they recognize that the high productivity levels are limited to the urban fringe where green patches represent a larger proportion of the landscape. In fact, Imhoff et al. (2000) found evidence of this relationship by analyzing the NPP for the 48 contiguous states. They found total NPP to be higher in urban areas than in wildlands, both in cities located in arid environments and in low-density development (Imhoff et al. 2000, 2004).

Although the relationship between habitat productivity and diversity is generally unimodal, the strength, shape, and sign of the relationship vary with scale of observation (Mittelbach et al. 2001). In urban regions, we can expect that the productivity-diversity relationship will vary both with biophysical characteristics as well as with urban form (Shochat 2006). Complex interactions and especially feedback between diversity and productivity (Naeem et al. 1996, Tilman et al. 1996) have not been studies in urban setting. And since different mechanisms operate at multiple scales, studies of urban ecology must incorporate a hierarchical approach (Cleargeau et al. 2006). These investigations are essential to better understand the role that humans play in mediating this relationship.

## Biotic interactions

Urbanization affects ecosystem function by altering the way species are distributed and interact with each other (Marzluff 2001, Hansen et al. 2005). For example, both native and nonnative predators may increase near human settlements, a change that may then affect other native species. Species found at the urban fringe are edge-adapted generalists who are able to use most effectively a variety of natural and human-generated resources for their survival (McKinney 2002). Higher densities of nest predators explain high rates of nest predation of migratory songbirds in suburban woodlots in

Maryland (Wilcove 1985). Another mechanism that lowers diversity is competition. Species often colonize urban areas within regions where they would not normally thrive. Examples include the grey-headed flying foxes (*Pteropus poliocephalus*) in Melbourne, Australia (Parris and Hazell 2005) and the house gecko (*Hemidactylus frenatus*) in Hawaii (Petren and Case 1996). As such species become more abundant, the native urban species, usually good adapters, could become extinct locally (Shochat et al. 2006).

How different are the key processes governing the diversity of an urbanizing ecosystem from the compensatory processes of colonization and extinction that are postulated by island biogeography theory (MacArthur and Wilson 1963, 1967)? For cities, the high turnover of species is primarily affected by the introduction, dispersal, and local extermination mediated by human activity (Rebele 1994). Marzluff (2005) developed a graphical model to form a series of testable hypotheses about how extinction and colonization are affected by urbanization in determining local diversity. Colonization and extinction, the fundamental processes governing diversity, respond to organisms' rates of survival, reproduction, and dispersal (Marzluff and Dial 1991, Bolger 2001). In human dominated ecosystems, colonization and extinction are affected by changes in land cover, the removal of barriers to dispersal, the introduction of new species, and the action of new selective forces caused by changes in climatic regimes, predators, competitors, and diseases (Marzluff 2005, Husté and Boulinier 2007). And while diversity still emerges as the balance between extinction and colonization, species invasion plays a prominent role (Olden and Poff 2003, 2004, Marzluff 2005).

## Trophic dynamics

Human activities in cities alter the food webs and trophic structure of biological communities. Ecologists have studied the factors that control the trophic structure and function of ecosystems for quite some time, but they have not thoroughly studied the trophic organization of urban areas (Faeth et al. 2005). Faeth et al. (2005) point out the importance for ecology to understand the structure and function of food webs in urban settings not only because urban habitats are increasingly important, but also because this knowledge is crucial to conservation efforts. Conservation planners and managers need to know what controls the number and diversity of trophic levels in urbanizing regions and how species in food webs interact through the processes of competition, predation, parasitism, and mutualism.

Urban habitats affect both above- and belowground food-webs (Bramen et al. 2002). Studies of the Sonoran Desert, developed as part of the Central Arizona-Phoenix LTER (CAP LTER), reveal some surprising human-induced modifications with respect to factors controlling trophic dynamics, according to Faeth et al. (2005). They found that species composition was radically altered (e.g., generalist species increased), and that resource subsidies caused by people increased and stabilized productivity (i.e., via modified water availability). This supports their hypothesis that the absence or reduction of predators and the increased abundance and predictability of resources in urban areas may cause a shift in control from top-down to bottom-up. Based on the Phoenix study, they propose that urbanization caused shifts from a system that is resource-based or controlled from the bottom up—typical of the Sonoran Desert—to a combined bottom-up and top-down model.

## Spatial and temporal heterogeneity

The diversity of species in urbanizing regions is greatly affected by the quality of habitat and template of resources. These factors are the results of biophysical processes operating at multiple scales and mediated by human actions. Ecosystem processes are heterogeneous and highly related to species distributions (Turner and Chapin 2005). The *habitat heterogeneity* hypothesis states that greater spatial variation in physical or environmental conditions allows for greater niche differentiation and, hence, more species (MacArthur and MacArthur, 1961). Humans affect habitat quality and resource availability by changing habitat heterogeneity in space and time. In urbanizing regions both spatial and temporal heterogeneity are influenced by human and biophysical processes associated with high fragmentation of land use and management.

The effect of human action on spatial heterogeneity in urbanizing regions is very well documented; however, how we know less about how this heterogeneity varies with scale, partly because studies have tended to focus primarily on aggregated measures (Band et al. 2005). At the scale of meters or below, urbanization may reduce the heterogeneity of land cover, but at the patch level, it may introduce highly heterogeneous new bio-physical conditions as the varied behaviors of land owners result in fragmented management patterns. As the scale increases, we may observe a further reduction in heterogeneity due to consistent patterns of urban development and habitat fragmentation.

Changes in temporal variability in urban ecosystems are driven by both human structures and high inputs of resources. A good example of change in temporal heterogeneity is the buffering effect that microclimatic changes associated with urbanization can have on habitat: In temperate cities, heat islands can extend the growing season while in desert cities, they can cause thermal stress and extend droughts. Shochat et al. (2006) report that the heat island in Phoenix, Arizona, has increased the stress on cotton plants (*Gossypium hirsutum*). Highly managed green areas in temperate cities such as Seattle, and irrigation in semi-arid cities such as Salt Lake City, provide water for plants throughout the year with subtle effects on wildlife.

Urban management and the built infrastructure can artificially reduce the variation—in both space and time—of resource availability, thus altering seasonal variations and dampening temporal variability. Some species thrive when they have less variation to endure, and their urban populations rise. A well-known example is the grey-headed flying-fox (*Pteropus poliocephalus*), a large, nomadic bat from eastern Australia that became established in Melbourne, Australia, when a heat island effect led to long-term climatic changes. Parris and Hazell (2005) found that human activities have increased temperatures and effective precipitation in central Melbourne, creating a more suitable climate for camps of the grey-headed flying-fox. Changed habitat, due to the high availability of water and continuous availability of food, interacts with biotic, trophic, and genetic processes and may help some species adapt to urban environments.

Heterogeneity in urban ecosystems is driven simultaneously by natural and human agents and processes. Urbanization tends to increase spatial and temporal heterogeneity on some scales and reduce it on others. For example, urban development enhances spatial variability by fragmenting the land cover. At the same time, the built landscape and infrastructure tend to decrease heterogeneity within patches and at larger scales. As a result, carbon, water, nutrient, and energy cycling are highly modified (Band et al. 2005). Increases in heterogeneity can also favor certain species and penalize others, changing biotic interactions and community composition. Furthermore, spatial heterogeneity affects disturbance regimes, and is another fundamental mechanism that links urban patch dynamics to ecosystem process and function.

## Disturbance regimes

Ecosystem disturbances are events that affect the pathways by which matter or energy flows in an ecosystem (Pickett et al. 1999). Disturbances affect resource availability (i.e., water and nutrient), ecosystem productivity, and species diversity. Changes in resources and related ecosystem processes caused by anthropogenic disturbances affect plants and animals over time, and ultimately successional dynamics. Urbanization modifies existing disturbance regimes (e.g., through fire and flood management) and creates novel disturbances (e.g., through new or disrupted dispersal pathways or species introduction). Cardinale et al. (2006) suggest that disturbance can moderate relationships between biodiversity and ecosystem functioning in two ways. It can increase the chance that diversity will generate unique system properties (i.e., emergent properties), or it can suppress the probability of ecological processes being controlled by a single taxon (i.e., the selection-probability effect).

As urbanization changes disturbance regimes, it affect species diversity (Rebele 1994). Species diversity is high at intermediate frequency or intensity of disturbance (Connell 1978, Pickett and White 1985). Disturbance impacts directly species interaction by precluding competitive dominance (Poff et al. 1997) and Cardinale et al. (2000) articulates the mechanisms by which disturbance mediate the relationship between species diversity and ecosystem function by focusing on variation of species abundance across scale. More recently Cardinale and Palmer (2002) use a laboratory experiment to show that indeed disturbance mediates the response of stream ecosystems to species richness.

## Evolutionary processes

Humans influence evolutionary processes by changing speciation and extinction patterns (Palumbi 2001). For example, humans are challenging bacteria with antibiotics, poisoning insects, rearranging and exchanging genes, creating and dispersing thousands of synthetic compounds, and fishing selectively. By hunting, moving around the globe, and massively reconfiguring the planet's surface, humans have increased species extinction rates to levels 1,000 to 10,000 times higher than those resulting from nonhuman causes, through resulting changes in predation and competition (Pimm et al. 1994, Vitousek et al. 1997, Flannery 2001). The combined effect of changing speciation and extinction is rapid evolutionary change (Palumbi 2001).

Urban environments may facilitate speciation by bringing together species previously isolated, or isolating populations through habitat destruction, as well as by introducing new exotic species, phenomena of speciation are much more likely to occur (Sax and Gaines 2003). Chances of extinction, on the other hand, are increased by changes in habitat and selective forces (Ledig 1992). In urban environments, selective changes are caused by eliminating variation in resource availability (i.e., food and water) and modifying biotic interactions (i.e., predation). Humans in cities also create new selective forces affecting the genetic structure and diversity of urban ecosystems (Yeh and Price 2004).

The evolution of behavioral flexibility and adaptive phenotypic plasticity in response to spatial and temporal variation in species interactions in urban environments can facilitate the success of organisms in novel habitats, and potentially contributes to genetic differentiation and speciation (Agrawal

2001). Adaptation generally does not occur in a relatively short period of ecological time. But in urban environments, resource availability and lower risks of predation may facilitate persistence until genetic change occurs. Wood and Yezerinac (2006) hypothesize that song sparrows have changed the frequency of their notes to adapt to the noisy urban environment. Another example is the dark-eyed junco (*Junco hyemalis*) reveals how this species has adapted its tail feathers in San Diego, California (Yeh and Price 2004). At the same time, the extreme turnover in biological communities might prevent the genetic differentiation of urban populations and impede evolutionary responses to the novel selective forces associated with urbanization (Shochat 2006).

## 8.4 The Intermediate Hypothesis: A Case Study in the Puget Sound

The intermediate disturbance hypothesis is a nonequilibrium ecological hypothesis that provides an explanation for the coexistence of species in ecological communities based on complex coexistence mechanisms (Wilson 1994, Dial and Roughgarden 1998, Buckling et al. 2000). It also suggests that in a situation of intermediate disturbance, more species can be expected at a given instant in time, whether or not that diversity is maintained over the long term (Roxburgh et al. 2004).

To test this hypothesis in urbanizing regions, researchers have studied many taxa including mammals (Racey and Euler 1982), birds and butterflies (Blair and Launer 1997, Blair 2001), ants (Nuhn and Wright 1979), and plants (Kowarik 1995). Birds are excellent indicators of the effects that urbanization has on ecosystems since they are highly mobile and respond rapidly to changes in landscape configuration, composition, and function (Marzluff et al. 1998). Urbanization affects birds directly as the ecosystem processes change, along with habitat and food supply. In fact, the percentage of land cover that is vegetation is a good predictor of the number of bird species. Birds are also affected indirectly as urbanization influences other factors, especially predation, interspecies competition, and diseases (Marzluff et al. 1998). Urbanization increases the number of introduced species and drastically reduces the number of native species (Marzluff 2001), thus altering the composition of urban avian communities. Populations of native species decline because their natural habitats are reduced and they cannot tolerate human disturbances (Beissinger and Osborne 1982, Blair and Walsberg 1996).

As part of a Biocomplexity project, we examined the interactions and feedback between urban development and land cover change and the effects on bird diversity in the Central Puget Sound region (Alberti et al. 2006a). Led by Marzluff, a team of wildlife biologists are studying the response of birds to the complex landscape changes produced by urbanization. Marzluff (2005) found that songbird diversity peaks in landscapes with 50% to 60% forest cover, because such areas gain more synanthropic birds (those that benefit from contact with humans) and more successional species; Meanwhile, some of the native forest species tend to leave such areas. The birds' dynamic response to changing land cover allows us to demonstrate how one component of biological diversity might respond to urbanization.

## Study area and avian surveys

The bird study area is the Puget Sound lowland (<500 m above sea level) (3,200 km$^2$) of temperate, moist forest surrounding Seattle, Washington. 139 study sites of 1 km$^2$ were selected and characterized according to their land use. Of the 139 sites, 119 were characterized as single-family residential (SFR), 13 as mixed use/commercial/industrial, and seven as forested ("control") sites with minimal development (Marzluff et al. 2001). The team also randomly selected 126 single-family residential and forested control sites along several axes of urbanization including: 1) urbanization intensity, 2) mean patch size of urban land cover, 3) similarity of adjacent areas, and 4) development age. The sites defined as suburban contained >70% urban land cover, and <15% forest land cover. The exurban sites contained <50% urban land cover, and >40% forest land cover (Alberti et al. 2004). During the springs and summers of 1998 through 2005, trained observers conducted 6437 fixed-radius (50 m) point-count surveys of breeding birds at 992 locations within the 139 study landscapes. (For a full description of the study see Donnelly and Marzluff 2004a, 2004b, 2006, Blewett and Marzluff 2005, and Hepinstall et al. in press).

## Observations

Marzluff (2005) shows that birds respond to human settlement along an urban-to-rural gradient in a complex way. The number of bird species in a 1 km$^2$ landscape made up of single-family housing and fragments of native coniferous forest was strongly correlated with the percentage of forest. The relationship is non linear (Richness = 17.7 + 46.03 (% forest) − 42.9 (% forest)$^2$; $F_{2,56}$ = 22.3, P < 0.0001) as shown in Figure 8.2. The quadratic relationship accounted for nearly half of the variation in bird species richness ($R^2_{adjusted}$ = 43.2%) (Marzluff 2005).

The richness peaked at the ratio of approximately 50% forest in the landscape. The richness of bird species is determined by the balance between two factors: the retention of native forest birds and the addition of synanthropic and early successional species. Species gain was quadratically related to the amount of forest. The number of native bird species that remained decreased only slightly as the amount of forest increased. Thus, bird communities in landscapes of 50% forest have high species diversity because they support rich mixes of native forest birds, early successional species that use grasslands and forest openings, and synanthropic species that benefit from human activities. Bird communities in more urban areas are impoverished because only about ten synanthropic species live on the mostly paved landscapes, along with fewer than five native species. Likewise, communities in the mostly forested areas are impoverished because they are composed almost entirely of the fifteen native forest birds.

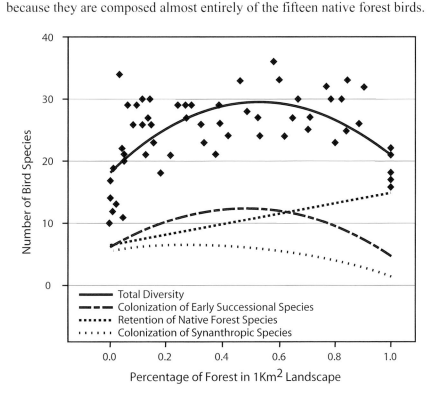

Figure 8.2. Impact of urbanization on songbirds. Measures of total species richness, which peaked at approximately 50% forest cover, reflect the combined effect of a steady decline in native species with decreasing percentage of forest cover and an increase in early successional and synanthropic species, which benefit from human activities with decreasing forest. The lower species richness at very low levels of forest cover (i.e., more heavily paved urban areas) reflects the dominance of a smaller number of more competitive synanthropic and remnant native species. Similarly, lower richness measures at higher levels of forest cover reflect the presence of predominantly native bird species (Marzluff 2005, p. 166).

Statistical models

To predict future patterns of richness and relative abundance of bird species, we developed a coupled model of land cover change and bird diversity by linking the land cover change model (LCCM) and a series of statistical models of birds' responses to change in landscape composition and configuration (Hepinstall et al. in press). Marzluff et al. (2001) used point-count data from the 139 study landscapes to develop separate models of species richness for all the species and the three development-sensitive guilds (Marzluff et al. 2001). The study involved a land cover character-ization of each 1 km$^2$ bird study area based on land cover data from 2002 to calculate the percent forest, percent urban, and aggregation index (Fragstats 3.3, McGarigal et al. 2002) of the forest, as well as the number of patches of forest, and the number and mean size of the urban patches. To better characterize the study sites, Hepinstall et al. (in press) used land use information derived from 2002 parcel data to calculate the percent patch density, and aggregation index of residential parcels, and the development age of all the parcels within each study area.

Marzluff and Hepinstall developed two *a priori* models of species richness and relative abundance based on previous studies (e.g., Donnelly and Marzluff 2004a, 2004b, 2006, Hepinstall et al., in press), on landscape measures relevant to urban planners, and on variables available as output from our LCCM. A first simple model (SM) included: 1) percentage of forest (in linear and quadratic form); 2) aggregation of residential land use; and 3) development age of parcels within a 1-km$^2$ window. A more complex model (FM) added seven more variables: 1) percentage of grass and agriculture; 2) forest aggregation index; 3) number of unique patches of forest land cover; 4) number of unique patches of urban land cover; 5) mean patch size of unique patches of urban land cover; 6) percent of residential land use; and 7) patch density of residential land use.

Applying the parameter estimates from models of species richness and relative abundance to the future landscapes generated by the land cover change model, Hepinstall et al. (in press) predicted the total guild- and subguild-specific species richness, and relative abundance for all species.

## Future land cover and avian diversity

Our LCCM predicts that in the next two decades we will observe a decline in mature forest types (deciduous, mixed, and coniferous) from 60% of the study area to 38%, and an increase in developed land (heavy, medium, and low urban classes) from 17% to 34%. According to our model, the proportion of land in grass and agriculture will decrease from 14% to 10% of the area, while the proportion of clearcut and regenerating forest will increase from 9% in 2003 to 18% in 2027 (Alberti et al. 2006a, Hepinstall et al. in press). This substantial reduction in forest cover and increase in developed land will affect the region's avian populations. Although, overall, only three to five species are expected to be lost, the loss in ecological resilience due to forest loss and fragmentation will make bird diversity significantly more vulnerable to future loss of forest. In addition, the observed pattern of avian diversity along the gradient of urbanization has an influence on the overall avian population dynamic.

The spatially explicit land cover change predictions can be mapped to explore the local effects of landscape change. We expect that changes in species richness will be concentrated in those regions of the study area where land cover change is most dramatic, primarily in the transition development zone surrounding the present heavy urban core where forest loss and aging of developments are more pronounced. We see similar patterns of loss in richness for the total species, the native forest species, and the early successional species. We predict a species loss at any specific locale up to 23 species, and up to nine species lost respectively from the native forest and early successional guilds. The model predicts a gain for the synanthropic species of two to four species in the transition zone.

Changes in the landscape pattern in the central Puget Sound region will affect the diversity of the transition zone. Currently, avian diversity peaks at the transition zone and exurban zones, but we expect that in the near future as the transition zone shifts, avian communities will gradually become more diverse the farther they are from development, rather than peaking at an intermediate level of settlement. (Figure 8.3).

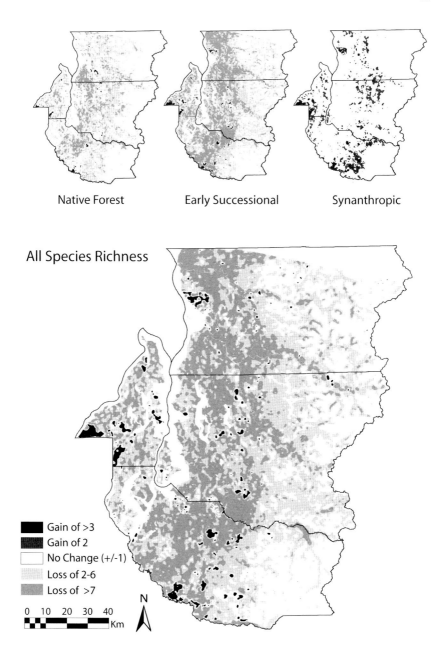

Native Forest          Early Successional          Synanthropic

All Species Richness

■ Gain of >3
▨ Gain of 2
□ No Change (+/-1)
▒ Loss of 2-6
▓ Loss of >7

0  10  20  30  40
▭▭▭▭ Km

N

Figure 8.3. Change in bird diversity in Central Puget Sound. The maps illustrate locational changes in abundance of native forest, early successional, synanthropic, and total bird species in the central Puget Sound region. Areas exhibiting significant losses in species richness reflect highly urbanized regions (e.g., the greater Seattle metropolitan area, center) mostly dominated by small numbers of synanthropic species, whereas gains in (mostly synanthropic) species along the Cascade range to the east of Seattle reflect areas exhibiting significant conversion of forest to other land uses (e.g., development, forest harvesting, etc.) (Hepinstall et al. in press).

# Chapter 9

# FUTURES OF URBAN ECOSYSTEMS

## 9.1 The Challenges: Complexity, Heterogeneity, and Surprise

Planning agencies in urbanizing regions face unprecedented challenges: Rapid environmental change places enormous pressure on their ability to support urban populations while maintaining a healthy ecosystem. Agencies must devise policies to guide urban development and to make decisions about where and how to invest in infrastructure that is economically viable while simultaneously minimizing environmental impact. In regions that are becoming urbanized, the form and pace of urbanization mediate the complex interactions between humans and ecological processes. In turn, urbanization places increasing pressures and constraints on natural resources. Planning decisions, especially those about urban growth and infrastructure, can influence the directions of urban development and determine the sustainability of our urban planet. To make sound decisions, it is crucial to assess the effectiveness of infrastructure choices and the robustness of urban planning strategies under alternative future scenarios.

Strategic decisions about urban infrastructure and growth management are based on our assessment of the past and our expectations for the future. How we think about the future has important consequences for how we define the problems to be addressed and how we search for solutions. Traditional approaches to planning and management typically rely on predictions of probable futures extrapolated from past trends. Alternatively, planners and managers have developed participatory processes to imagine desirable futures based on a set of shared community values and goals. However, with respect to long-term trends, complexity and uncertainty in coupled human-natural systems make their future increasingly unpredictable. Planners and managers need to rely on a much broader and diverse knowledge of the past to build a view of what Stewart Brand (1999) calls the "long now." Expanding on this concept, Steve Carpenter (2002, 2069), in his MacArthur lecture entitled "Building an Ecology of the Long Now," noted that in many cases of environmental decision-making, what ecologists cannot predict is at least as important as what can be predicted. Thinking about the future, however, is challenging. Scientists and managers are constrained by their assumptions about how the world works and what drives change. In combination, all these conditions tend to limit planners' imagination to a default set of scenarios, and thus limit their ability to deal with surprise.

For example, climate variability and change are expected to signifycantly impact essential human services (i.e., supplies of water and energy) and ecological functions (i.c., primary production) within urban areas. Increasingly, policy makers and planners must balance the need to provide critical services to the urban population while maintaining important ecological functions. While climate change may be inevitable, strategies can be implemented to make urban systems more resilient to potential changes in climate, and simultaneously maintain human and ecological functions. To identify and prioritize urban planning and management strategies, we must explore future scenarios of climate change and the ways they interact with other drivers of change. Further, we must also identify vulnerable systems, and assess the effectiveness of alternative strategies in reducing risk under each scenario.

For environmental decision-makers, it has become crucial to assess alternative strategies and take action in the face of irreducible uncertainties. Predictive models that are designed to provide accurate assessments of future conditions can only partly account for the interactions between highly uncertain drivers of change and the surprising, but plausible, futures over the long term. While important progress has been made in complex modeling, and improved simulation and computer power have allowed us to process quite astonishing amounts of data, our models are constrained by our limited knowledge, assumptions, and mindsets. How can we articulate and explore possible futures when coupled human-natural systems are so complex, and their futures so uncertain? How can we challenge the assumptions and mindsets that both scientists and managers hold about the future? And how can we build institutions that take a "long now" perspective?

In this chapter, I discuss the challenges of predicting future dynamics of coupled human-ecological systems and of building models of urban ecosystems to assess the ecological future of urbanizing regions. Aiming at predicting the future by estimating probabilities on the basis of past trends is not sufficient. Coupled human-natural systems exhibit complex dynamics that can cause surprising behaviors (Holling 1996, Scheffer et al. 2001 Carpenter 2001, 2002). I propose to link predictive modeling with scenario building in order to explore systematically and creatively plausible futures. Strategists at Royal Dutch/Shell originally proposed this approach in the 1970s, and it was recently applied in the Millennium Ecosystem Assessment (MEA 2005). By focusing on key drivers, complex interactions, and irreducible uncertainties, scenario building generates the narratives within which we can use predictive models to test hypotheses and develop adaptive management strategies. This approach, of linking models and scenarios, is relevant to both science and policy. It provides a basis for developing an integrated understanding of the processes and mechanisms that govern urban ecosystem dynamics; at the same time, it provides tools to predict and assess the impact of future urban growth and to evaluate alternative urban planning and growth-management strategies.

## 9.2 Complexity and Predictability

In modeling the interactions between human and ecological systems, we need to consider that multiple factors operate simultaneously at various scales. Coupled human-ecological systems are very complex. Complexity emerges as interacting agents engage in simple behaviors in systems that are nonlinear (i.e., strongly coupled) or open (driven from equilibrium by external factors). Change and evolution are inherent in these systems. The feedback mechanisms that operate between ecological and human processes can amplify or dampen changes and thus regulate the system's response to external pressures. In the Millennium Ecosystem Assessment summary report, the authors conclude that "there is established but incomplete evidence that changes being made in ecosystems are increasing the likelihood of nonlinear and potentially high-impact, abrupt changes in physical and biological systems that have important consequences for human well-being" (MEA 2005).

In urban ecosystems, urban development controls the ecosystem structure in complex ways. Land use decisions affect species composition directly by introducing species and altering land cover, and indirectly by modifying the agents that naturally cause disturbances. Human choices about what and how much to produce and consume determine the amounts of resources that are extracted and the amounts of emissions and wastes. Decisions about investing in infrastructures or adopting control policies may either mitigate or exacerbate these effects. But there are also important feedbacks. Ecological productivity affects the regional economy. Thus interactions between decisions and ecological processes at the local scale can result in large-scale environmental change.

Emerging properties in complex systems are aggregate behaviors that we cannot infer from studying the system components that have generated them (Parrott 2002). Emergent properties make complex systems inherently unpredictable. But while highly unpredictable, complex systems can be studied with the objective of gaining knowledge on their behavior and dynamics. Holling (2001, 391) explains that "Complexity does not emerge from a random association of a large number of interacting factors." Rather, it emerges from a smaller number of controlling processes. These systems are self-organized; a small set of critical processes creates and maintains this self-organization and drives the systems' evolution (Levin 1999). Holling (2001) argues that there is an essential simplicity behind complex systems that can lead us to understand their dynamics. The approach for studying such systems should be "as simple as possible but no simpler" than is required for understanding and communication. It should be dynamic and prescriptive, not static and descriptive, so we can connect policies and actions to the evaluation of different futures. And it should embrace uncertainty and unpredictability because change and surprise are inevitable in coupled human-natural systems.

In science, all predictions involve some level of uncertainty. Although in some cases uncertainty may be great, scientists most often assume that it is quantifiable. Complex systems are inherently unpredictable, thus the uncertainty cannot be completely quantified. In coupled human-natural systems, future dynamics are dependent on multiple drivers, such as population growth and climate change, that have very different degrees of uncertainty. The probability distribution of predictions for coupled human-natural system depends on the distributions of such drivers. But since future driver distributions may be unknown, the uncertainty in such predictions cannot be calculated (Carpenter 2002).

In coupled human-natural systems, uncertainty and unpredictability can be generated by surprising interactions among the driving forces and the reflexive interactions between human behavior and their anticipated knowledge for the environmental change their action can have. As Steven Carpenter (2002, 2080) points out: "Even the uncertainties are uncertain, because we do not know the set of plausible models for the dynamics of the probability distributions."

## Dynamic urban ecosystems

Human systems and natural systems both change over space and time. One major problem in modeling their relationships is that they operate at very different spatial and temporal scales. The lag time between a human decision and its impact on the environment complicates any attempt to model their interactions. Human impacts on ecosystems may become apparent only after irreversible changes have already taken place and caused undesirable consequences. The delayed response to human-induced changes in ecological systems has been known for a long time. The depletion of fish stocks is an example of a lagged environmental response to over-fishing (Jackson et al. 2001). Another example is climate change (Burkett et al. 2005). It took a long time for the climate to respond to the concentrations of greenhouse gases in the atmosphere, and it can take a long time before ecosystems respond to climate change.

Time-lags between human and ecosystem function are exacerbated by scale mismatches. People and ecosystems work at very different time scales (Cumming et al. 2006). In coupled human-natural systems, drivers of change differ considerably in their response times (MEA 2005), and the speed at which a driver reacts has an important impact on the system's ability to adjust. The extinction of species due to habitat loss or climate change can be considered irreversible on a human time scale. Failure of people or society to detect or recognize such impacts adds to the lag in response. In an urban region, the extraction of groundwater may exceed the system's capacity for recharge long before the costs of extraction begin to reflect that depletion. The economic costs associated with long-term ecological effects are often not reflected in market prices for goods and services.

Complex spatial dynamics are also affected by boundaries and scale mismatches, and human impacts may be widely distributed and distanced over space. Changes in upstream catchments affect water availability and ecological conditions in downstream regions. Emissions of atmospheric pollutants affect those living downwind. Changes in landscape structure in one part of a watershed has consequences on disturbance regimes on a larger scale. Global climate change affects regions unequally. The spatial units of human influence and institutional response (through land-use regulations and policies) do not match those that govern biophysical processes and ecosystem functioning.

Both spatial and temporal lags have important consequences on the ability to manage coupled human-natural systems. In fact, the temporal and spatial separation between cause and effect make it extremely difficult to assess costs and benefits and to fully appreciate the distributional consequences of human action. For people to take responsibility for the environmental consequences of their behaviors in space and time requires a much more expanded perspective in time and space.

We must treat time and space explicitly if we are to accurately represent the dynamics of urban and environmental change. Spatially-explicit models are increasingly being implemented in modeling coupled human-natural system dynamics, taking advantage of the advancement in geographic information science technology, available spatial data, and computer power (Goodchild 2003). When building models of urban ecosystems, time can be represented as either a discrete or as a continuous variable. While treating time as continuous is certainly daunting, we can use multiple time steps for the different processes; this approach represents an important improvement over stationary models. Today, most operational urban models are based on a cross-sectional, aggregate, equilibrium approach; we could improve on them by representing time explicitly as a discrete variable, and by explicitly identifying slow- and fast-changing variables.

## Multiple adaptive agents

We also need to challenge the implicit assumption of most urban models that decisions are made by one single decision-maker at one point in time. Urban development is the outcome of dynamic interactions among the choices of many actors, including households, businesses, developers, and governments (Waddell 1995). These actors make decisions that determine and alter the patterns of human activities and ultimately affect ecosystem change. Their decisions are interdependent; for example, employment activity affects housing location, which affects retail activity and infrastructure, which in turn affect housing development. Urban decision-makers are a broad and diversified group of people who make a series of relevant decisions over time.

In order to model urban ecological interactions, we must explicitly represent the location, production, and consumption behaviors of these multiple actors. This approach requires a highly disaggregated representation of human agents (i.e., households and businesses) and ecological agents (i.e., species and populations). We can disaggregate economic sectors by using a revised version of the input-output model methodology. Microsimulation may also help us address the difficult tradeoffs that households and businesses make between location, production, and consumption preferences. Agent-based models allow us to implement multi-attribute utility functions that correspond to the different agents and allow them to place different values to the different attributes (Monticino et al. 2007).

## Feedback mechanisms

In modeling urban ecological systems, we also need to consider the feedback mechanisms that connect the natural and human systems. These are control elements that can amplify or dampen a given output. For example, ecologists describe negative feedback in the biosphere: A homeostatic integration of biotic and physical processes keeps the amount of carbon dioxide in the air relatively constant. But we do not completely understand the feedback loops—both positive and negative—between human and environmental systems. We know that human decisions leading to the burning of fossil fuels and land use change affect the carbon cycle and in turn, the associated climate changes will affect human choices, but the nature of these interactions remains controversial. In particular, the feedback of environmental change on human decisions is difficult to represent: Environmental change affects all people independently of who has caused the environmental impact in the first place, while the impact of each individual decision-maker on the environment depends on the choices of others (Ostrom 1991).

Interactions between ecological and human functions involve several feedback mechanisms. Within urban development, for example, real estate markets involve the feedback mechanisms of buyers and sellers adjusting their prices in reaction to the relative abundance or scarcity of real estate. Feedback mechanisms can be negative, or dampening forms that tend to stabilize systems—such as real estate markets. Feedback can also be positive, accelerating adjustments and leading to unstable conditions that change catastrophically as in the case of ecological succession or the extinction of species. The shift between these multiple states is often abrupt, and systems respond to perturbation in ways that are complex and highly nonlinear. The process becomes nonlinear as multiple agents, such as natural vegetation and urban development, interact and compete for space. The characteristic response shows strong hysteresis; that is, when an ecosystem shifts from the vegetation state to the sprawl state, it becomes highly resistant to switching back.

## 9.3 Spatial and Temporal Heterogeneity

Variability in time and space is a defining characteristic of ecosystems—on all scales. Individual organisms, populations, and communities vary among them, and over space and time (Hewitt et al. 2007). Sources of heterogeneity in urban ecosystems are both natural and human. Natural sources of heterogeneity include biological and physical agents, disturbance regimes such as storms and earthquakes, and stresses such as droughts and flooding (Pickett and Rogers 1997). Humans also increase heterogeneity: They introduce exotic species, modify landforms and develop drainage networks, control or modify natural disturbance agents, and build extensive infra- structure (Pickett et al. 1997). They also break existing landscapes into smaller patches; landscape ecologists have started to explore the impact that the various, dynamic arrangements of patch structures have on ecosystems (Godron and Forman 1982, Turner 1989, Forman 1995, Collinge 1996).

Ecological scholars have long-recognized the importance of spatial and temporal heterogeneity and the consequences on scaling across hetero- geneous systems (O'Neill et al. 1986, Wiens 1989, Levin 1992). But empirical studies often still assume stationarity. This assumption creates important limitations on what ecologists can infer and how much they can generalize (Wagner and Fortin 2005, Hewitt et al. 2007).

Hewitt et al. (2007) point out several ways that heterogeneity challenges extrapolation. First, the variance in ecological systems tends to increase with spatial and temporal extent (Schneider 1994). Second, large-scale variations can dominate the dynamics of small-scale processes (i.e., demographic or biotic) and potentially confound small-scale experiments (Schneider et al. 1997). A third major problem is generalizing the results from studies conducted at one or a few locations or times, or at one spatial and temporal scale. The results may change dramatically at another scale, even shifting the direction of ecological responses (Hewitt et al. 2007).

Spatial heterogeneity created by human-natural interactions (i.e., patchiness of the urban landscape) has important consequences for the ability to model coupled systems. Spatial heterogeneity violates assumptions of parametric tests by creating autocorrelation in the error structure and reducing the degree of freedom (Wagner and Fortin 2005). Spatial autocorrelations imply that characteristics and dynamics of nearby land cover patches tend to be more spatially clustered and similar than expected due to random chance. In urban ecosystems, positive spatial autocorrelation and spatial dependence are driven by urban development.

To model complex coupled human-ecological systems, we must explicitly address the effect of spatially and temporally heterogeneous processes across multiple scales. The challenge is to identify these heterogeneities, and to take them into account in model building. A number of approaches and techniques can be used to identify and address spatial and temporal non-stationarity. The most straightforward approach is model segmentation. Wagner and Fortin (2005) suggest wavelet analysis as a promising alternative to characterizing and partitioning landscapes in the presence of multiple, overlapping processes. Integrating discrete and continuous data models in representing landscapes may be an effective approach for identifying and modeling heterogeneity (Cova and Goodchild 2002). In general, the integration of multiple approaches and methods that combine statistical analysis with natural history and experiments is critical to allow for generalizations (Hewitt et al. 2007).

## 9.4 Threshold, Discontinuity, and Surprises

Thresholds are transition points between alternate states or regimes (Liu et al. 2007). A regime shift between alternate stable states occurs when a controlling variable in a system reaches a threshold, modifying its dynamics and feedbacks (Walker and Meyers 2004). Subtle environmental change can set the stage for large, sudden, surprising, and sometimes irreversible, changes in ecosystems. Regime shifts depend not only on the perturbation, but also on the size of the basin of attraction (Holling 1973, Sheffer et al. 2001, Figure 9.1). In systems with multiple stable states, gradually changing conditions may reduce the size of the basin of attraction around a state. This is what Holling (1973) defines a loss of ecological resilience. This is typically described using the heuristic of the fate of a ball in a landscape of hills and valleys. As represented in Figure 9.1, a small perturbation or external event may be enough to cause a shift to an alternative stable state. However, this loss of resilience makes the system more fragile, in the sense that the system can easily be tipped into a contrasting state by stochastic events.

Recent studies have provided empirical evidence that alternative stability domains exist in a variety of ecosystems such as lakes, coral reefs, oceans, forests, and arid lands (Scheffer et al. 2001). Walker and Meyers (2004) describe a database documenting thresholds in ecological and socio-ecological systems that drive system-shifts. In coupled human-natural systems the effects of environmental change on human function and well-being may not be apparent until ecological changes reach a threshold. Complex feedbacks between natural and ecosystem thresholds can generate regime shifts (Walker and Meyers 2004). Regime shifts in ecosystems are

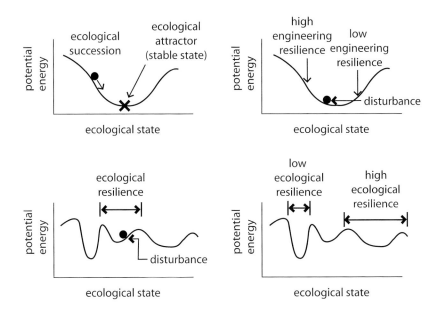

Figure 9.1. Ecological resilience. The four diagrams describe the difference between ecological and engineering resilience according to Holling (1973). Engineering resilience is the rate at which a system returns to a single steady state following a perturbation. In the diagrams, the steepness of the sides of a stability pit. The deeper a pit the more stable it is. Ecological resilience is a measure of the amount of change that is required to transform a system from being maintained by one set of mutually reinforcing processes and structures to a different set of processes and structures. Ecological resilience is a measure of the regional topography of a stability landscape. In the diagram, ecological resilience of a system corresponds to the width of its stability pit. (Peterson 2004).

difficult to predict (Scheffer and Carpenter 2003). There is, however, increasing evidence that ecosystem dynamics become more variable prior to some regime shifts (Berglund and Gentz 2002, Brock and Carpenter 2006, Carpenter and Brock 2006). For example, by studying variability around predictions of a simple time-series model of lake eutrophication, Carpenter and Brock find that rising standard deviation (SD) could signal impending shifts about a decade in advance. Brock et al. (2006) explain how this can occur for one-dimensional systems. Carpenter and Brock (2006) showed that the variance component related to an impending regime shift could be separated from environmental noise using methods that required no knowledge of the mechanisms underlying the regime shift.

The presence of alternative stable states has profound implications for our response to environmental change. In urbanizing regions, multiple steady and unstable states exist simultaneously (Alberti and Marzluff 2004). Eventually urban sprawl leads the ecosystem to shift from a natural steady state of abundant and well-connected natural land cover to a second steady state of greatly reduced and highly fragmented natural land cover (Figure 9.2). The exact form of the natural "steady" state depends on natural disturbance regimes. The sprawl state is a forced equilibrium that relies on incomplete information regarding the full ecological costs of providing human services to low-density development. Sprawl is an unstable stable because it is based on importing ecosystem services from other areas.

As I showed in Chapter 1, the state of an urban ecosystem is driven between the natural and sprawl states by the amount of urbanization. As we replace ecological functions with human functions in urbanizing regions, the processes supporting the ecosystem may reach a threshold and drive the system to collapse (Figure 9.2). An incomplete view of the relationship between urbanization, ecological functions, and human functions assumes that ecological and human functions are independent (Figure 9.3, p. 236). In contrast, a view of urban ecosystems as coupled human-natural systems indicates that there may be a threshold in the relationship between urbanization and ecological conditions. Ecosystem function directly supports the human population in non-urbanized areas and indirectly supports human function in urbanized areas. Urbanization degrades ecological conditions to a level in which ecological functions collapse. Eventually human function in urbanized areas declines as the ecosystem functions are reduced by urbanization.

To assess the resilience of urban ecosystems, we must first understand how interactions between humans and ecological processes affect the

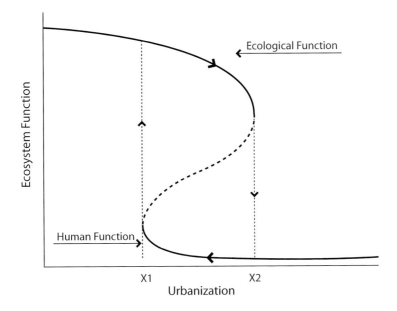

Figure 9.2. Multiple states in urban ecosystems. The graph illustrates alternate stable states in urban systems. Structural changes in the urban landscape (e.g., Chapter 1, Figure 1.8) result in a shift in system's dynamics.

resilience of the inherently unstable equilibrium points between the natural ecosystem attractor and the sprawl attractor. In other words, we must understand how best to balance human and ecosystem functions in urban ecosystems. Ecosystem functions are the ecological processes and conditions that sustain humans and other species (Daily 1997). Human functions in urban areas, such as housing, water supply, transportation, waste disposal, and recreation depend on ecosystem functioning and productivity over the long term. Urban areas also depend on the ecosystem's ability to act as a sink to absorb emissions and waste. Ecosystems provide other important functions to the urban population: They regulate climate, control flooding, and absorb carbon, to mention a few (Ehrlich and Mooney 1983, Daily 1997, Costanza et al. 1997).

Using as a framework the resilience hypothesis proposed in Chapter 1, consider how the built infrastructure in an urban ecosystem modifies hydrological functions. As an area becomes urbanized, humans tend to

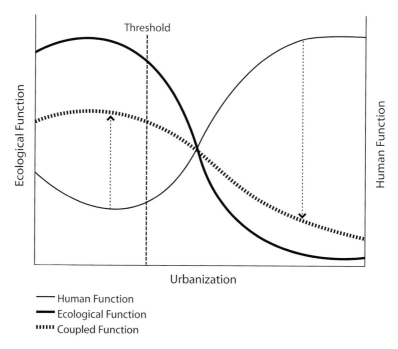

Figure 9.3. Interaction between human and ecological functions in urban ecosystems. The two solid lines represent a partial view of the relationship between urbanization, ecological functions, and human functions. Urbanization is measured as the amount of developed land per unit area. The heavy black line indicates the decline in ecological function that results from urbanization. The lighter black line indicates human services replacing ecological services. This view however is flawed since it assumes that ecological and human functions are independent. In contrast, a view of urban ecosystems as coupled human-natural systems indicate that there may be a threshold in the relationship between urbanization and ecological conditions.

replace the natural hydrological functions with built infrastructure; doing so lets them control the water flow, extract and distribute water for human uses, and purify water before it returns to the natural water bodies. In this process, urbanization decreases the amount and quality of natural hydrological functions, and replaces them with the built infrastructure that supports human functions.

At first glance, the human and natural functions in an urban ecosystem may seem to be operating independently, but in reality they are highly coupled. For example, the infrastructure's ability to serve multiple uses depends on the size, availability, and recharge capacity of the clean water supply. But the decline of the natural hydrological function may constrain that supply. As a result of human pressure, the coupled hydrological function (both human and natural) may decline as ecosystem functions, both local and global, are reduced. The functional form of the relationships between ecosystem function, human function, and urbanization depends on the specific ecosystem functions being considered; it may also depend on alternative future conditions caused by complex interactions between drivers of change (i.e., climate or technology), as I discuss later in this chapter.

Making it even more difficult to model interactions between drivers and predict potential shifts in system behaviors is the fact that their impacts are both cumulative and synergistic. In general, environmental disturbances have an important impact when the factors causing them are grouped so closely in space or time that they overwhelm the natural system's ability to remove or dissipate those impacts (Clark 1986). Human stresses in cities may cross thresholds beyond which the stresses may irrevocably damage important ecological functions. In most ecological systems, processes occur step-wise rather than progressing smoothly (Holling 1986), and sharp shifts in behavior are natural. These related properties of ecosystems require us to consider resilience: the amount of disturbance a system can absorb without changing its structure or behavior.

Modeling urban ecological systems will require us to pay special attention to uncertainty, which can be caused by many factors: We may not sufficiently understand a given phenomenon; we may make systematic and random errors or subjective judgments; natural systems can change abruptly and in discontinuous ways; and characterizing the responses of the system function will involve thresholds and multiple domains of stability. Because the knowledge of environmental systems is always incomplete and uncertain, surprise is inevitable (Holling 1996). For all these reasons, urban planners must explicitly characterize and analyze uncertainty.

## 9.5 Scenario Planning and Adaptive Management

Coupled human-natural systems challenge our traditional assumptions and strategies for planning and managing natural resources and the environment (Liu et al. 2007). The success or failure of many policies and management practices is based on their ability to take into account the complexities and uncertainty of these systems. For instance, many decisions that do not consider cumulative impacts, cross-boundary effects, threshold, and uncertainty may result in unexpected and undesirable environmental consequences. When assumptions regarding climate variability and extreme events do not take uncertainty into account, it is harder to prepare and respond effectively (consider Hurricane Katrina). Furthermore, when policies aim at stabilizing the ecological system or eliminating its variability, the outcome is inevitably its collapse (Carpenter and Gunderson 2001).

Authors of recent studies of coupled human-natural systems indicate that to face the challenges described above, planners and managers must take several factors into account: emergent properties, reciprocal effects, nonlinearity, and surprises (Liu et al. 2007). Because humans' predictive ability is inherently limited, we must incorporate uncertainty into decision-making. Scenario building is a strategy to explicitly consider what futures are plausible in the face of irreducible uncertainties (Shwartz 1991). We need strategies that enhance the adaptive capacity of systems while preserving key aspects of their structure and functioning, because coupled human-natural systems will inevitably change in response to various exogenous stressors, and we cannot predict their internal dynamics (Carpenter 2002).

Scenario building is both a systemic method and a framework to expand our ability to think creatively about the future by focusing on complexity and uncertainty (Peterson et al. 2003). Rather than focusing on accurately predicting a single outcome, scenarios let us examine the interactions of various key uncertain factors creating alternative futures. If we focus only on what is likely or predictable within some reasonable confidence interval, we will not be able to identify as broad a range of possible risks and surprises. Scenario building requires an open mind, and a willingness to explore uncharted territory.

Several authors have examined alternative approaches to future studies and have discussed the benefits of building scenarios in light of increasing uncertainty (Amara 1981, Marien 2002, Börjeson et al. 2005). Peterson et al. (2003) identify three major benefits of using scenarios in ecological conservation: (1) Planners can better understand key uncertainties; (2) they can incorporate alternative perspectives into conservation planning; and (3) they can formulate decisions that become more resilient to surprises. In the rest of this chapter, I discuss scenario building as an approach for planning and managing in urban ecosystems.

## What are scenarios?

Scenarios are narratives about alternative environments in which the participants can play out their decisions about planning and management strategies (Ogilvy and Schwartz 1998). They are not predictions or visions. Instead, they are hypotheses about different futures designed to highlight the risks and opportunities involved in specific strategic issues. To clarify this distinction, we can represent predictions, visions, and scenarios in terms of probability distributions (Figure 9.4). Peterson et al. (2003) define a scenario

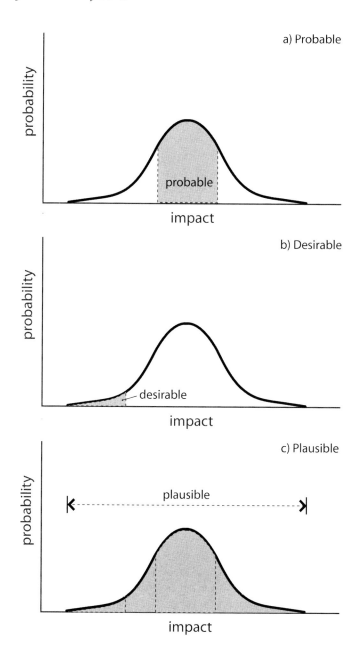

Figure 9.4. Probability distributions. This graph compares predictive models (a) probable, visions (b) desirable, and scenarios (c) plausible using probability distributions.

as a "structured account of a possible future." Scenario planning is a method for learning about the future by exploring how the potential impact of the most unpredictable and important driving forces can shape the future.

Scenarios are written as plausible stories—not probable ones. Traditional approaches to planning and management rely on predictions based on probabilities and quantified uncertainties. A prediction is understood to be the best possible estimate of future conditions (Peterson et al. 2003). By comparing scenarios to predictions, Peterson et al. (2003) emphasize that ecological predictions assume that we know the probability distribution of specified ecological variables at a specified time in the future; the accuracy of this prediction will depend on current conditions, specified assumptions about drivers, the measured probability distributions used in model parameters, and the measured probability of the model itself being correct (Clark et al. 2001).

Predictive models are effective over short time frames and under stable conditions (Lindgren and Bandhold 2003). To build such models, we generally need to know significant amounts about the mechanisms that drive the behavior of a phenomenon, and to have a substantial amount of historical data. Predictive models generate probabilities (Peterson et al. 2003). However, such models do not work as well with complex, nonlinear processes or thresholds. As I point out at the beginning of this chapter, all predictions are approximations of what may happen, and they involve some level of uncertainty. But what makes future dynamics of coupled human-natural systems inherently unpredictable is the interactions of multiple drivers with very different degrees of uncertainty. The probability distribution of coupled human-natural system predictions is unknown (Carpenter 2002).

Scenarios are also distinct from what planners have traditionally called visions. Visions are generated through a participatory process: Planners use a variety of techniques to engage stakeholders in imagining a desirable future. The objective of a visioning process is to identify a set of goals and strategies to guide a planning effort. Through a visioning process, stakeholders collectively imagine possible futures rather than exploring what might occur. Although visioning has its important role in the planning process, the lack of systematic considerations of what we know does not provide an effective tool to assess alternative futures.

Scenarios are intended to go beyond predictions, using uncertainty and complexity to provoke the participants' imaginations and provide a more comprehensive view of risk, so that the results can be embedded in critical strategic decisions. Scenarios first emerged following World War II as a method for military planning. In the 1960s, Herman Kahn, refined scenarios as a tool for what he called business prognostication: predicting according to present indications or signs. In the 1970s, Pierre Wack, working for Royal Dutch/Shell, used scenario building to analyzed two plausible energy futures: a realistic one in which oil prices would increase dramatically, and a less-realistic one in which they did not.

Approaches to scenario planning vary, but they follow a similar structure. Scenario planning focuses on key drivers, uncertainties, and interactions that might shape plausible alternative futures. Scenario planning generally involves eight key steps: (1) Identify the focal issue; (2) identify the driving forces; (3) rank the drivers by their importance and uncertainty; (4) select the scenario logics; (5) flesh out the scenarios; (6) select indicators for monitoring; (7) assess the impacts for different scenarios; and (8) evaluate alternative strategies (Schwartz 1991, Peterson et al. 2003, Lindgren and Bandhold et al. 2003).

Scenario planning has many purposes. The list below can be used to evaluate the effectiveness of a scenario planning process:

- Expand perspectives.
- Help decision-makers think about the future.
- Challenge assumptions (in both science and planning).
- Develop and test strategies and plans.
- Synthesize and communicate complex information to decision-makers.
- Provide insight into drivers of change.
- Reveal the implications of potential future trajectories.
- Illuminate options for action.
- Improve education and public awareness.
- Understand differences in perspectives among stakeholders and jointly explore the consequences.

Scenario planning should not be seen as an alternative to traditional planning approaches such as predictions and visions; rather, scenarios complement other planning tools. Essential in long-term planning, scenarios work best when the trajectories of driving forces are highly uncertain, when multiple interactions among these drivers may occur, and when looking far into the future. To succeed, however, scenario planning requires the integration of science and imagination. Participants need to keep an open mind and be willing to challenge their assumptions and mindsets. By making use of scientific facts, models, and imagination, scenarios can be effectively integrated with predictive models to test hypotheses on the possible impacts of alternative urban strategies. They also can serve to make communication among scientists, policymakers, and the public more effective.

## 9.6 Hypothetical Scenarios of Urban Ecosystem Functions

To explore how scenarios can be used to assess alternative planning strategies, I use a hypothetical example of how changes in urban hydrological function can interact with technology and human choice. Such interaction can potentially reach a threshold in the human and ecosystem functions and generate a system shift between multiple stable states under climate change. How resilient are alternative patterns of urban development to environmental change? This is a fundamental question in urban planning. As I discussed in Chapter 1, in urbanizing regions, urban sprawl leads the ecosystem to move from a natural steady state to a second steady state of greatly reduced ecosystem function (McDonnell et al. 1997, Vitousek et al. 1997b, Dobson et al. 2001, Costanza et al. 2002, Hansen et al. 2002). Human functions may also be compromised in sprawling developments because of higher infrastructure costs and the effect of reduced ecosystem function on human function in the long term (Frank 1989). Time lags in system feedbacks can result in a rapid collapse of system dynamics. Feedback can take decades to appear, and it often appears in unexpected forms that decision-makers do not see as connected to the original cause. For example, most people living in the suburban periphery do not appreciate how much it costs for the municipality to provide them with public services (e.g., utilities) (Ottensmann 1977). Often, such provision subsidizes sprawling development, because the price of services does not reflect their real cost and distance from central facilities (Ewing 1994, 1997).

I use this hypothetical example to illustrate how scenarios can effectively help in exploring plausible futures. In the example, the interaction between natural and human hydrological function may vary under different scenarios, but we do not know the probability distribution of the scenario emerging from such interaction. The probability distribution of such outcome depends on the distributions of multiple driving forces. In the example, climate change and technology drive human and ecosystem function. Since the distributions for these factors can take different forms, their interaction is unpredictable. As I show in Figures 9.5, 9.6, and 9.7, not only may the mean impact shift under alternative scenarios, but also the shift in variance may offset our ability to calculate the distribution of the outcome, given alternative hypotheses about thresholds in human and ecological functions under alternative climate and technological scenarios.

For example, the ability of an urbanizing watershed to perform hydrological functions essential to humans and other species depends on various scenarios of climate change, as well as various alternative technological scenarios. In the hypothesized example, climate change and its impact range from high to low (based on IPCC scenarios). I also hypothesize alternative technological trajectories based on the type of innovation that emphasizes reactive versus proactive and self-regulating end points. Once we consider the various possibilities, we can then hypothesize alternative thresholds under different scenarios that may result from the interactions between two driving forces: climate change and technological innovation (Table 9.1, Figure 9.8).

In response to the human, ecological, and economic costs of sprawling development, the field of urban planning has attempted to stabilize inherently unstable states—by balancing the conversion of natural land cover with the development needed to support human services. The assumption behind such planned development is that the development pattern affects ecological conditions, as well as the maintenance of ecosystem and human functions. In the phase of reorganization and renewal of adaptive cycles, urban ecosystems have a chance to change their trajectory and begin to develop self-organizing processes of interacting ecological and socioeconomic functions. This forced equilibrium is inherently unstable, as it has to balance the tension between providing for human functions and ecosystem functions as shown in Figure 9.2. Alternative patterns of development may have an impact on the ecosystem functions under alternative scenarios. Human and ecosystem functions are interdependent. Alternative urbanization patterns have different levels of resilience, measured as their capacity to simultaneously support ecological and human functions (Figure 9.9).

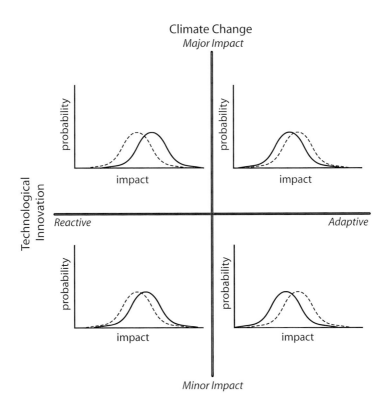

Figure 9.5. Probability distributions under alternative scenarios. The probability distributions of impact may shift in mean under different scenarios.

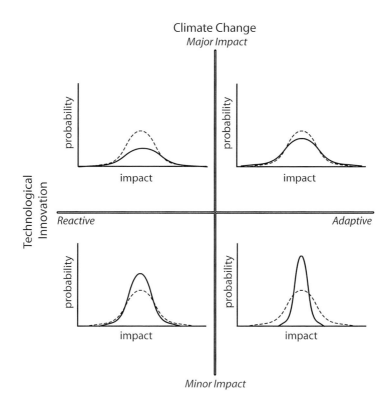

Figure 9.6. Probability distributions under alternative scenarios. The probability distribution of impact may shift in variance under different scenarios.

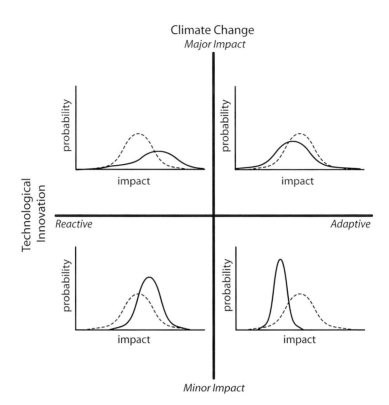

Figure 9.7. Probability distributions under alternative scenarios. The probability distribution of impact may shift in both mean and variance under different scenarios.

Table 9.1. Scenario descriptions.

### Climate Change

*Major Impact*: Large magnitude of climate impacts, as described in Scenario A1 of the IPCC scenarios (IPCC 2000). High sea level rise, glacial melting, temperature increase, summer droughts, winter flooding.

*Minor Impact*: Limited climate impacts, as illustrated in the IPCC scenario B1 (IPCC 2000). Few regional effects from changes in temperature and hydrology.

### Technological Innovation

*Adaptive*: Technological innovation solutions are proactive, mimic natural cycles, reflect context and variability, aim at resource efficiency, flexibility, and self-reliance.

*Reactive*: Technological innovation solutions are reactive, aim at controlling natural cycles, are rigid, and depend highly on resources inputs.

| Scenario | Hydrological function | Infrastructure |
|---|---|---|
|  | System collapse, reaching an early threshold due to droughts & flooding and reactive & rigid infrastructure. | Stepwise change provides minor improvements and are unable to substitute for lost hydrological function. |
|  | Overall function decline is dampened. Thresholds delayed due to adaptive technologies and proactive approach. | Self-regulating infrastructure adapts to climate impacts and results in a sustainable level of service. |
|  | Decline is significantly delayed by reduced impacts of climate change. Threshold is reached in third quarter. | Minor climate impacts leave more resources to expand on infrastructure. More improvements, but less effective. |
|  | Optimum hydrological scenario adaptive technologies increase hydrological resliency and delay impacts. | Greater gains in efficiency coupled with minor disturbances produce high service levels. |

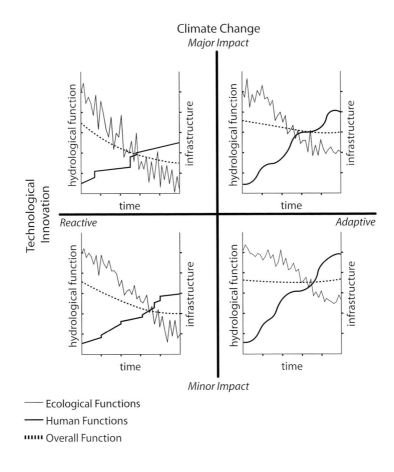

Figure 9.8. Alternative trajectories of hydrological functions under four scenarios.

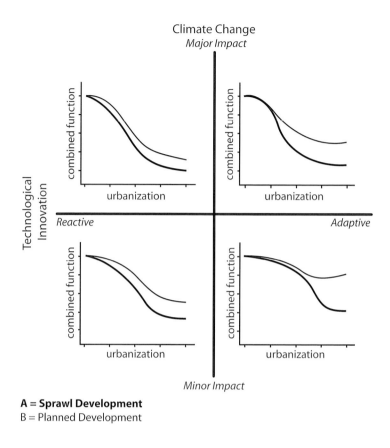

Figure 9.9. Relationships between urban development patterns and combined ecosystem and human functions under alternative scenarios. The graphs represent the hypothesized relationship between urban development patterns and combined human and natural ecosystem function under alternative scenarios to emphasize the potential role that urban form can play in minimizing environmental impact and increasing urban ecosystem resilience. Under all scenarios, sprawling development (A) will lead to reduced ecosystem function and, ultimately, will affect human function. Planned development (B) enables urban ecosystems to support increasing levels of urbanization. The effectiveness of planned development is dependent on the alternative scenarios.

How can we plan in the face of complexity, uncertainty, and hetero-geneity? What can we learn from this hypothetical example? Six principles for planning and management emerge:

- *Resilience.* Focus on increasing system resilience instead of aiming to control system dynamic and/or to eliminate change.
- *Diversity.* Maintain diverse development patterns to enhance system resilience and to support a diversity of species and ecosystem functions.
- *Integration.* Minimize resource use and diversify resource supplies (e.g., water and energy), and invest in infrastructure that supports integration.
- *Learning.* Create buffers for error and opportunities for experimenting, updating, and learning about system function and thresholds.
- *Flexibility.* Create flexible policies that mimic the variability of environmental processes and the heterogeneity of human commun-ities, and their evolution over time.
- *Adaptation.* Plan as designing a set of experiments, monitor progress, evaluate outcomes, and systematically adapt strategies.

# Chapter 10

# URBAN ECOLOGY: A SYNTHESIS

*When we speak of Nature it is wrong to forget that we are ourselves a part of Nature. We ought to view ourselves with the same curiosity and openness with which we study a tree, the sky or a thought, because we too are linked to the entire universe.*

— Henri Matisse

## 10.1 A Hybrid Ecology

Scholars working at the interface between ecology and the social sciences have started to articulate the opportunities and challenges for ecology to fully and productively integrate the complexity and global scale of human activity into ecological research (Liu et al. 2007). The future of Earth's ecosystems is increasingly influenced by human action, particularly the pace and pattern of urbanization. An ecology that does not include humans in its theories and experiments will rapidly evolve into paleoecology.[1] Meanwhile, urban scholars will need to expand their approach to fully appreciate that the ecology of a region and its biophysical processes shape the human habitat and the city just as much as do human action and perceptions. They have yet to write the "natural history of urbanization" that Mumford (1956) called for half a century ago, and for the same reason: "only a small part of the preliminary work has been done" (387). It is critical that we understand how human and ecological systems have coevolved over time to generate the present urban world, if we are to anticipate how environmental change will shape the cities of the future.

---

[1] Paleoecology studies the ecosystems of the past based on data from fossil and sub-fossil records. Most paleoecological research studies fossil organisms—their life cycle, their living interactions and their natural environment—over the last two million years (known as the Quaternary period), and more precisely the Holocene epoch (the last 10,000 years), or the last glacial stage of the Pleistocene epoch (from 50,000 to 10,000 years ago). This reference to paleoecology is metaphorical. Paleoecology and evolutionary biology are essential to ecology and to urban ecology. Both must inform the ecology of the present and of the future.

Cities are hybrid phenomena. We cannot understand them fully just by studying their component parts separately. Cities are not simply the combination of human and ecological systems. As Walker et al. (2006) point out, coupled human-ecological systems are a "different thing altogether." In conceptualizing coupled human-ecological systems, Westley et al. (2002) suggest that humans are not embedded in ecological systems; neither are ecosystems embedded in human systems. The study of urban ecosystems entails the study of hybrid systems emerging from interactions between human and ecological systems. Cities are the result of simultaneously occurring human and ecological processes in time and in space and the legacy of the simultaneous processes of the past. As Mumford (1956, 388) reminds us, "whether one looks at the city morphologically or functionally, one cannot understand its development without taking in its relationship to earlier forms of cohabitation that go back to non-human species." Thus urban ecology is the study of the coevolution of human-ecological systems, not the separate studies of the human habitat and of the ecosystems upon which humans depend.

In this book, I have argued that we need a theory of urban ecology, a hybrid between urban theory and ecological theory. Emerging models of urban ecology still cannot effectively take into account the complex interactions between humans and ecology. In Chapter 1 I reviewed several conceptual models that have attempted to integrate human and ecological systems to understand the dynamics of urbanizing regions. Emerging urban ecological studies place a different emphasis on one of several approaches. Referring to the traditional distinction in ecology between population-community and process-functional studies, Grimm et al. (2000) distinguish between an "ecology *in* cities" (that primarily focuses on the study of habitats or organisms within cities) and an "ecology *of* cities" (that studies urban areas from an ecological systems perspective). Others emphasize a complex systems approach (Wu and David 2002, Alberti and Marzluff 2004), seeing urban ecosystems as complex dynamic systems of many interacting agents. Similarly to other ecosystems described by Scheffer et al. (2001), shifts in such systems from one relatively stable state to another can be triggered either by the action of slowly changing variables or by relatively discrete shocks. In urban ecosystems these shifts may be controlled by complex interactions between human and ecological processes (Alberti et al. 2006b). For example, the ecological conditions of an urban stream can change from good to poor as a result of an incremental loss or degradation of riparian vegetation, or by substantially paving the drainage basin to make way for a large development or built infrastructure, or by both of these changes occurring simultaneously. Over the long term, what controls the ability of an urban ecosystem to support both its ecological and human function can be affected by slow-changing variables (i.e., climate change) or discrete shocks (i.e., hurricane Katrina) that can force the system over a threshold.

Scholars of urban ecosystems have made important progress in studying interactions between human and ecological processes in such systems (Grimm et al. 2000, Pickett et al. 2001, Alberti et al. 2003), yet we are just beginning to understand their organization and behavior. While several authors have addressed the relationships between urbanization and eco-system function (Collins et al. 2000, Grimm et al. 2000, Pickett et al. 2001), few have directly asked how human and ecological patterns emerge from these interactions. Nor have they investigated how these patterns control the distribution of energy, materials, organisms, and information in human-dominated ecosystems at the local and global scales, and how human-dominated ecosystems differ from their nonhuman-dominated counterpart. We lack an understanding of the mechanisms linking emerging urban ecosystem patterns to ecosystem processes and controlling their dynamics.

To achieve such an understanding we need to change the way we pose questions and search for answers. The ecologists and social scientists who have begun to investigate questions at the interface between humans and ecological processes have done so within their own academic disci-plines (Redman et al. 2000). Furthermore, these studies have privileged the empirical testing of hypotheses developed within these disciplines over the more challenging task of developing hypotheses that explore the inter-actions. But if we remain within the traditional disciplinary boundaries, we will not make progress towards a theory of urban ecosystems as coupled human-ecological systems, because no single discipline can provide an integrated perspective without bias. Questions and methods of inquiry specific to our traditional domains yield partial views that reflect different epistemologies and understandings of the world. Simply linking these views is not enough to achieve the level of synthesis required to see the urban ecosystem as a whole.

The challenge for scholars of coupled human-ecological systems is to collaborate in generating new research questions, not simply to integrate the findings of disciplinary studies. We need a theory that builds on multiple world views to develop testable hypotheses about the mechanisms that govern coupled human-natural systems. Such a theory requires that scholars of both urban systems and ecology be willing to challenge the assumptions and world views within their disciplines. But that alone is not enough; we must also engage in this process many other social and natural scientists who study human and natural systems from various perspectives.

More importantly, we need to change the way we train the new generation of scholars, especially ecologists, planners, and economists. In graduate school, we are taught to break problems apart into manageable parts. That works for relatively simple problems, but not for complex ones. We cannot reassemble the knowledge that different experts produce separately and expect to understand complex problems—because each expert sees different things and no one expert can see all the connections that allow us to understand the whole. Trying to reconstruct the whole from a fragmented knowledge is a futile exercise as Peter Senge reminds us in *The Fifth Discipline* (1990, 3) by quoting physicist David Bohm: "it is like trying to reassemble the fragments of a broken mirror to see a true reflection."

## 10.2 Toward a Theory of Urban Ecology

In ecology, theory consists of heuristics used to construct models that describe the interaction of living systems with their environments (Sarkar 2006). In ecology, we have no unified theory, or even general principles comparable to Schrodinger's standing wave equations in quantum physics that describe matter in physical space. Ecological theory builds on principles developed within multiple sub-fields: population ecology, metapopulation ecology, community ecology, metacommunity ecology, physiological ecology, functional ecology, behavioral ecology, and the ecologies of conservation, ecosystems, landscapes, evolution, and conservation (Roughgarden et al. 1989, Sarkar 2006). Similarly, urban theory builds on concepts and principles from urban economics, sociology, geography, political science, anthropology, and urban planning to explain urban systems and their functioning (Short 2006).

A theory of urban ecology will have to build on a plurality of concepts from multiple disciplines to address questions about the mechanisms that govern the behavior of urban ecosystems. These questions are complex and diverse: How do human populations and other organisms come to be distributed in time and space? How do energy and material fluxes emerge from the interactions between humans and ecological processes? How do human populations and activities interact with ecological processes at the levels of the individual, population, and community to determine system stability (i.e., resilience)? How do human populations, interacting with ecological processes, generate emergent system-level behaviors? How do landscape-scale organizations of structures and processes arise? How are they maintained, and how do they evolve? (Alberti et al. 2003).

The challenge for urban ecology is to articulate the discrepancies between ecological theory and observations in urban ecosystems in a set of testable hypotheses and, by exploring differences and communalities with nonhuman-dominated systems, to gain clearer insights into how ecosystems work. With my colleagues on the urban ecology team at the University of Washington, I suggest that integrating humans into ecosystems would provide important opportunities for advancing ecosystem science (Alberti et al. 2003). We suggest that the study of urbanizing regions can lead to fundamental changes in niche theory; it would be possible to distinguish realized niches from fundamental ones on the basis of human interaction. If we redefine the realized niche as an organism's hypervolume of occurrence in the presence of a gradient of human domination, we would be able to identify and quantify the multiple and diverse ways that humans force the population-level ecological functions that structure communities. Understanding how niche assembly functions in human-dominated ecosystems would allow ecologists to directly test the effects of competitors, predators, diseases, and other biophysical changes induced by humans on community organization (Alberti et al. 2003).

Empirical studies conducted at the two Urban Long-Term Ecological Sites (LTER) in Phoenix, AZ and Baltimore, MD, indicate that urban ecosystems may deviate from conventional expectations based on theoretical models of non-human-dominated systems (Alberti et al. 2003, Kaye et al. 2006). Kaye et al. (2006) indicate that urban biogeochemical cycles might be distinct from nonurban systems as a result of human-controlled energy and element fluxes. Concentrated human populations not only alter nutrient sources by changing atmospheric deposition rates and by importing fertilizer and food, but also influence nutrient fluxes in plants and soils through hydrological changes (Kaye et al. 2006). In urban areas, stormwater detention basins, ditches, gutters, and lawns can also be hotspots for denitrification as nitrates and organic matter accumulate (Kaye et al. 2006).

Similarly, urban ecosystems may diverge from nonhuman-dominated systems in what controls the food web dynamics. Trophic dynamics in urban ecosystems are influenced by complex human social processes and feedback mechanisms (Shochat 2004, Faeth et al. 2005). Urbanization in Phoenix, for example, causes shifts from a bottom-up (resource-based) system typical of the Sonoran Desert to a system driven simultaneously by bottom-up and topdown processes (Faeth et al. 2005). In such an environment, predation becomes increasingly important for some taxa as resources become abundant and predictable (Faeth et al. 2005), although these resources are variable across the urban landscape (Hope et al. 2003). Complex trophic dynamics are not predictable based only on knowledge of species composition.

By integrating the complex human interactions with the food web in our studies of urban ecosystems, we can also shed light on another scientific controversy in ecology—the influence of biological diversity on ecological stability (Alberti et al. 2003). Humans make the food web more complex, but despite the complexity of trophic interactions this may not necessarily increase the stability of the ecological or human systems. By decoupling diversity and stability in human-dominated ecosystems, we can explore the importance of other factors such as species identity, rather than simply species richness, to community stability and ecosystem functioning. Investigating the changing relationship between diversity and stability along a gradient of human domination can also clarify when diversity produces stability, when diversity simply means redundancy, and when diversity leads to instability (e.g., diversity caused by invasive exotics).

Marzluff (2005) has tried to resolve a key puzzle for "island biogeography" in human dominated ecosystems by developing formal hypotheses of the processes governing biological diversity in urbanizing landscapes. Expanding on work by Blair (1996, 2001), Marzluff (2005) suggests that in a human-dominated ecosystem, diversity is an emergent property of extinction and colonization forces, but the actions of invading species take on greater relevance (Olden and Poff 2004). By studying bird communities in the Seattle, WA, metropolitan area, Marzluff (2005) has shown that bird diversity peaks at intermediate levels of human settlement primarily because intermediately disturbed forests are being colonized by early successional native species. His research aims to determine the relative importance of colonization versus extinction to bird communities in Seattle and derive testable hypotheses about how extinction and colonization are affected by urbanization to determine avian diversity in urban habitat islands.

From an evolutionary perspective, humans also influence natural selection by changing the genetic fitness of organisms and reconfiguring the physical environment and biogeochemistry (Pimm et al. 1994,Vitousek et al. 1997, Flannery 2001). By changing speciation and increasing extinction, humans are causing evolutionary change (Palumbi 2001). Environmental change resulting from human action can in turn force human cultural and natural selection and fitness. Marzluff and Angell (2005) go further to propose an additional synergy between human culture and the environment—a coevolution of humans and birds–by providing evidence from studying crows and ravens. Populations evolve as genetic and learned information is transferred through time by genetic and cultural inheritance. As people interact with their environment they change natural and cultural fitness. Reciprocal changes and feedbacks between humans and their environment lead to coevolved human and natural cultures (Marzluff and Angell 2005).

The scientific community is just beginning to address the philosophical and methodological implications of defining a theory of urban ecology. Some scholars have suggested that we need a distinct theory of urban ecology to understand ecological patterns and processes in the urban ecosystem because humans are qualitatively different from other organisms (Trepl 1995). Trepl (1995) points out that a theory of urban ecology should explicitly address what distinguishes urban ecosystems from other types, and that it should outline the systematic relationships among these characteristics. He suggests a set of hypotheses concerning integration, succession, and invasion. First, in urban ecosystems the degree of integration among urban habitat patches and communities (i.e., organization or connectivity) is low, the systems are not in equilibrium, and stochastic processes predominate over deterministic ones (Trepl 1994). Second, succession is strongly linked to site history and is relatively unpredictable since urban ecosystems are not deterministically directed by functional dynamics. Moreover, cities are open to invasions by unknown numbers of alien species.

Others have argued that humans are like other organisms, and therefore ecological theory can be extended to encompass human-dominated environments (Niemala 1999). Urban ecosystems, according to Niemala (1999), differ from nonhuman-dominated ones in the degree of influence of human activities (Gilbert 1989, Sukopp and Numata 1995, Walbridge 1997). Niemala (1999) argues that the basic ecological patterns and processes in human- and nonhuman-dominated ecosystems are similar, and there is no need for a distinct theory of urban ecology. Thus, we can successfully study urban ecosystems using existing ecological theories, such as metapopulation theory (Niemala 1999).

Do we need a distinct theory of urban ecology to understand ecological patterns and processes in the urban ecosystem (Trepl 1995), or can we instead extend ecological theory to encompass human-dominated environments (Niemala 1999)? Collins et al. (2000) are more cautious in drawing a conclusion. Increasing evidence shows that the differences are not merely quantitative, but qualitative. There are also plausible explanations for leaving the question open. Collins et al. (2000) argue that several factors– including the influence of culture, the constraints and opportunities afforded by our institutions, and our ability to create strategies in response to anticipated selection pressures–mean that standard ecological and evolutionary theories and principles might apply only imperfectly to human populations.

The answer, I think, depends on how we define "urban ecosystem." If we define it as a coupled human-ecological system, I propose that neither urban theory nor ecological theory can fully explain how urban ecosystems work within their separate disciplinary domains. As several authors have suggested, we cannot define coupled human-ecological systems either as humans embedded in an ecological system or as ecosystems embedded in human systems (Westley et al. 2002, Walker et al. 2004, 2006). How, then, can we synthesize our existing knowledge into a set of hypotheses that can inform the development of a theory of urban ecosystems as coupled human-ecological systems?

Building on previous work that applies complex system and hierarchy theory to coupled human-ecological systems (Levin 1998, Wu and David

2002, Holling 2001, Holling et al. 2002a, Gunderson and Holling 2002, Walker et al. 2004), I propose a framework to identify significant properties that govern the functioning of urban ecosystems. I propose that urban eco-systems are hybrid, multi-equilibria, hierarchical systems, in which patterns at higher levels emerge from the local interactions among multiple agents interacting among themselves and with their environments. They are proto-typical complex adaptive systems, which are open, nonlinear, and highly unpredictable (Hartvigsen et al. 1998, Levin 1998, Portugali 2000, Folke et al. 2002, Gunderson and Holling 2002). Disturbance is frequent and intrinsic (Cook 2000). Change has multiple causes, can follow multiple pathways, and is highly dependent on historical context (Allen and Sanglier 1978, 1979, McDonnell and Pickett 1993). Agents are autonomous and adaptive, and change their rules of action based upon new information.

We can use a set of heuristics to describe these characteristics of urban ecosystems, and articulate formal hypotheses about their functioning and dynamics. Based on an extensive review of the literature on coupled human-natural systems, in this book I have identified eight elements that characterize urban ecosystems: (1) hierarchies, (2) emergent properties, (3) multiple equilibria, (4) non-linearity, (5) discontinuity, (6) spatial heterogeneity, (7) path-dependency, and (8) resilience.

*Hierarchies.* Urban ecosystems can be described as near-decomposable and nested spatial hierarchies, in which hierarchical levels correspond to structural and functional units operating at distinct spatial and temporal scales (Levin 1992, Reynolds and Wu 1999, Wu and David 2002). In the urban landscape, the lowest hierarchical level and the smallest land-scape spatial unit vary with socioeconomic and biophysical processes from households and buildings to habitat patches or remnant ecosystems. At a coarser spatial scale land parcels and habitat patches interact with each other to create a new functional level and unit such as a neighborhood or sub-basin. Neighborhoods and sub-basins initiate and are constrained by regional economic and biophysical processes. Since landscapes are non-linear systems, they can simultaneously exhibit instability at lower levels and complex meta-stability at broader scales (Wu 1999, Burnett and Blaschke 2003). Near decomposability is a key tenet of hierarchy theory; ecological systems can be simplified based on the principle of time-space to simplify the complexity of nature, yet retain its essence (Wu, 1999). Urban landscapes are also hierarchically organized. At the higher levels, processes occur on a larger spatio-temporal scale and define the boundary conditions in which the system functions; at the lower levels, processes are faster and local and act as initiating conditions. In applying hierarchical theory to urban ecosystems, holons (horizontal structure) can be represented by patches (the ecological unit) and parcels (the economic unit). Through loose horizontal and vertical coupling, patches and parcels interact with and between other patches and parcels at the same, and at higher and lower, levels of organization (Wu 1999).

*Emergent properties.* Urban ecosystems exhibit emergent properties: properties that do not belong to any of their component parts. In emergent phenomena, a small number of rules or laws can generate complex systems and behaviors through local-scale interactions. Ecosystem stability and resilience emerge from the self-organized interaction of many different ecological processes occurring at different scales (Peterson et al. 1998). Similarly, self-organizing principles can also be applied to spatial economies to understand the clustering of land uses (Krugman 1995). The economics of urban systems and their spatial competition may lead to a clustering of similar land uses, while monopolistic competition can lead to their spatial dispersal (Parker et al. 2001). A more complex question is how to identify the emergent properties of coupled systems. Urban landscape patterns emerge from local-scale interactions among variables such as human preferences for residential location, individual mobility patterns, transportation infrastructure, and real estate markets, but also regional climate, hydrology, and topography (Torrens and Alberti 2000).

*Multiple equilibria.* No ecosystem has a single equilibrium; instead, multiple equilibria define functionally different states and contribute to their persistence and their self-monitoring and self-correcting capacity. Multiple equilibria are an emergent property of the coupled human-natural system. Urbanization can drive urban ecosystems to a state of sprawl. As urbanization increases, the system moves between natural vegetation attractors and sprawl attractors. The system moves away from the natural vegetation attractor toward the sprawl attractor and beyond, until increasing urbanization reduces the system's ability to support the human population. In urbanizing regions, as human services replace ecosystem services, the ecosystem reaches a threshold where it is likely to collapse. If the ecosystem collapse has reduced settlement enough to allow substantial natural vegetation to regrow, this process drives the system back toward the natural vegetation attractor.

*Nonlinearity.* An essential aspect of complex systems is nonlinearity (Levin 1998). A system is considered nonlinear if its outputs are not proportional to its inputs across the range of the inputs. In nonlinear systems, a very small change in some parameters can cause great qualitative differences in the resulting behavior. In complex systems, the system parts may interact strongly, leading to the emergent properties. In urban ecosystems interactions between human and ecological systems may lead to sharp shifts in behaviors when an unstable equilibrium threshold is crossed. Traffic flow patterns, urban development and decay, and sprawl are examples of nonlinear system behaviors.

*Path dependency.* Current driving forces can only partially explain current landscape patterns. Most changes in complex adaptive systems are reinforced by local chance events, such as mutation and environmental variation, leading to potential alternative developmental pathways (Levin 1998). Landscape change may depend on initial conditions, and small random events may lead to very different outcomes. Nonlinearity leads to path dependency. That is, the local rules of interaction change as the system evolves and develops (Levin 1998). One example of path dependency is the effect of transportation infrastructure on the pattern of development, through both increased development and changes in the real estate market, that in turn affect further infrastructure development (Turner et al. 1995).

*Discontinuity.* In ecosystems, change is neither continuous and gradual nor consistently chaotic. Rather it is episodic; periods of slow movement are punctuated by sudden change (Holling et al. 2002b). Moreover, events vary widely in the way they occur over space and scale (Holling 1973, O'Neill et al. 1988, Levin 1992). Discontinuities arise in an endogenous way, i.e., from within ecosystems (Holling 1973). An example is the spruce budworm (*Choristoneura fumiferana*) and its outbreaks in Canadian forests which periodically defoliate and kill large areas of mature balsam fir (Holling 1978b, Ludwig et al. 2002). Multiple equilibria dynamics and discontinuities occur also in coral reefs (Hughes 1994) and kelp forests (Estes and Duggins 1995). Discontinuities have also been documented in coupled human-ecological systems (Carpenter et al. 2002, Holling 2001, Holling et al. 2002a, Rosser 2006). Examples are fishery collapses, discontinuities in forestry, and eutrophic shallow lakes (Schindler 1990, Scheffer 1998, Carpenter et al. 2002, Wagener 2003).

*Spatial heterogeneity.* Events are not uniform over space (Holling 1978a). Landscape patchiness is a well known phenomenon, although as Turner and Chapin (2005) remind us, ecology still lacks a theory of ecosystem function that is spatially explicit. We do know that spatial variations in biophysical factors (e.g., topography) and disturbance determine the natural matrix of spatial variability in ecosystems (Holling 1992). Species patchiness, for example, can be caused by one of two different phenomena: the positive spatial autocorrelation of the ecological spatial processes of individual organisms (e.g., dispersal, competition) or spatial dependence due to species responses to underlying environmental conditions (Wagner and Fortin 2005). In urban ecosystems the sources of heterogeneity are multiple, and are generated by both human decisions and ecological processes (Band et al. 2005).

*Resilience.* Resilience is the capacity of a system to absorb shocks without reorganizing around a new set of structures and processes. More precisely, Holling (1973) defines resilience as the amount of disturbance a system can absorb without shifting into an alternate regime. In urban ecosystems, resilience depends on the ability to simultaneously support human and ecological functions (Alberti and Marzluff 2004). When thresholds are exceeded, the system shifts to a new regime that may be reversible, irreversible, or effectively irreversible—that is, not reversible on human time scales (Scheffer et al. 2001, Carpenter 2003).

Urban ecosystems provide unique opportunities to test hypotheses about interactions between humans and ecological processes. Using the heuristics proposed above, I suggest we can develop and test several hypotheses to study urban ecosystems and understand complex phenomena such as sprawl. The synthesis I have proposed in this book of the rapidly growing empirical work by multiple research teams on urban ecology points to five major themes: 1) Urban ecosystems are dynamic, hierarchically structured, patch mosaics resulting from local interactions between human and biophysical agents. 2) Urban ecosystems are likely driven between multiple states by the amount and pattern of urbanization (Alberti and Marzluff 2004). 3) Spatial interactions between socioeconomic and biophysical patterns and processes in urban ecosystems lead to emergent properties (e.g., sprawl). 4) Emergent landscape patterns affect ecological and socioeconomic processes in nonlinear ways (e.g., the intermediate disturbance hypothesis). 5) Ecosystem functions (both ecological and human) are moving targets with multiple and unpredictable futures; thus, policies that aim to achieve fixed goals cause a loss of resilience and are destined to fail.

## 10.3 Building Integrated Models

*Complex systems theory* (CAS, Levin 1998, 1999, Gunderson and Holling 2002) provides the theoretical framework for linking the local interactions between human dynamics and ecological processes to the overall structure and dynamics of urban landscapes. Drawing on *hierarchy theory*, urban landscapes can also be modeled as nested hierarchies with vertical (levels) and horizontal (holons) structures (O'Neill et al. 1986, Wu 1999, Wu and David 2002). Wu and David (2002) hypothesize that while the dynamics of urban landscapes are primarily driven by bottom-up processes, top-down constraints and hierarchical structures are also important for predicting these dynamics. I propose that a *multi-agent modeling* system can provide a platform for integrating these approaches and modeling the agents, the environment through which agents interact, and the rules that define both the relationships among agents and the relationships between agents and the environment.

Important progress has been made in modeling dynamic multi-agent human and ecological systems, but no one has formally tested hypotheses about the interacting emergent behaviors of coupled human and ecological systems. There are several research challenges. First, how can we simulate emergent behavior in ways that reasonably capture the patterns observed in urban landscapes? Second, how can we explicitly represent the human and biophysical agents at a level of disaggregation that allows us to explore the mechanisms linking patterns to processes (Portugali 2000)? Third, in modeling the interactions between human and natural systems, we find that many factors operate simultaneously at different levels of organization (Alberti 1999a). Simply linking these models in an additive way may not adequately represent the behavior of the coupled systems because interactions may occur at hierarchical levels that are not represented (Pickett et al. 1994). Additionally, since urban landscapes are spatially heterogeneous, changes in driving forces may be relevant only at certain scales or in certain locations (Levin 1992, Turner et al. 1995). At present, however, we understand too little about the interactions between spatial scales. To simulate the behavior of urban ecological systems we will need to explicitly consider the temporal and spatial dynamics of these systems, and also identify the interactions between human and ecological agents across the different temporal and spatial scales at which various processes operate (Alberti 1999a).

To address the inherent complexity of urban systems, we will need to integrate many complementary research strategies. Urban ecosystems are conceptualized as nested hierarchies where individual domains (e.g., scales in landscapes) occasionally interact with domains at higher and lower levels; the strongest and most frequent interactions occur within one level (Allen and Starr 1982). Domains in the hierarchy are separated by different characteristic rates of processes and thresholds (abrupt changes in system processes; i.e., Meentemeyer 1989, Wiens 1989). At higher levels, we observe slower rates and larger entities; at lower levels we see faster rates and smaller entities (Wu 1999). The theory of *patch dynamics* provides a framework to address spatial heterogeneity and explicitly represent the structure, function, and dynamics of patchy systems (Levin and Paine, 1974, Pickett and White 1985, Wu and Levin 1994, 1997, Pickett and Rogers 1997). An *agent-based* structure makes it possible to model decision-making processes and integrate the different approaches into a coherent modeling framework.

I propose that we use pattern-oriented modeling (POM, Grimm et al. 2005) to test alternative models of underlying processes and structures, thus making agent-based modeling of urban ecosystems more rigorous. Patterns contain information about essential underlying processes and structures, and the strategy of POM can decode the information about the internal organization of the system. Users of this approach begin by observing patterns at multiple scales and testing hypotheses about agent behaviors and interactions across scales. The assumption is that for complex systems, a single pattern observed at a specific scale and hierarchical level is not sufficient to understand a system or to reduce the uncertainties in model structure and parameters (Grimm et al. 2005). Pattern-oriented modeling can reduce uncertainty in model parameters, both by making models structurally realistic, which usually makes them less sensitive to parameter uncertainty, and by making parameters interact in ways that resemble the interactions of real mechanisms. It is therefore possible to fit all the calibration parameters by finding values that reproduce multiple patterns simultaneously (Grimm et al. 2005, Wiegand et al. 2003).

## 10.4 A Research Agenda for Urban Ecology

While various schools of urban ecology have developed alternative models to integrate human and ecological systems, they all point to the same need: we must redefine the set of questions that will guide the next generation of urban ecological inquiry. Redman et al. (2001) ask: "How did the social-ecological system develop into its current state, and how will it change in the future?" The focus here is the system: the nature of feedback, the rates of change, system components, and the specifics of resource use and production. Three associated questions further focus their inquiry. 1) How have the characteristics of ecological systems in the region under study influenced the social patterns and processes that have emerged? 2) How have social patterns and processes influenced the use and management of ecological resources? 3) How are these interactions changing over time and what does this mean for the state of the social-ecological system? (Redman et al. 2001).

   With my colleagues in the UW Urban Ecology Program (2003, 1176), I emphasize a slightly different focus: "How do humans interacting with their biophysical environment generate emergent collective behaviors (of humans, other species, and the systems themselves) in urbanizing landscapes?" We ask four questions: 1) How do socioeconomic and biophysical variables influence the spatial and temporal distributions of human activities in human-dominated ecosystems? 2) How do the spatial and temporal distributions of human activities redistribute energy and material fluxes and modify disturbance regimes? 3) How do human populations and activities interact with processes at the levels of the individual (birth, death, dispersal), the population (speciation, extinction, cultural or genetic adaptation), and the community (competition, predation, mutualism, parasitism) to determine the resilience of human-dominated systems? 4) How do humans respond to changes in ecological conditions, and how do these responses vary regionally and culturally (Alberti et al. 2003)?

## Urban landscape patterns as emergent phenomena

A new research project at the University of Washington (UW) and Arizona State University (ASU), funded by the National Science Foundation (NSF) as part of the Biocomplexity Program, investigates the complex coupled human-natural system dynamics of the Seattle and Phoenix metropolitan areas (Alberti et al. 2006b).[2] The study aims to empirically test hypotheses about how the interactions of human agents, real estate markets, built infrastructure, and biophysical factors drive current patterns of development and how these patterns affect human and ecological function in these two different bioregions. The study employs a pattern-oriented hierarchical approach to model how complex agent-based interactions generate landscape patterns at multiple times and spatial scales. We address four overarching questions: 1) How do dynamic landscape systems evolve to generate the emergent patterns that we see in urban landscapes? 2) What nonlinearities, thresholds, discontinuities, and path dependencies explain the divergent trajectories of urban landscapes? 3) How do emergent urban landscape patterns influence biodiversity and ecosystem functioning? 4) How can urban planning integrate this knowledge to develop sustainable urban landscape patterns? (Alberti et al 2006).

This project is one of several that have started to articulate key research questions so we can begin to test hypotheses and develop a theory of urban ecosystem dynamics that is crucial if urban ecology is to advance as a science. Our project is still at its beginning stage, and after completion we expect that it will only start to shed some light on these fundamental questions, but we believe that the questions we pose provide a useful starting point to develop a research agenda for urban ecology.

---

[2] BE/CNH: Urban Landscape Patterns: Complex Dynamics and Emergent Properties (Alberti PI: BCS 0508002). The project is a joint effort by the UW Urban Ecology Research Lab (www.urbaneco.washington.edu) and the ASU Global Institute of Sustainability (http://sustainable.asu.edu/gios/).

*1) How do coupled human-ecological systems evolve to generate emergent patterns that we see in urban landscapes?*

What agents, processes, hierarchies, and interactions govern the emergence of these patterns? How do spatial interactions among the human and biophysical processes lead to emergent properties? In the Biocomplexity Project we hypothesize that distinctive urban landscape patterns are associated with alternative states of urban ecosystems (Alberti and Marzluff 2004). These patterns can be characterized as highly developed vs. undeveloped land cover, clustered vs. dispersed development, specialized vs. mixed land use, and high vs. low level of urban infrastructure. Operationalizing the questions posed above into a set of testable hypotheses requires explicitly modeling 1) the agents, 2) the interactions among and between agents and their environment through time and space, and 3) the dynamic changes resulting from the interactions. We hypothesize that urban landscapes are spatially nested hierarchies in which hierarchical levels correspond to structural and functional units. Using a hierarchical modeling approach, we aim to identify the structural and functional units at distinct spatial and temporal scales of human and biophysical processes and specify the agents and rates of processes that characterize and distinguish the levels in the hierarchy.

*2) What nonlinearities, thresholds, discontinuities, and path dependencies explain divergent trajectories of urban landscapes?*

What are the multiple equilibria in such systems? Dynamic complex systems are typically characterized by two or more possible system states (defined either at a specific point in time or as a developmental trajectory) to a given set of inputs and boundary conditions (Levin 1999, Gunderson 2000, Gunderson and Holling 2002). In a *state* space defined by all the variables or components of a system, we define the region of the space to which all the evolutionary trajectories are drawn as an *attractor* or *basin of attraction*. We hypothesize that urban landscapes are likely driven between the natural vegetation and sprawl states by the amount and pattern of urbanization (Alberti and Marzluff 2004). We propose that the emerging pattern mediates the relationship between urbanization and movement between states.

What are the sources of nonlinearities, thresholds, and discontinuities in the relationships between human and biophysical systems that explain the pattern in the landscape? A system is considered nonlinear if its outputs and inputs are not proportional across the range of the inputs. Urban landscapes exhibit characteristics of nonlinear systems. In this project we explore the sources of nonlinearities, including thresholds, self-reinforcing and self-limiting processes, self-organization, hysteresis and the multiple adjustments in the interactions between human and biophysical processes that lead to emergent properties.

How are urban landscape trajectories shaped by prior conditions? Current driving forces can only partially explain current landscape patterns. Landscape change may depend on initial conditions, and small random events may lead to very different outcomes. An example of path dependency is the way transportation infrastructure affects the pattern of development by leading to increased development and changes in the real estate market. The resulting development also feeds back into further infrastructure development (Turner et al. 1995). We hypothesize that trajectories of landscape change result from the phenomenon of "lock-in systems" (Turner et al. 1995).

*3) How do emergent urban landscape patterns influence human and ecosystem functioning?*

Current NSF studies in three major urban regions-Seattle, WA, Phoenix, AZ, and Baltimore, MD–are starting to articulate hypotheses about some unique characteristics of human-dominated ecosystems and their functioning including their biogeochemistry and trophic dynamics. Urban ecosystems exhibit properties that might be distinct from nonurban systems as a result of fragmentation of natural habitats, altered hydrological systems, human-controlled energy flow and nutrient cycles, and their consequences on trophic interactions. Urbanization favors some species, but selects against others so that the composition of urban communities differs from those found in native environments. However, we do not know how patterns emerging in urbanizing regions affect biodiversity, since empirical tests of mechanisms controlling ecosystem functions have been primarily conducted in non-human-dominated ecosystems.

The challenge for urban ecology is to start to formalize hypotheses that link patterns to processes in urbanizing regions. Marzluff and his colleagues find the highest bird diversity in Puget Sound landscapes that have intermediate levels of disturbance in the form of urban development and a mosaic of forested landscapes (Alberti and Marzluff 2004, Marzluff 2005, Hansen et al. 2005). These results are consistent with the "intermediate disturbance" hypothesis (Blair 1996, 2004), but we still do not understand the specific mechanisms that cause these patterns of diversity to emerge. In our Biocomplexity Project, Marzluff and his team aim to explore how patterns in avian demographics (survival, reproduction, and dispersal) emerge as across the urban-to-rural gradient.

Since urban ecosystems are characterized by both ecological and human functions, we can expect important feedback mechanisms between ecological and human processes to control ecosystem dynamics. Ecological changes at local and regional scales affect human well-being and preferences as well as the decisions people make. Assessing the resilience of urban ecosystems requires understanding how interactions between human and ecological processes affect human functions such as housing, water supply, transportation, waste disposal, and recreation. Ecosystems provide important services to the urban population: they regulate climate, control flooding, and absorb carbon, to mention a few (Ehrlich and Mooney 1983, Daily 1997, Costanza et al. 1997). One important trajectory of future research is to articulate how emerging patterns in urban ecosystems affect household preferences and land development choices.

*4) How can urban planning integrate emerging knowledge about human-ecological systems to develop resilient urban landscapes?*
   The questions that motivate urban ecology research are important to public policy because of the multiple challenges facing policy makers: they must plan and manage urban ecological systems in ways that minimize the ecological impacts on ecosystems while sustaining economically and so-cially viable urban communities. We aim to generate empirical knowledge and to develop tools that can inform decision-making and support the assessment of alternative strategies and investment decisions in the processes of urban development and ecological conservation.

## 10.5 Implications for Urban Planning

A systematic understanding of the relationships between human and ecological processes in urban landscapes is central to urban design and planning. In response to the costs of sprawling development patterns, urban planning has attempted to stabilize inherently unstable states in urbanizing regions by devising plans and strategies that aim to achieve a balance between the conversion of natural land cover and the maintenance of ecological conditions that support human services (Alberti and Marzluff 2004). The assumption behind planned development and smart growth is that urban development patterns affect the ability of the natural processes and built infrastructure to support human and ecological function in urban areas. An understanding of how alternative development patterns can simultaneously support ecological (i.e., bird diversity, water quality) and human function (housing and water supply) seems essential to guide planning practice, especially given that urban patterns are being further decentralized.
   It is also critical to understand how coupled human-ecological systems work if we are to more effectively target questions that are relevant to policy decisions. More than ever, urban policymakers are challenged by the task of redirecting urban growth towards a more sustainable course. To do so, they

expect scholars of urban ecology to answer fundamental questions about the ecological resilience of alternative urban patterns. However, as we have seen throughout this book, the study of coupled human-ecological systems in urbanizing regions is still too fragmented to let us answer such questions—and it lacks a fully integrated theoretical framework.

The challenge, as I pointed out at the beginning of this chapter, is for both ecology and urban planning. It implies the development of a hybrid theory of the urban phenomena. While urban analysts have been interested in the question of appropriate urban form for more than a century, only since the 1950s have they recognized the need for a theory of urban form. Kevin Lynch and Lloyd Rodwin (1958) were the first to stress the importance of an analytical framework to link human goals to city form, and then to sketch the elements of one (Alberti 1999c). In their incisive article "A Theory of Urban Form," they developed analytical categories to explore the relationships between elements of form and basic values such as health, survival, growth, and adaptability. Although they developed a general model that would apply to various human values, as the values would be continuously redefined, so would the analytical system (Alberti 1999c).

Lynch's *Good City Form* (1981) is the first complete theoretical exploration of how urban patterns perform in relation to specific human values. Since then, society's goals have changed profoundly as scientists have learned far more about human interactions with the environment. We now recognize that human and environmental systems interact in very complex, often nonlinear ways, on multiple scales of time and space (Holling et al. 2002a). But this knowledge is not reflected in urban theory and practice. If we are to analyze the city as a complex system that evolves in response to changes in both socioeconomic and biophysical forces, we will need not only to extend our current approaches but also to integrate modes of inquiry combining historical, comparative, and experimental approaches (Alberti 1999c).

I suggest six implications of coupled human-natural systems. First, urban planning must fully appreciate that coupled human-ecological systems are dynamic, open, and non-equilibrial. This has implications for developing and evaluating urban planning strategies and the ability of the ecosystem to maintain or recover ecological function after development. Instead of aiming at achieving a specific condition (e.g., fixed urban density or distance of a development from a stream, as set in most planning regulations), planning must aim at maintaining the characteristics of the system that simultaneously support ecological and human function (i.e., resilience). Furthermore, if variability rather than consistency characterizes ecological conditions, multiple urban patterns might be "desirable" under different ecological conditions as opposed to a single "optimal" one.

Second, we need to recognize that change in coupled human-ecological systems can occur abruptly and discontinuously. We can characterize this response by drawing on thresholds and multiple domains of stability. As Holling (1996) suggests, knowledge of the system is always incomplete and surprise is inevitable. This perspective on environmental change requires a new framework for both understanding and including surprise as we explore and plan for resilient urban patterns. Typically planning relies primarily on predictive models, but complexity and uncertainty of coupled human-ecological systems make their interactions highly unpredictable. By focusing on key drivers, complex interactions, and irreducible uncertainties, scenario planning generates plausible futures within which predictive models can be used to test hypotheses and develop adaptive management strategies.

Third, we must see biophysical processes as drivers of urban change. Since human and ecological systems are interdependent—humans affect ecosystems functions, and ecosystem change simultaneously affects human well-being. Therefore we will have to extend urban theory to include an understanding of how urban systems respond to changes in the biophysical structure. The idea of a city being interdependent with its regional natural resources is not new to planning theory (Geddes 1905). What is new is considering the urban ecosystem as a coupled human-ecological system that evolves through the dynamic interactions between human and ecological functions. Ecosystems provide essential services to urban areas. When ecosystems change—watersheds are contaminated, biodiversity is lost, or climate changes—human well-being is affected over the long term.

A fourth implication is the importance of investigating mechanisms and thresholds. Where significant relationships exist between urban patterns and ecosystem functions, we must investigate the mechanisms that explain the relationships and explore whether the functional relationship indicates the existence of thresholds. John Marzluff and I (Alberti and Marzluff 2004) hypothesize that patterns of urbanization are critical in balancing the tension between providing human and ecological services and maintaining the unstable equilibrium created by planned development. We must learn more about the dynamics of these relationships so we can understand the factors that determine such thresholds of changing patterns.

A fifth implication concerns the consideration of scale. The relationship between urban patterns and ecological function depends on scale. Of course the study of urban patterns and ecological resilience must apply to the scales of both the city and metropolitan area—but urban patterns are relevant to environmental processes operating at multiple scales. Scale considerations include both the resolution of a given urban pattern measurement and the geographic extent or boundary of the area being considered. To study the relationships between urban patterns and human and ecological functions, we will have to cross spatial and hierarchical scales. Therefore, we need a nested approach.

A sixth implication is that the unpredictability of today's urban ecosystems challenges traditional planning and management assumptions and strategies for natural resources and environmental conservation. Planning and management strategies that aim to achieve a stable state are likely to make the system less resilient and reduce the options for sustaining human and ecological functions simultaneously. For urban planning and management practices to succeed, they must take complexity and uncertainty into account and redirect strategies toward building flexibility, adaptability, and resilience (Gunderson and Holling 2002). The challenge is to develop an adaptive capacity to learn and incorporate such knowledge in managing change.

## 10.6 A Final Note

In this book I have intentionally not resolved one key dilemma: whether we need to develop new ecological and urban theories, or whether we can extend current theories to describe how urban ecosystems work. We must learn much more before we can resolve this question; in fact, I leave it to my students and their students. It will take another generation of thought and scholarship before we will understand what kinds of dual and hybrid knowledge we need to achieve an effective synthesis between humans and nature in cities. Several scholars have tried to resolve the dilemma by proposing that the need for fundamentally novel theories to study urban ecosystems does not exclude disciplinary perspectives from playing a valuable role. But, as several others have started to articulate, a successful theory of urban ecology will require a number of specialists to think in interdisciplinary and multidisciplinary ways (Collins et al. 2000).

The task, I think, goes beyond the natural and social sciences to include the arts. Cities are the product not only of natural history and human activity. They are also the product of human imagination. In Italo Calvino's *Invisible Cities* (1974) Marco Polo describes the city of Fedora to Kublai Khan. The city's museum contains crystal globes that hold miniature representations of the city as individual inhabitants imagined it might have become but did not. As sociologist Howard Becker (2002) points out, Calvino's dialogues with Kublai Khan have important epistemological implications for our theories about the world. Her reading of Calvino's methodology is highly relevant here. A unified theory of urban ecology has to find room for both the "true" Fedora and the little Fedoras in the glass globes. "Not because they are equally real, but because they are all only assumptions. The one contains what is accepted as necessary when it is not yet so; the others, what is imagined as possible and, a moment later, is possible no longer" (Calvino 1974, 32). Marco Polo suggests that in order to understand general rules that apply to the city, "we must exclude [from the number of imaginable cities] those whose elements are assembled without a connecting thread, an inner rule, a perspective, a discourse" (Calvino 1974, 43-44). But Calvino (44) warns us that "Cities also believe they are the work of the mind or of chance, but neither the one nor the other suffices to hold up their walls." That is, neither is a sufficient explanation of how they work (Becker 2002). For that reason perhaps a "unified" theory of urban ecology will never exist, and many will argue that aiming at one will defy the mysteries of how urban ecosystems actually work and evolve.

# GLOSSARY

**Agent:** An autonomous decision-maker. The word "agent" is derived from the Greek "agein," which means to drive or lead, and from Latin "agere," which means to act. The term agent describes something that is acting. Originator of the process of acting and interacting with, other agents and their environments.

**Agent Based Modeling (ABM):** Method for testing the collective effects of individual actions by simulating dynamic interactions and feedbacks at both the individual and group levels.

**Biodiversity** (contraction of biological diversity): Biodiversity reflects the number, variety and variability of living organisms. Biodiversity may refer to the diversity within species (genetic diversity), between species (species diversity), or between ecosystems (ecosystem diversity) (MEA 2005).

**Complex system:** System that has properties not fully explained by an understanding of its component parts, that is characterized by nonlinear behavior, and that exhibits structural and functional characteristics emerging from interactions of its constituent parts (as opposed to arising via external or centralized organizational processes) (Goldenfeld and Kadanoff 1999).

**Disturbance:** Discrete event in time that causes a temporary change in average environmental conditions. A disturbance is any relatively discrete event in time that disrupts ecosystems, communities, or population structure and change resources, substrate availability, or the physical environment (Pickett and White 1985).

**Ecosystem function:** Ability to dynamically sustain life through inter actions among biotic and abiotic components and processes (e.g., primary production, nutrient cycling, decomposition etc.).

**Emergent property:** Phenomenon that is not evident in the constituent parts of a system, but that appears when they interact in the system as a whole.

**Gradient analysis:** Empirical analytical method based on the theory that environmental variation is ordered in space, and that spatial environmental patterns govern the corresponding structure and function of ecological systems, be they populations, communities, or ecosystems (Whittaker 1967).

**Hierarchy theory:** Evolution of general systems theory that has emerged as part of the general science of complexity. Hierarchy theory focuses upon levels of organization and scale. These nested levels emerge according to the dominant spatiotemporal scales at which system elements operate, causing them to interact most strongly with other elements of the same type; less strongly with elements that are dissimilar; and more weakly with patterns, processes and elements operating at disparate scales. Hierarchies occur both in social systems and ecological systems (Ahl and Allen 1996).

**Human function:** Socioeconomic, political, and residential/interpersonal processes, institutions and interactions that sustain and maintain human populations (Alberti and Marzluff 2004).

**Hydrologic function:** Biophysical processes and interactions that maintain or affect dynamics of the hydrologic cycle, as well as system dynamics that rely on water resources.

**Landscape:** Area that is spatially heterogeneous in at least one factor of interest (Turner et al. 2001).

**Natural habitat:** The collection of biotic and abiotic elements within a given location (predominantly unaltered by human activities) that provides elements for survival (food and shelter) for a species.

**Net primary productivity:** The amount of solar energy converted to chemical energy through the process of photosynthesis (production minus respiration) measuring the energy flowing into an ecosystem.

**Network theory:** Network theory has evolved in mathematics and physics as part of graph theory. It has been applied in many fields including computer science, biology, epidemiology, economics, and sociology. A network consists of a set of nodes joined by links. While initially represented by random graphs, real networks display some organizing principles (Albert and Barabasi 2002).

**Nutrient cycling:** The processes by which elements are extracted from their mineral, aquatic, or atmospheric sources, recycled from their organic forms, and converted to ionic form where biotic uptake occurs, ultimately returning them to the atmosphere, water, or soil (MEA 2005).

**Patch:** Discrete area of relatively homogeneous environmental conditions that are relevant to a given organism or ecological phenomenon (Pickett and White 1985).

**Pattern metrics:** Analytic measures describing characteristics of spatial heterogeneity. Pattern metrics describe three levels of spatial pattern: 1) landscape-level metrics describe the composition and configuration of the mosaic of classes across the entire landscape; 2) class-level metrics describe the spatial configuration of specific class types within the landscape; and 3) patch-level metrics describe the spatial configuration and characteristics (e.g., size, shape, edge length) of discrete patches in the landscape (McGarigal and Mark 1995).

**Resilience:** The size and shape of the basin of attraction around a stable system state, which defines the maximum perturbation that can be tolerated by a system without causing a shift to an alternative stable state (Holling 1973).

**Scenario:** Plausible description of how the future may develop, based on a coherent and internally consistent set of assumptions about the interactions of key uncertain and important driving forces (e.g., rate of technology change, prices) (MEA 2005).

**Sprawl:** Growth of an urban area that is unplanned and uncontrolled. Sprawl usually results in loss of rural areas and terrestrial habitats. Sprawl is typically characterized as poorly planned low density development on the edge of cities and towns, which is land-consumptive, auto-dependent, and designed without respect to the surroundings.

**Urban system:** Built environment with a high human population density. Urban systems are operationally defined as human settlements with a minimum population density commonly in the range of 400 to 1,000 persons per square kilometer, minimum size of typically between 1,000 and 5,000 people, and maximum agricultural employment usually in the vicinity of 50–75% (MEA 2005).

**Urban gradient:** Continuum in the level of development (and associated patterns of land use and land cover) ranging from the urban core(s) to the rural fringe of development, and the concomitant change in human and biophysical processes along that continuum (McDonnell and Pickett 1993).

**Urban heat island:** Region of elevated temperatures surrounding an urban area caused by geometric effects of the city and heat absorbed by structure and pavement.

**Urban pattern:** Spatial characteristics and interrelationships associated with the mosaic of urban land use types, and the resultant heterogeneity in land cover characteristics.

**Watershed:** All of the land area, determined largely by topographic flow patterns, that contributes surface run-off to the water supply of a body of water such as a river, stream, or lake (USGS).

# REFERENCES

Abel, T., and J. R. Stepp. 2003. A new ecosystems ecology for anthropology. Conservation Ecology 7(3):12.

Aber, J. D., K. J. Nadelhoffer, P. Steudler, and J. M. Melillo. 1989. Nitrogen saturation in northern forest ecosystems. BioScience 39:378–386.

Abrams, P. A. 1995. Monotonic or unimodal diversity–productivity gradients: What does competition theory predict? Ecology 76:2019–2027.

Agrawal, A. 2001. Phenotypic plasticity in the interactions and evolution of species. Science 294:321–326.

Ahl, V., and T. F. H. Allen. 1996. Hierarchy Theory: A Vision, Vocabulary, and Epistemology. Columbia University Press, New York.

Albert, R. 2006. General network theory. Pages 3–20 in S. J. Hwang, W. J. Sullivan, and A. R. Lommel (eds.), LACUS Forum 32: Networks. Linguistic Association of Canada and the United States, Houston.

Albert, R., and A. L. Barabasi. 2002. Statistical mechanics of complex networks. Reviews of Modern Physics 74(1):47–97.

Alberti, L. B. 1485. In Ten Books on Architecture, ed., J Rykwert. New York Transatlantic Arts, 1966.

Alberti, M. 1999a. Modeling the urban ecosystem: A conceptual framework. Environment and Planning B 26:605–630.

Alberti, M. 1999b. Urban form and ecosystem dynamics: Empirical evidence and practical implications. Pages 84–96 in K. Williams, E. Burton, and M. Jenks (eds.), Achieving Sustainable Urban Form. Routledge, London.

Alberti, M. 1999c. Urban patterns and environmental performance: What do we know? Journal of Planning Education and Research 19(2):151–163.

Alberti, M. 2001. Quantifying the urban gradient: Linking urban planning and ecology. In J. M. Marzluff, R. Bowman, R. McGowan, R. Donnelly, (eds.), Avian Ecology in an Urbanizing World. Kluwer, New York.

Alberti, M. 2005. The Effects of Urban Patterns on Ecosystem Function. International Regional Science Review. 28(2):169–192.

Alberti, M. 2007. Ecological signatures: Seeking a more scientific understanding of sustainable urban forms. Places. (In Press)

Alberti, M., D. Booth, K. Hill, C. Avolio, B. Coburn, S. Coe, and D. Spirandelli. 2007. The impact of urban patterns on aquatic ecosystems: An empirical analysis in Puget lowland sub-basins. Landscape and Urban Planning. Volume 80, Issue 4.

Alberti, M., E. Botsford, and A. Cohen. 2001. Quantifying the urban gradient: Linking urban planning and ecology. Pages 89–115 in J. M. Marzluff, R. Bowman, and R. Donnelly (eds.), Avian Ecology and Conservation in an Urbanizing World. Kluwer Academic Publishers, Boston.

Alberti, M., P. Christie, J. Marzluff, and J. Tewksbury. 2007. Interactions between natural and human systems in Puget Sound. Sound Science: Synthesizing Ecological and Socioeconomic Information about the Puget Sound Ecosystem. January 2007.

Alberti, M., and J. Hepinstall. Forthcoming. Modeling land-use land-cover change in urbanizing landscapes.

Alberti, M., J. Hepinstall, P. Waddell, J. Marzluff, and M. Handcock. 2006a. Modeling Interactions Among Urban Land-Cover Change, and Bird Diversity. Final Report. National Science Fundation, Biocomplexity project, 2001–2006 (BCS-0120024).

Alberti, M., and J. Marzluff. 2004a. Ecological resilience in urban ecosystems: Linking urban patterns to human and ecological functions. Urban Ecosystems 7:241–265.

Alberti, M., and J. Marzluff. 2004b. Resilience in urban ecosystems: Linking urban patterns to human and ecological functions. Urban Ecosystems 7:241–265.

Alberti, M., J. Marzluff, E. Shulenberger, G. Bradley, C. Ryan, and C. Zumbrunnen. 2003. Integrating humans into ecology: Opportunities and challenges for studying urban ecosystems. BioScience 53(12):1169–1179.

Alberti, M., K. Puruncajas, L. Hutyra et al. Forthcoming Detecting landscape signatures in urbanizing ecosystems. Ecology Research Lab. Seattle, WA.

Alberti, M., C. Redman, J. Wu, J. Marzluff, M. Handcock, J. Anderies, P. Waddell, D. Fox, and H. Kautz. 2005. BE/CNH: Urban Landscape Patterns: Complex Dynamics and Emergent Properties (Alberti PI: BCS 0508002).

Alberti, M., C. Redman, J. Wu, J. Marzluff, M. Handcock, J. Anderies, P. Waddell, D. Fox, and H. Kautz. 2006b. Urban landscape patterns and global environmental change (GEC): Complex dynamics and emergent properties. IHDP Update. Newsletterr of the International Human Dimension Programme on Global Environmental Change 2:5–6.

Alberti, M., and L. Susskind. 1996. Managing urban sustainability: Introduction to the special issue. EIA Review 16(4–6):213–221.

Alberti, M., and P. Waddell. 2000. An integrated urban development and ecological simulation model. Integrated Assessment 1:215–227.

Alberti, M., R. Weeks, and S. Coe. 2004. Urban land cover change analysis for the Central Puget Sound: 1991–1999. Photogrammetric Engineering and Remote Sensing 70:1043–1052.

Alexandridis, K., Pijanowski, B.C. 2007. Assessing multiagent parcelization Performance in the MABEL simulation model using Monte Carlo replication experiments. Environment and Planning B: Planning and Design 34(2):223–244.

Alig, R. J., J. D. Kline, and M. Lichtenstein. 2004. Urbanization on the U.S. landscape: Looking ahead in the 21st century. Landscape and Urban Planning 69:219–234.

Alig, R. J., A. Plantinga, S. Ahn, and J. D. Kline. 2003. Land use changes involving forestry for the United States: 1952 to 1997, with projections to 2050. USDA Forest Service, Pacific Northwest Research Station, Portland, OR. General Technical Report 587.

Allan, J. D. 2004. Landscapes and riverscapes: The influence of land use on stream ecosystems. Annual Review of Ecology and Systematics 35:257–284.

Allan, J., D. Erickson, and J. Fay. 1997. The influence of catchment land use on stream integrity across multiple spatial scales. Freshwater Biology 37:149–161.

Allen, P. M., and M. Sanglier. 1978. Dynamic models of urban growth. Journal of Social Biological Structure 1:265–80.

Allen, P. M., and M. Sanglier. 1979a. A dynamic model of growth in a central place system. Geographical Analysis 11:156–272.

Allen, P. M., and D. M. Sanglier. 1979b. Dynamic models of urban growth II. Journal of Social and Biological Structure 2:269–278.

Allen, T. F. H., and T. B. Starr. 1982. Hierarchy: Perspectives for Ecological Complexity. University of Chicago Press, Chicago.

Alonso, W. 1964. Location and Land Use. Harvard University Press, Cambridge, MA.

Amara, R. 1981. The futures field: searching for definitions and boundaries. The Futurist 15(1):25–29.

Anas, A. 1983. Discrete choice theory, information theory and the multinomial logit and gravity models. Transportation Research 17B:13–23.

Anas, A. 1986. From physical to economic urban models: The Lowry framework revisited. Pages 163–172 in B. G. Hutchinson and M. Batty (eds.), Advances in Urban Systems Modelling. North-Holland, Amsterdam.

Anas, A. 1987. Modeling in Urban and Regional Economics. Harwood Academic Publishers, Chur, Switzerland.

Anas, A., R. Arnott, and K. Small. 1998. Urban spatial structure. Journal of Economic Literature 36(3):1426–1464.

Anas, A., and I. Kim. 1996. General equilibrium models of polycentric urban land use with endogenous congestion and job agglomeration. Journal of Urban Economics 40(2):232–256.

Anderies, M. J., B. H. Walker, and A. P. Kinzig, 2006. Fifteen weddings and a funeral: Case studies and resilience-based management. Ecology and Society 11(1): 21. [online] URL: http://www.ecologyandsociety.org/vol11/iss1/art21/.

Anderson, H. R., R. W. Atkinson, J. L. Peacock, L. Marston, and K. Konstantinou. 2004. Meta-analysis of time-series studies and panel studies of particulate matter (PM) and ozone ($O_3$). World Health Organization, Copenhagen.

Andersson, C., K. Frenken, and A. Hellervik. 2006. A complex networks approach to urban growth. Environment and Planning A 38:1941–1964.

Andrews, E. D., and J. M. Nankervis. 1995. Effective discharge and the design of channel maintenance flows for gravel-bed rivers. Pages 151–164 in J. E. Costa, A. J. Miller, K. W. Potter, and P. R. Wilcock (eds.), Natural and Anthropogenic Influences in Fluvial Geomorphology. AGU Geophysical Monograph 89. American Geophysical Union, Washington, DC.

Araújo, M. B. 2003. The coincidence of people and biodiversity in Europe. Global Ecology and Biogeography 12:5–12.

Armstrong, R. A. 1976. Coexistence of two competitors on one resource. Journal of Theoretical Biology 56:499–502.

Arnfield, A. J. 1990. Canyon geometry, the urban fabric and nocturnal cooling: A simulation approach. Physical Geography 11:220–239.

Arnfield, A. J. 2001. Micro- and mesoclimatology. Progress in Physical Geography 25:560–569.

Arnfield, A. J., and C. S. B. Grimmond. 1998. An urban canyon energy budget model and its application to urban storage heat flux modelling. Energy and Buildings 27:61–68.

Arnfield, A. J., and Mills, G. M. 1994. An analysis of the circulation characteristics and energy budget of a dry, asymmetric, east–west urban canyon. I. Circulation characteristics. International Journal of Climatology 14:119–134.

Arnold, C. L., P. J. Boison, and P. C. Patton. 1982. Sawmill Brook: An example of rapid geomorphic change related to urbanization. Journal of Geology 90:155–166.

Arnold, C. L., and C. J. Gibbons. 1996. Impervious surface coverage: The emergence of a key environmental indicator. Journal of the American Planning Association 62:243–258.

Arrhenius, O. 1921. Species and area. Journal of Ecology 9:95–99.

Atkinson, B. J. 1985. The Urban Atmosphere. Cambridge University Press, Cambridge, UK.

Audirac, I., A. H. Shermyen, and M. T. Smith. 1990. Ideal urban form and visions of the good life: Florida's growth management dilemma. Journal of the American Planning Association 56:470–482.

Auerbach, F. 1913. Das gesetz der bevölkerungskonzentration. Petermanns Geographische Mitteilungen 59:74–76.

Austin, M. P. 1987. Models for the analysis of species response to environmental gradients. Vegetation 69:35–45.

AWWA. 1999. Water Rate Structures and Pricing, Second Edition. American Water Works Association, Denver, CO.

Ayres, R. U., and U. E. Simonis. 1994. Industrial Metabolism: Restructuring for Sustainable Development. United Nations University Press, NewYork.

Bailey, R. A., S. A. Harding, and G. L. Smith. 1989. Cross-validation. Pages 39–44 in S. Kotz, and N. L. Johnson (eds.), Encyclopedia of Statistical Sciences. J Wiley, New York.

Bak, P. 1996. How Nature Works: the Science of Self-Organized Criticality. Copernicus, New York, NY.

Bak, P., K. Chen, and M. Creutz. 1989. Self-organized criticality in the "game of life." Nature 342:780–782.

Bak, P., C. Tang, and K. Wiesenfeld. 1987. Self-organized criticality: an explanation of 1/f noise. Physical Review Letters 59:381.

Baker, L. A., Y. Xu, D. Hope, L. Lauver, and J. Edmonds. 2001. Nitrogen balance for the Central Arizona-Phoenix (CAP) ecosystem. Ecosystems 4:582–602.

Baker, V. R. 1977. Stream channel response to floods with examples from central Texas. Geological Society of America Bulletin 88:1057–1071.

Baldassare, M. 1992. Suburban communities. Annual Review of Sociology 18:475–494.

Balling, R. C., and R. S. Cerveny. 1987. Long-term associations between wind speeds and the urban heat island of Phoenix, Arizona. J. Clim. Appl. Meteorol. 26:712–716.

Balmford, A., J. L. Moore, T. Brooks, N. Burgess, L. A. Hansen, P. Williams, and C. Rahbek. 2001. Conservation conflicts across Africa. Science 291:2616–2619.

Band, L. E., M. L. Cadenasso, C. S. B. Grimmond, J. M. Grove, S. T. A. Pickett. 2005. Heterogeneity in urban ecosystems: Patterns and process. In G. Lovett, C. G. Jones, M. G. Turner, and K. C. Weathers (eds.), Ecosystem Function in Heterogeneous Landscapes. Springer-Verlag, New York.

Band, L. E., D. Mackay, I. Creed, R. Semkin, and D. Jeffries. 1996. Ecosystem processes at the watershed scale: Sensitivity to potential climate change. Limnology and Oceanography 4:928–938.

Band, L. E., C. L. Tague, P. M. Groffman, and K. Belt. 2001. Forest ecosystem processes at the watershed scale: Hydrological and ecological controls of nitrogen export. Hydrological Processes 15(10):2013–2028.

Barabási, A. L. 2005. Taming complexity. Nature Physics 1:68–70.

Barabasi, A.-L., R. Albert, and H. Jeong. 1999. Mean-field theory for scale-free random networks. Physica A: Statistical Mechanics and its Applications 272:173–187.

Bastin, L., and C. D. Thomas. 1999. The distribution of plant species in urban vegetation fragments. Landscape Ecology 14:493–507.

Batten, D. F. 2001. Complex landscapes of spatial interaction. Annals of Regional Science, 35:81–111.

Batty, M. 1997. Cellular automata and urban form: a primer. Journal of the American Planning Association 63:266–269.

Batty, M. 1998. Urban evolution on the desktop: Simulation using extended cellular automata. Environment and Planning A 30:1943–1967.

Batty, M. 2005. Cities and Complexity: Understanding Cities with Cellular Automata, Agent-Based Models, and Fractals. The MIT Press, Cambridge, MA.

Batty, M., and B. Hutchinson, (eds.) 1983. Systems Analysis in Urban Policy-Making and Planning. Plenum Press, New York.

Batty, M., and P. Longley. 1994. Fractal Cities: A Geometry of Form and Function. Academic Press.

Batty, M., and P. Torrens. 2001. Modeling complexity: The limits to prediction. CyberGeo, 201.

Batty, M., and Y. Xie. 1994. From cells to cities. Environment and Planning B: Planning and Design 21:S31–S48.

Beatley, T., and K. Manning. 1997. The Ecology of Place: Planning for Environment, Economy, and Community. Island Press. Washington, DC.

Becker, S. H. 2002. Italo Calvino as Urbanologist. L'Année Sociologique.

Beissinger, S., and D. Osborne. 1982. Effects of urbanization on avian community organization. Condor 84:75–83.

Bell, M. L., F. Dominici, and J. M. Samet. 2005. A meta-analysis of time-series studies of ozone and mortality with comparison to the national morbidity, mortality, and air pollution study. Epidemiology 16:436–445.

Bell, M. L., R. Goldberg, C. Hogrefe, P. L. Kinney, K. Knowlton, B. Lynn, J. Rosenthal, C. Rosenzweig, and J. A. Patz. 2007. Climate change, ambient ozone, and health in 50 US cities. Climatic Change 82:61–76.

Bell, M. L., A. McDermott, S. L. Zeger, J. M. Samet, and F. Dominici. 2004. Ozone and short-term mortality in 95 US urban communities, 1987–2000. Journal of the American Medical Association 292:2372–2378.

Bell, S. 1999. Landscape: Pattern, Perception and Process. E.& F.N. Spon, London.

Bender, D. J., and L. Fahrig. 2005. Matrix structure obscures the relationship between interpatch movement and patch size and isolation. Ecology 86:1023–1033.

Benenson, I., and P. M. Torrens. 2004. Geosimulation: Automata-Based Modeling of Urban Phenomena. London, Wiley.

Bennett, E. M., S. R. Carpenter, and N. F. Caraco. 2001. Human impact on erodable phosphorus and eutrophication: A global perspective. BioScience 51(3):227–234.

Berger, T., and D. C. Parker. 2002. Examples of specific research–introduction. Pages 26–25 in D. C. Parker, T. Berger, and S. M. Manson (eds.), Agent-Based Models of Land Use/Land Cover Change. LUCC Report Series 6, Louvain-la-Neuve.

Berger, T., and C. Ringler. 2002. Trade-offs, efficiency gains and technical change: Modeling water management and land use within a multiple-agent framework. Quarterly Journal of International Agriculture 41(1/2):119–144.

Berkowitz, A. R., C. H. Nilon, and K. S. Hollweg. 2002. Understanding Urban Ecosystems: A New Frontier for Science and Education. Springer-Verlag, New York.

Berry, B. J. L. 1990. Urbanization. Pages 103–119 in B. L. Turner II, W. C. Clark, R. W. Kates, J. F. Richards, J. T. Mathews, and W. B. Meyer (eds.), The Earth as Transformed by Human Action. Cambridge University Press. Cambridge, UK.

Berry, M. W., and K. S. Minser. 1997. Distributed Land-Cover Change Simulation Using PVM and MPI. Land Use Modeling Workshop, USGS EROS Data Center, Sioux Falls, SD, 5–6 June 1997.

Bjorklund, A., C. Bjuggren, M. Dalemo, and U. Sonesson. 1999. Planning biodegradable waste management in Stockholm. Journal of Industrial Ecology 3(4):43–58.

Black, P. E. 1991. Watershed Hydrology. Prentice Hall, Englewood Cliffs, NJ.

Blair, R. 1996. Land use and avian species diversity along an urban gradient. Ecological Applications 6:506–519.

Blair, R. B. 1999. Birds and butterflies along an urban gradient: Surrogate taxa for assessing biodiversity? Ecological Application 9:164–170.

Blair, R. B. 2001. Birds and butterflies along urban gradients in two ecoregions of the U.S. Pages 33–56 in J. L. Lockwood, and M. L. McKinney (eds.), Biotic Homogenization. Kluwer, Norwell, MA.

Blair, R. B., and A. E. Launer. 1997. Butterfly diversity and human land use: Species assemblages along an urban gradient. Biological Conservation. 80:113–125.

Blair, J. M., R. W. Parmelee, and P. Lavelle. 1995. Influences of earthworms on biogeochemistry. Pages 127–158 in P. F. Hendrix (ed.), Earthworm Ecology and Biogeography in North America. Lewis Publishers. Boca Raton, FL.

Blair, W., and G. Walsberg. 1996. Thermal effects of radiation and wind on a small bird and implications for microsite selection. Ecology 77:2228–2229.

Blewett, C. M., and J. M. Marzluff. 2005. Effects of urban sprawl on snags and the abundance and productivity of cavity-nesting birds. Condor 107:677–692.

Bock, C. E., J. H. Bock, and B. C. Bennett. Songbird abundance in grasslands at a suburban interface on the Colorado High Plains. Pages in J. Herkert, ed. Ecology and Conservation Grassland Birds in the Western Hemisphere. Ornithological Society. Berkeley, CA.

Bohlen, P., R. V. Pouyat, V. Eviner, and P. M. Groffman. 1996. Short and long term effects of earthworms on nitrous oxide fluxes in forest soils. Supplemental Bulletin of the Ecological Society of America 77:43.

Bojkov, R. D. 1986. Surface ozone during the second half of the nineteenth century. Journal of Climate and Applied Meteorology 25: 343–352.

Bolger, D. T. 2001. Urban birds: Population, community, and landscape approaches, p. 155–177. In J. M. Marzluff, R. Bowman, and R. E. Donnelly (eds.), Avian Ecology and Conservation in an Urbanizing World. Kluwer Academic, Norwell, MA.

Bolger, D., A. Alberts, R. Sauvajot, P. Potenza, C. McCalvin, D. Tran, S. Mazzoni, and M. Soule. 1997. Response of rodents to habitat fragmentation in coastal Southern California. Ecological Applications 7:552–563.

Bolger, D., A. Alberts, and M. Soulé. 1991. Occurrence patterns of bird species in habitat fragments: Sampling, extinction, and nested species subsets. American-Naturalist 137(12):155.

Bolliger, J., J. C. Sprott, and D. J. Mladenoff. 2003. Self-organization and complexity in historical landscape patterns. Oikos 100:541–553.

Bonan, G. 2002. Ecological Climatology – Concepts and Applications. Cambridge University Press. New York.

Bookchin, M. 1980. Towards an Ecological Society. Black Rose Books, Montreal.

Booth, D. B. 1990. Stream channel incision following drainage-basin urbanization. Water Resources Bulletin 26:407–417.

Booth, D. B. 2000. Forest Cover, Impervious-Surface Area, and the Mitigation of Urbanization Impacts in King County, Washington. Report to King County Water and Land Resources Division. Center for Urban Water Resources Management, University of Washington, Seattle, WA, USA.

Booth, D., and C. J. Jackson. 1997. Urbanization of aquatic systems-degradation thresholds, stormwater detention, and the limits of mitigation. Water Resources Bulletin 33:1077–1090.

Booth, D. B., J. R. Karr, S. Schauman, C. P. Konrad, S. A. Morley, M. G. Larson, and S. J. Burges. 2004. Reviving urban streams: Land use, hydrology, biology, and human behavior. Journal of the American Water Resources Association 40:1351–1364.

Booth, D. B., J. R. Karr, S. Schauman, C. P. Konrad, S. A. Morley, M. G. Larson, P. Henshaw, E. Nelson, and S. J. Burges. 2001. Urban Stream Rehabilitation in the Pacific Northwest. Final report to U. S. EPA, grant no. R82-5284-010. Center for Urban Water Resources, University of Washington, Seattle, Washington, USA.

Booth, D. B., D. R. Montgomery, and J. Bethel. 1996. Large Woody Debris in Urban Streams of the Pacific Northwest. Pages 178–197 in L. A. Roesner (ed.), Effects of Watershed Development and Management on Aquatic Ecosystems. American Society of Civil Engineers, New York, NY, USA.

Börjeson, L., M. Höjer, K.-H. Dreborg, T. Ekvall, and G. Finnveden. 2005. Scenario types and techniques: Towards a user's guide. Futures 38(7): 723–739.

Borman, F. H., and G. E. Likens. 1979. Pattern and Process in a Forested Ecosystem. Springer-Verlag, New York.

Bornstein, R. D. 1968. Observation of the urban heat island effect in New York city. Journal of Applied Meteorology 7:575–582.

Bornstein, R., and M. LeRoy. 1990. Urban barrier effects on convective and frontal thunderstorms. Proceedings of the Fourth American Meteorological Society Conference on Mesoscale Processes, Boulder, CO. American Meteorological Society, 120–121.

Bowman, R., and G. E. Woolfenden. 2001. Nest success and the timing of nest failure of Florida Scrub-Jays in suburban and wildland habitats. Pages 385–404 in J. M. Marzluff, R. Bowman, and R. E. Donnelly (eds.), Avian Ecology and Conservation in an Urbanizing World. Kluwer Academic, Norwell, MA.

Boyden, S., S. Millar, K. Newcombe, and B. O'Neill. 1981. The Ecology of a City and Its People: The Case of Hong Kong. Australian National University Press, Canberra.

Boyce, D. E., M. R. Lupa, M. Tatineni, and Y. He. 1993. Urban activity location and travel characteristics: Exploratory scenario analyses. SIG1 Seminar Paper on Environmental Challenges in Land Use Transport Coordination, Blackheath, Australia, December 6B10. Urban Transportation Center, University of Illinois, Chicago.

Bradshaw, A. D. 2003. Natural ecosystems in cities: A model for cities as ecosystems. Pages 77–94 in A. R. Berkowitz, C. H. Nilon, and K. S. Hollweg (eds.), Understanding Urban Ecosystems. Springer-Verlag, New York.

Bramen, S. K., A. F. Pendley, and W. Corley. 2002. Influence of comercially available wildflower mixes on beneficial arthropod abundance and predation in turfgrass. Environmental Entomology 31:564–572.

Brand, S. 1999. The Clock of the Long Now: Time and Responsibility. Basic Books, New York.

Brazel, A., P. Gober, S. J. Lee, S. Grossman-Clarke, J. Zehnder, B. Hedquist, and E. Comparri. 2007. Determinants of changes in the regional urban heat island in metropolitan Phoenix (Arizona, USA) between 1990 and 2004. Climate Research 33(2):171–182

Breheny, M. (ed.), 1992. Sustainable Development and Urban Form. Pion, London.

Breuse, Jurgen, Feldmann, Hildegard, and Uhlmann, Ogarit. 1998. Urban Ecology. Springer, Germany.

Britton, D. L., J. A. Day, and M. P. Henshallhoward. 1993. Hydrochemical response during storm events in a South African mountain catchment – the influence of antecedent conditions. Hydrobiologia 250(3):143–157.

Brock, W. A., and S. R. Carpenter. 2006. Variance as a leading indicator of regime shift in ecosystem services. Ecology and Society 11(2):9. [online] URL: http://www.ecologyandsociety.org/vol11/iss2/art9/.

Brooks, A. P., G. J. Brierley, and R. G. Millar. 2003. The long-term control of vegetation and woody debris on channel and flood-plain evolution: Insights from a paired catchment study in southeastern Australia. Geomorphology 51(1–3):7–29.

Brothers, T., and A. Spingarn. 1992. Forest fragmentation and alien plant invasion of central Indiana old-growth forests. Conservation Biology 6:91–100.

Brown, J. H. 1995. Macroecology. University of Chicago Press, Chicago.

Brown, J. H. 1995. Organisms as engineers: A useful framework for studying effects on ecosystems. Trends in Ecology and Evolution 10:51–52.

Brown, J. H. 1995. Organisms and species as complex adaptive systems: Linking the biology of populations with the physics of ecosystems. Pages 16–28 in C. G. Jones, and J. H. Lawton (eds.), Linking Species and Ecosystems. Chapman and Hall, New York.

Brown, J. H., and E. G. Heske. 1990. Control of a desert-grassland transition by a keystone rodent guild. Science 250:1705–1707.

Brunekreef, B., and S. T. Holgate. 2002. Air pollution and health. Lancet 360:1233–1242.

Buckling, A., R. Kassen, G. Bell, and P. B. Rainey. 2000. Disturbance and diversity in experimental microcosms. Nature 408:961–964.

Burchell, R. W., G. Lowenstein, W. R. Dolphin, C. C. Galley, A. Downs, S. Seskin, K. Gray Still, and T. Moore. 2002. Costs of Sprawl 2000. TCRP Report 74, National Academy Press, Washington, DC.

Burges, S. J., M. S. Wigmosta, J. M. Meena, 1998. Hydrological effects of landuse change in a zero-order catchment. Journal of Hydrologic Engineering 3:86–97.

Burgess, E. W. 1925. The growth of the city: An introduction to a research project. Pages 47–62 in R. E. Park, E. W. Burgess, and R. D. McKenzie (eds.), The City. University of Chicago Press, Chicago.

Burnett, C., and T. Blaschke. 2003. A multi-scale segmentation/object relationship modeling methodology for landscape analysis. Ecological Modelling 168:233–240.

Burton M. L., L. J. Samuelson, S. Pan. 2005. Riparian woody plant diversity and forest structure along an urban-rural gradient. Urban Ecosystems 8:93–106.

Cadenasso, M., M. Traynor, and S. T. A. Pickett. 1997. Functional location of forest edges: Gradients of multiple physical factors. Canadian Journal of Forest Resources 27:774–782.

Cadenasso, M. L., S. T. A. Pickett, and J. M. Grove. 2006. Dimensions of ecosystem complexity: Heterogeneity, connectivity, and history. Ecological Complexity 3:1–12.

Cadenasso, M. L., S. T. A. Pickett, and K. Schwarz. 2007. Spatial heterogeneity in urban ecosystems: Reconceptualizing land cover and a framework for classification. Frontiers in Ecology and Evolution 5: 80–88.

Calvino, I. 1974. Invisible Cities. Harcourt, Inc., Orlando, FL.

Cambridge Systematics, Inc. and Parsons Brinkerhoff Quade and Douglas, Inc. 1992.

Cardelino, C. A., and W. L. Chameides. 1990. Natural hydrocarbons, urbanization and urban ozone. Journal of Geophysical Research 95:13971–13979.

Cardinale, B. J., H. Hillebrand, and D. F. Charles. 2006. Geographic patterns of diversity in streams are predicted by a multivariate model of disturbance and productivity. Journal of Ecology 94:609–618.

Cardinale, B. J., K. Nelson, and M. A. Palmer. 2000. Linking species diversity to the functioning of ecosystems: On the importance of environmental context. Oikos 91:175–183.

Cardinale, B. J., M. A. Palmer, and S. L. Collins. 2002. Species diversity increases ecosystem functioning through interspecific facilitation. Nature 415:426–429.

Cardinale, B. J., M. A. Palmer, C. M. Swan, S. Brooks, and N. L. Poff. 2001. The influence of substrate heterogeneity on biofilm metabolism in a stream ecosystem. Ecology 83:412–422.

Carlson, T. N., and S. T. Arthur. 2000. The impact of land use/land cover changes due to urbanization on surface microclimate and hydrology: A satellite perspective. Global and Planetary Change 25:49–65.

Carpenter, S., B. Walker, J. M. Anderies, and N. Abel. 2001. From metaphor to measurement: Resilience of what to what? Ecosystems 4:765–781.

Carpenter, S. R. 2001. Alternate states of ecosystems: Evidence and its implications. Pages 357–383 in M. C. Press, N. Huntly and S. Levin (eds), Ecology: Achievement and Challenge. Blackwell, London.

Carpenter, S. R. 2002. Ecological futures: Building an ecology of the long now (Robert H. MacArthur Award Lecture). Ecology 83(8):2069–2083.

Carpenter, S. R. 2003. Regime shifts in lake ecosystems: Pattern and variation. Excellence in Ecology Series, Volume 15. Ecology Institute, Oldendorf/Luhe, Germany.

Carpenter, S. R., B. J. Benson, R. Biggs, J. W. Chipman, J. A. Foley, S. A. Golding, R. B. Hammer, P. C. Hanson, P. T. J. Johnson, A. M. Kamarainen, T. K. Kratz, R. C. Lathrop, K. D. McMahon, B. Provencher, J. A. Rusak, C. T. Solomon, E. H. Stanley, M. G. Turner, M. J. Vander Zanden, C. -H. Wu, and H. Yuan. 2007. Understanding regional change: A comparison of two lake districts. Bioscience 57(4):323–335.

Carpenter, S. R., and W. A. Brock. 2006. Rising variance: A leading indicator of ecological transition. Ecology Letters 9:311–318.

Carpenter, S. R., W. Brock, and D. Ludwig. 2002. Collapse, learning and renewal. in L. H. Gunderson and C. S. Holling (eds.), Panarchy: Understanding Transformations in Human and Natural Systems. Island Press, Washington, DC.

Carpenter, S. R., N. F. Caraco, D. L. Correll, R. W. Howarth, A. N. Sharpley, and V. H. Smith. 1998. Nonpoint pollution of surface waters with phosphorus and nitrogen. Ecological Applications 8:559–568.

Carpenter, S.R. and L. H. Gunderson. 2001. Coping with collapse: Ecological and social dynamics in ecosystem management. BioScience 51:451–457.

Carpenter, S. R., R. C. Lathrop, P. Nowak, E. M. Bennett, T. Reed, and P. A. Soranno. 2006. The ongoing experiment: Restoration of Lake Mendota and its watershed. Pages 236–256 in J. J. Magnuson, T. K. Kratz, and B. J. Benson (eds.), Long-Term Dynamics of Lakes in the Landscape: Long-Term Ecological Research on North Temperate Lakes. Oxford University Press, New York.

Carpenter, S. R., and P. R. Leavitt. 1991. Temporal variation in paleo-limnological record arising from a trophic cascade. Ecology 72:277–85.

Carpenter, S. R., D. Ludwig, and W. A. Brock. 1999. Management of eutrophication for lakes subject to potentially irreversible change. Ecological Applications 9:751–771.

Carreiro M. M., K. Howe, D. F. Parkhurst, and R. V. Pouyat. 1999. Variations in quality and decomposability of red oak litter along an urban-rural land use gradient. Biology and Fertility of Soils 30:258–268.

Cassin, J., R. Fuerstenberg, L. Tear, K. Whiting, D. St. John, B. Murray, J. Burkey. 2005. Development of Hydrological and Biological Indicators of Flow Alteration in Puget Sound Lowland Streams. King County Water and Land Resources Division. Seattle, Washington.

Cayan, D. R., and A. V. Douglas. 1984. Urban influences of surface temperatures in the southwestern United States during recent decades. J. Climate and Applied Meteorology 23:1520–1530.

Cervero, R., and R. Gorham. 1995. Commuting in transit versus automobile neighborhoods. Journal of the American Planning Association 61(2):210–225.

Cervero, R., and K.-L. Wu. 1995. Polycentrism, community and residential location in the San Francisco Bay Area. Working Paper 640, Institute of Urban and Regional Development, University of California, Berkeley.

Chandler, T. J. 1976. Urban climatology and its relevance to urban design. Technical Note 149. Geneva: World Meteorological Organization.

Chapin, F. S., III, O. E. Sala, I. C. Burke, J. P. Grime, D. U. Hooper, W. K. Lauenroth, A. Lombard, H. A. Mooney, A. R. Mosier, S. Naeem, S. W. Pacala, J. Roy, W. L. Steffen, and D. Tilman. 1998. Ecosystem consequences of changing biodiversity: Experimental evidence and a research agenda for the future. Bioscience 48:45–52.

Chapman, R. E., J. Wang, and A. F. G. Bourke. 2003. Genetic analysis of spatial foraging patterns and resource sharing in bumble bee pollinators. Molecular Ecology 12:2801–2808.

Charbonneau, R., and G. M. Kondolf. 1993. Land use change in California: Nonpoint source water quality impacts. Environmental Management 17:453–460.

Chen, J., J. Franklin, and T. Spies. 1992. Vegetation responses to edge environments in old-growth Douglas-fir forests. Ecological Applications 2:387–396.

Chown, S. L., B. J. van Rensburg, K. J. Gaston, A. S. L. Rodrigues, and A. S. van Jaarsveld. 2003. Energy, species richness, and human population size: Conservation implications at a national scale. Ecological Applications 13:1233–1241.

Churcher, P. B., and J. H. Lawton. 1987. Predation by domestic cats in an English village. Journal of Zoology 212:439–455.

Clark, G. L., M. P. Feldman, and M. S. Gertler. (eds.). 2000. The Oxford Handbook of Economic Geography. Oxford University Press, Oxford, UK.

Clark, J. S., S. R. Carpenter, M. Barber, S. Collins, A. Dobson, J. A. Foley, D. M. Lodge, M. Pascual, J. R. Pielke, W. Pizer, C. Pringle, W. V. Reid, K. A. Rose, O. Sala, W. H. Schlesinger, D. H. Wall, and D. Wear. 2001. Ecological forecasts: An emerging imperative. Science 293:657–660.

Clark, W. C. 1986. Sustainable development of the biosphere: Themes for a research program. Pages 5–48 in W. C. Clark and R. E. Munn (eds.), Sustainable Development of the Biosphere. Cambridge University Press, Cambridge.

Clarke, K., S. Hoppen, and L. Gaydos. 1997. A self-modifying cellular automaton model of historical urbanization in the San Francisco Bay Area. Environment and Planning B 24: 247–261.

Clarke, R., and J. King. 2004. The Water Atlas: A Unique Visual Analysis of the World's Most Critical Resource. The New Press, New York.

Clergeau, P., J. Jokimaki, and R. Snep. 2006. Using hierarchical levels for urban ecology. Trends in Ecology & Evolution 21:660–661.

Clergeau, P., G. Mennechez, A. Sauvage, and A. Lemoine. 2001. Human perception and appreciation of birds: A motivation for wildlife conservation in urban environments of France. Pages 69–86 in J. M. Marzluff, R. Bowman, and R. Donnelly (eds.), Avian Ecology in an Urbanizing World. Kluwer: Norwell, MA.

Coleman, B. D. 1981. On random placement and species-area relations. Mathematical Biosciences 54:191–215.

Collinge, S. 1996. Ecological consequences of habitat fragmentation: Implications for landscape architecture and planning. Landscape and Urban Planning 36:50–77.

Collins, J. P., A. Kinzig, N. B. Grimm, W. F. Fagan, D. Hope, J. Wu, and E. T. Borer. 2000. A new urban ecology. American Scientist 88:416–425.

Collins, S. L., and S. M. Glenn. 1997. Intermediate disturbance and its relationship to within- and between-patch dynamics. New Zealand Journal of Ecology 21:103–110.

Connell, J. H. 1978. Diversity in tropical rain forests and coral reefs. Science 199:1302–1310.

Connell, J. H. 1979. Tropical rainforests and coral reefs as open non-equilibrium systems. Pages 243–252 in R. M. Anderson, B. D. Turner, and L. R. Taylor (eds.), Population dynamics. Blackwell Scientific Publications, Oxford, UK.

Connor, E. F., and E. D. McCoy. 1979. The statistics and biology of the species-area relationship. American Naturalist 113:791–833.

Conway, R. 1990. The Washington projection and simulation model: A regional interindustry econometric model. International Regional Science Review 13:141–165.

Cook, R. E. 2000. Do landscapes learn? Ecology's "New Paradigm" and design in landscape architecture. Pages 115–132 in M. Conan (ed.), Environmentalism in Landscape Architecture. Dumbarton Oaks Trustees for Harvard Library, Washington DC.

Costa, J. E., and J. E. O'Connor. 1995. Geomorphically effective floods. pp. 45–56 in J. E. Costa, A. J. Miller, K. W. Potter, and P. R. Wilcock (eds.), Natural and anthropogenic influences in fluvial geomorphology. AGU Geophysical Monograph 89. American Geophysical Union, Washington, DC.

Costanza, R., R. d'Arge, R. de Groot, S. Farber, M. Grasso, B. Hannon, K. Limburg, S. Naeem, R. V. O'Neill, J. Paruelo, R. G. Raskin, P. Sutton, M. van den Belt. 1997. The value of the world's ecosystem services and natural capital. Nature, 387(6230):255.

Costanza, R., A. Voinov, R. Boumans, T. Maxwell, F. Villa, L. Wainger, and H. Voinov. 2002. Integrated ecological economic modeling of the Patuxent River Watershed. Ecological Monographs 72:203–231.

Costanza, R., L. Wainger, and N. Bockstael. 1995. Integrated ecological economic systems modeling: Theoretical issues and practical applications. In Milon, J. W., and Shogren, J. F. Integrating Economics and Ecological Indicators. Praeger, Westport, CT.

Cova, T. J., and M. F. Goodchild. 2002. Extending geographical representation to include fields of spatial objects. International Journal of Geographical Information Science 16(6):509–532.

Cressie, N. A. C. 1993. Statistics for Spatial Data. Revised edition. John Wiley and Sons, New York.

Cronon, W. 1991. Nature's metropolis: Chicago and the Great West. Norton, London.

Cronon, W. 1996. The trouble with wilderness: or, getting back to the wrong nature. Pages 69–90 in W. Cronon (ed) Uncommon ground: rethinking the human place in nature. W.W. Norton and Co., New York, NY.

Crooks, K. R., A. V. Suarez, and D. T. Bolger. 2004. Avian assemblages along a gradient of urbanization in a highly fragmented landscape. Biological Conservation 115:451–462.

Csillag, F., and S. Kabos. 2002. Wavelets, boundaries, and the spatial analysis of landscape pattern. Ecoscience 9:177–190.

Cullinan, V. I., M. A. Simmons, and J. M. Thomas. 1997. A Bayesian test of hierarchy theory: Scaling up variability in plant cover from field to remotely sensed data. Landscape Ecology 12(5):273–285.

Cumming, G. S., D. H. M. Cumming, and C. L. Redman. 2006. Scale mismatches in social-ecological systems: Causes, consequences, and solutions. Ecology and Society 11(1):14. [online] URL: http://www. ecologyandsociety.org/vol11/iss1/art14/.

Dahl, T. E. 1990. Wetland losses in the United States 1780s to 1980s. United States Department of Interior, Fish and Wildlife Service. Washington, DC.

Dahlgren, R. A. 2006. Biogeochemical processes in soils and ecosystems: From landscape to molecular scale. Journal of Geochemical Exploration 88(1–3):186–189.

Daily, G. C (ed.), 1997. Nature's Services: Societal Dependence on Natural Ecosystems. Island Press, Washington, DC.

Dale, M. R. T., and M. Mah. 1998. The use of wavelets for spatial pattern analysis in ecology. Journal of Vegetation Science 9(6):805–814.

D'Antonio, C. M., and P. M. Vitousek. 1992. Biological invasions by exotic grasses, the grass/fire cycle, and global change. Annual Review of Ecology and Systematics 23:63–87.

Daszak, P., A. A. Cunningham, and A. D. Hyatt. 2000. Emerging infectious diseases of wildlife: Threats to biodiversity and human health. Science 287:443–449.

Davidson, C. 1998. Issues in measuring landscape fragmentation. Wildlife Society Bulletin 26:32–37

Davis, B. N. K. 1978. Urbanisation and the diversity of insects. Pages 126–138 in L. A. Mound, and N. Waloff (eds.), Diversity of Insect Faunas. Symposia of the Royal Entomological Society Number 9. Blackwell Scientific, Oxford, UK.

Davis, M. 1998. Ecology of fear: Los Angeles and the imagination of disaster. Metropolitan Books, New York.

Dawson, J. P., P. J. Adams, and S. N. Pandis. 2007. Sensitivity of ozone to summertime climate in the eastern USA: A modeling case study. Atmospheric. Environment 41:1494–1511.

Decker, E. H., S. Elliot, F. A. Smith, D. R. Blake, and F. S. Rowland. 2000. Energy and material flow through the urban ecosystem. Annual Review of Energy and the Environment 25:685–740.

DeFries, R. S., M. Hansen, R. G. Townshend, A. C. Janetos, and T. R. Loveland. 1999. Global 1km data set of percent tree cover derived from remote sensing. Available on-line [http://glcf.umiacs.umd.edu] from the Global Land Cover Facility, University of Maryland Institute for Advanced Computer Studies, College Park, Maryland.

Dendrinos, D. S. 1992. The Dynamics of Cities: Ecological Determinism, Dualism and Chaos. Routledge, London.

D'Eon, G. R., S. M. Glenn, I. Parfitt, and M. J. Fortin. 2002. Landscape connectivity as a function of scale and organism vagility in a real forested landscape. Conservation Ecology 6 http://www.consecol.org/vol6/iss2/art10

DeWalle, D. R., B. R. Swistock, T. E. Johnson, and K. J. McGuire. 2000. Potential effects of climate change on urbanization and mean annual streamflow in the United States. Water Resources Research 3(9):2655–2664.

Dial, R., and J. Roughgarden. 1998. Theory of marine communities: The intermediate disturbance hypothesis. Ecology 79:1412–1424.

Dickman, C. 1987. Habitat fragmentation and invertebrate species richness in an urban environment. Journal of Applied Ecology 24:337–351.

DiPasquale, D., and W. Wheaton. 1996. Urban Economics and Real Estate Markets. Princeton Hall, Englewood Cliffs, NJ.

Doak, D. F., P. C. Marino, and P. M. Kareiva. 1992. Spatial scale mediates the incluence of habitat of habitat fragmentation on dispersal success: Implications for conservation. Theoretical Population Biology 41(3): 315–336.

Dobson, A. P., J. P. Rodriguez, and W. M. Roberts. 2001. Synoptic tinkering: Integrating strategies for large-scale conservation. Ecological Applications 11:1019–1026.

Douglas, I. 1974. The impact of urbanization on river systems. Pages 307–317 in Proceedings of the International Geographical Union Regional Conference, New Zealand Geographic Society.

Douglas, I. 1983. The Urban Environment. Edward Arnold, Baltimore, MD.

Douglas, I. 1983. The Urban Environment. Edward Arnold, Baltimore/London.

Done, T. J. 1992. Phase shifts in coral reef communities and their ecological significance. Hydrobiology 247:121–132.

Donnelly, R., and J. M. Marzluff. 2004a. Importance of reserve size and landscape context to urban bird conservation. Conservation Biology 18:733–745.

Donnelly, R. E., and J. M. Marzluff. 2004b. Designing research to advance the management of birds in urbanizing areas. pp. 114–122 in W. W. Shaw, L. K. Harris, and L. Vandruff, (eds.), Proceedings of the 4th International Symposium on Urban Wildlife Conservation. May 1–5, 1999. University of Arizona Press, Tuscon.

Donnelly, R., and J. M. Marzluff. 2006. Relative importance of habitat quantity, structure, and spatial pattern to birds in urbanizing environments. Urban Ecosystems 9:99–117.

Dooling, S., G. Simon, and K. Yocom. 2007. Place-based urban ecology: A century of park planning in Seattle. Urban Ecosyst 9:299–321.

Dow, C. L., and D. R. DeWalle. 2000. Trends in evaporation and Bowen ratio on urbanizing watersheds in eastern United States. Water Resources Research 36(7):1835–1843.

Dublin, H., A. R. E. Sinclair, and J. McGlade. 1990. Elephants and fire as causes of multiple stable states in the Serengeti-Mara woodlands. Journal of Animal Ecology 59:1147–1164.

Dubos, R. 1968. So Human An Animal. Charles Scribner's Sons, New York.

Duncan, O. D. 1960. From Social System to Ecosystem. Sociological Inquiry. 31, 2:140–149.

Dunne, T., and L. B. Leopold. 1978. Water in Environmental Planning. W. H. Freeman & Company, San Francisco.

Duvigneaud, P. (ed.), 1974. Etudes Ecologiques de l'Ecosysteme Urbain Bruxellois. Ministere de L'Education Nationale Francais et Neerlandais.

Edwards, J. L., and J. L. Schofer. 1975. Relationships between transportation energy consumption and urban structure: Results of simulation studies. Department of Civil and Mineral Engineering, Minneapolis.

Ehrenfeld, J. G. 2003. Effects of exotic plant invasions on soil nutrient cycling processes. Ecosystems 6:503–23.

Ehrlich P. R., and A. H. Ehrlich. 1981. Extinction: The Causes and Consequences of the Disappearance of Species. Random House: New York.

Ehrlich, P. R., and H. A. Mooney. 1983. Extinction, substitution, and ecosystem services. Bioscience 33:248–254.

Ellefsen, R. 1990. Mapping and measuring buildings in the canopy boundary layer in ten U.S. cities. Energy and Buildings 15–16:1025–1049.

Ellickson, B. 1981. An alternative test of the hedonic theory of housing markets. Journal of Urban Economics 9:56–79.

Elton, C. 1927. Animal Ecology. Sidgwick and Jackson: London.

EPA. 2000. Draft EPA Guidelines for Management of Onsite/Decentralized Wastewater Systems. U.S. Environmental Protection Agency, Office of Wastewater Management, Washington, DC, USA. [online] http://www.epa.gov/owm/septic/pubs/septic_management_handbook.pdf (verified April 10, 2006).

Epstein, J.M. and R. Axtell. 1996. Growing Artificial Societies: Social Science from the Bottom Up. MIT Press/Brookings Institution Press, Cambridge, MA/Washington, DC.

Estes, J. A., and D. O. Duggins. 1995. Sea otters and kelp forests in Alaska – Generality and variation in a community ecological paradigm. Ecological Monographs 65:75–100.

Evans, K. L., and K. J. Gaston. 2005. People, energy and avian species richness. Global Ecology and Biogeography 14:187–196.

Evans, T. P., and H. Kelley. 2004. Multi-scale analysis of a household level agent-based model of landcover change. Journal of Environmental Management 72(1–2):57–72.

Ewing, R. 1994. Characteristics, causes, and effects of sprawl: A literature review. Environmental and Urban Issues 21:1–15.

Ewing, R. 1997. Is Los Angeles-style sprawl desirable? Journal of the American Planning Association 63:107–126.

Ewing, R., J. Kostyack, D. Chen, B. Stein, and M. Ernst. 2005. Endangered by Sprawl: How Runaway Development Threatens America's Wildlife. National Wildlife Federation, Smart Growth America, and NatureServe. Washington, DC.

Ewing, R., R. Pendall, and D. Chen. 2003. Measuring sprawl and its transportation impacts. Transportation Research Record 1831:175–183.

Færge, J., J. Magida, and F. W. T. Penning de Vries. 2001. Urban nutrient balance for Bangkok. Ecological Modelling 139(1):63–74.

Faeth, S. H., P. S. Warren, E. Shochat, and W. A. Marussich. 2005. Trophic dynamics in urban communities. BioScience 55:399–407.

Fahrig, L., and G. Merriam. 1985. Habitat patch connectivity and population survival. Ecology 66(6):1762–1768.

Fahrig, L., and J. Paloheimo. 1988. Effect of spatial arrangement of habitat patches on local population size. Ecology 69(2):468–475.

Filion, P., T. Bunting, and K. Warriner. 1999. The entrenchment of urban dispersion: Residential preferences and location patterns in the dispersed city. Urban Studies 36:1317–1347.

Finkenbine, J. K., D. S. Atwater, and D. S. Mavinic. 2000. Stream health after urbanization. Journal of the American Water Resources Association 36:1149–60.

Fischer-Kowalski, M. 1998. Society's metabolism: The intellectual history of material flow analysis. Part I, 1860–1970. Journal of Industrial Ecology 2(1):61–78.

Fischer-Kowalski, M., and W. Hüttler. 1998. Society's metabolism: The intellectual history of material flow analysis. Part II: 1970–1998. Journal of Industrial Ecology 2(4):107–137.

Fisher, S. G., L. J. Gray, N. B. Grimm, and D. E. Busch. 1982. Temporal succession in a desert stream ecosystem following flash flooding. Ecological Monographs 52:93–110.

Flannery, T. 2001. The Eternal Frontier. Atlantic Monthly Press, New York.

Fleming, T. H., and V. J. Sosa. 1994. Effects of nectarivorous bats and frugivorous mammals on the reproductive success of plants. Journal of Mammalogy 75(4):845–851.

Folke, C., S. Carpenter, T. Elmqvist, L. Gunderson, C. S. Holling, B. Walker, J. Bengtsson, F. Berkes, J. Colding, K. Danell, M. Falkenmark, L. Gordon, R. Kasperson, N. Kautsky, A. Kinzig, S. Levin, K. Maler, F. Moberg, L. Ohlsson, P. Olsson, E. Ostrom, W. Reid, J. Rockstrom, H. H. G. Savenije, and U. Svedin. 2002. Resilience and sustainable development: Building adaptive capacity in a world of transformation. Scientific Background Paper on Resilience for the process of The World Summit on Sustainable Development, Ministry of the Environment, Stockholm, Sweden.

Folke, C., T. Hahn, P. Olsson, and J. Norberg. 2005. Adaptive governance of social-ecological systems. Annual Review of Environment and Resources 30:441–473.

Folke, C., C. S. Holling, and C. Perrings. 1996. Biological diversity, ecosystems and the human scale. Ecological Applications 6:1018–1024.

Folke, C., A. Jansson, J. Larsson, and R. Costanza. 1997. Ecosystem appropriation by cities. Ambio 26:167–172.

Folke, C., N. Kautsky, H. Berg, A. Jansson, and M. Troell. 1998. The ecological footprint concept for sustainable seafood production: A review. Ecological Applications 8:63–71.

Forman, R. T. T. 1995. Land Mosaics: The Ecology of Landscapes and Regions. Cambridge University Press, Cambridge.

Forman, R. T. T. 2000. Estimate of the area affected ecologically by the road system in the United States. Conservation Biology 14:31–35.

Forman, R. T. T., and L. E. Alexander. 1998. Roads and their major ecological effects. Annual Review of Ecology and Systematics 29:207–231.

Forman, R., and M. Godron. 1986. Landscape Ecology. John Wiley, New York.

Fortin, M.-J., B. Boots, F. Csillag, and T. K. Remmel. 2003. On the role of spatial stochastic models in understanding landscape indices in ecology. Oikos 102:203–212.

Foster, D., F. Swanson, J. D. Aber, I. Burke, N. Brokaw, D. Tilman, and A. Knapp. 2003. The importance of land-use legacies to ecology and conservation. BioScience 53(1):77–88.

Foster, S. S. D., and P. J. Chilton. 2003. Groundwater: the processes and global significance of aquifer degradation, Philosophical Transactions Royal Society London B 358:1935–1955.

Frank, J. E. 1989. The Cost of Alternative Development Patterns: A Review of the Literature. Urban Land Institute, Washington, DC.

Frank, L., B. Stone, Jr., and W. Bachman. 2000. Linking land use with household vehicle emissions in the central Puget Sound: Methodological framework and findings. Tranportation Research D 5:173–196.

Franklin, J. F. 1992. Scientific basis for new perspectives in forests and streams. Pages 25–72 in R. J. Naiman (ed.), Watershed Management. Springer-Verlag, New York.

Frick, E. A., D. J. Hippe, G. R. Buell, C. A. Couch, E. H. Hopkins, D. J. Wangsness, and J. W. Garrett. 1998. Water quality in the Apalachicola-Chattahoochee-Flint River basin, Georgia, Alabama, and Florida, 1992–95. U.S. Geological Survey Circular 1164.

Fudali, E. 2001. The ecological structure of the bryoflora of Wroclaw's parks and cemeteries in relation to their localization and origin. Acta Societatis Botanicorum Poloniae 70:229–235.

Fujita, M. 1989. Urban Economic Theory. Cambridge: Cambridge University Press.

Fujita, M., and H. Ogawa. 1982. Multiple equilibria and structural transition in non-monocentric urban configurations. Regional Science and Urban Economics 12:161–196.

Gardner, R. H. 1998. Pattern, process and the analysis of spatial scales. pp. 17–34 in D. L. Peterson, and V. T. Parker (eds.), Ecological Scale: Theory and Applications. Columbia University Press, New York.

Gardner, R. H., B. T. Milne, M. G. Turner, and R. V. O'Neill. 1987. Neutral models for the analysis of broad-scale landscape pattern. Landscape Ecology 1:19–28.

Garreau, J. 1991. Edge City: Life on the New Frontier. Anchor Books, New York, NY.

Gaston, K. J. 2005. Biodiversity and extinction: Species and people. Progress in Physical Geography 29:239–247.

Geddes, P. 1905. Civics as concrete and applied sociology. Sociological Papers 2:57–119

Geddes, P. 1915. City in Evolution. Williams and Norgate Ltd., London, UK.

Gedzelman, S. D., S. Austin, R. Cermak, N. Stefano, S. Partridge, S. Quesenberry, and D. A. Robinson. 2003. Mesoscale aspects of the urban heat island around New York City. Theoretical and Applied Climatology 75:29–42.

Gerritse, R. G., J. A. Adeney, G. M. Dommock, and Y. M. Oliver. 1995. Retention of nitrate and phosphate in soils of the Darling Plateau in Western Australia: Implications for domestic septic tank systems. Australian Journal of Soil Research 33:353–367.

Gilbert, O. L. 1989. The Ecology of Urban Habitats. Chapman & Hall, New York.

Gill, D., and Bonnett, P. 1973. Nature in the urban landscape: A study of urban ecosystems. Baltimore. York Press.

Giuliano, G., and K. A. Small. 1991. Subcenters in the Los Angeles region. Journal of Regional Science and Urban Economics 21:163–182.

Glaab, C. N., and A. T. Brown. 1967. A History of Urban America. Macmillan, New York.

Godron, M., and R. T. Forman. 1982. Landscape modification and changing ecological characteristics. Pages 12–28 in H. A. Mooney and M. Godron (eds.), Disturbance and Ecosystems: Components of Response. Springer-Verlag, New York.

Godron, M., and R. T. T. Forman. 1985. Landscape Ecology New York: Wiley.

Goldenfeld, N., and L. P. Kadanoff. 1999. Simple lessons from complexity. Science 284(5411):87–89.

Goldman, M. B., P. M. Groffman, R. V. Pouyat, M. J. McDonnell, and S. T. A. Pickett. 1995. Methane uptake and nitrogen availability in forest soils along an urban to rural gradient. Soil Biology and Biochemistry 27:281–286.

Goodchild, M. F. 2003. Geographic information science and systems for environmental management. Annual Review of Environment and Resources 28:493–519.

Goodwin, B. J., and L. Fahrig. 2002. How does landscape structure influence landscape connectivity? Oikos 99(3):552–570.

Gordon, N. D., T. A. McMahon, and B. L. Finlayson. 1992. Stream hydrology. An introduction for ecologists. John Wiley and Sons, Chichester, UK.

Gordon, P., A. Kumar, and H. Richardson. 1989. The influence of metropolitan spatial structure on commuting time. Journal of Urban Economics 226:128–151.

Gordon, P., and H. W. Richardson. 1997. Are compact cities a desirable planning goal? Journal of the American Planning Association 63:95–106.

Gordon, P., H. Richardson, and H. Wong. 1986. The distribution of population and employment in a polycentric city: The case of Los Angeles. Environment and Planning A 18:161–173.

Graedel, T. E., and B. R. Allenby. 1995. Industrial Ecology. Prentice Hall, Englewood Cliffs, NJ.

Graedel, T. E., M. Bertram, A. Kapur, B. Reck, and S. Spatari. 2004. Multilevel cycle of anthropogenic copper. Environmental Science & Technology 38(4):1242–1252.

Graf, W. L. 1977. Network characteristics in suburbanizing streams. Water Resources Research 13(2):459–463.

Graybill, J., S. Dooling, V. Shandas, J. Withey, A. Greve and G. Simon. 2006. A rough guide to interdisciplinarity: Graduate student perspectives. BioScience 56:757–763.

Green, D. G. 1989. Simulated effects of fire, dispersal and spatial pattern on competition within vegetation mosaics. Vegetation 82:139–153.

Green, D. G. 1993. Emergent behaviour in biological systems. Pages 24–35 in D. G. Green and T. J. Bossomaier (eds.), Complex Systems – From Biology to Computation. IOS Press, Amsterdam.

Green, D. G. 1994. Connectivity and complexity in ecological systems. Pacific Conservation Biology 1:194–200.

Green, D. G., A. Tridgell, and A. M. Gill. 1990. Interactive simulation of bushfire spread in heterogeneous fuel. Mathematical and Computer Modelling 13: 57–66.

Green, D. G., and S. Sadedin. 2005. Interactions matter: Complexity in landscapes and ecosystems. Ecological Complexity 2(2):117–130.

Gregg, J. W., C. G. Jones, and E. Dawson. 2003. Urbanization effects on tree growth in the vicinity of New York City. Nature 424:183–187.

Gregory, K. J. 1977. Channel and network metamorphosis in Northern New South Wales. Pages 389–410 in K. J. Gregory (ed.), River Channel Changes. Wiley and Sons, Chichester, UK.

Gregory, K. J., R. J. Davis, and P. W. Downs. 1992. Identification of river channel change to due to urbanization. Applied Geography 12:299–318.

Gregory, S. V., F. J. Swanson, W. A. McKee, and K. W. Cummins. 1991. An ecosystem perspective of riparian zones: focus on links between land and water. BioScience 41:540–551.

Greve, A. 2007. Toward a more complex understanding of urban stream function: Assessing post-developmental recovery period and channel morphology and the relationship between urban built form, land cover pattern, and hydrologic flow regime. Ph.D. Dissertation, Department of Urban Design and Planning, University of Washington, Seattle, WA, USA.

Grime, J. P. 1997. Biodiveristy and ecosystem function. Science 277:1260–1261.

Grime, J. P., V. K. Brown, K. Thompson, G. J. Masters, S. H. Hillier, I. P. Clarke, A. P. Askew, D. Corker, and P. Kielty. 2000. The response of two contrasting limestone grasslands to simulated climate change. Science 289:762–765.

Grimm, N. B., J. M. Grove, S. T. A. Pickett, and C. L. Redman. 2000. Integrated Approaches to Long-Term Studies of Urban Ecological Systems. BioScience 50:571–584.

Grimm, N. B., R. W. Sheibley, C. L. Crenshaw, C. N. Dahm, W. J. Roach, and L. H. Zeglin. 2005. Nutrient retention and transformation in urban streams. Journal of the North American Benthological Society 24(3):626–642.

Grimm, V., E. Revilla, U. Berger, F. Jeltsch, W. M. Mooij, S. F. Railsback, H.-H. Thulke, J. Weiner, T. Wiegand, and D. L. DeAngelis. 2005. Pattern-oriented modeling of agent-based complex systems: Lessons from ecology. Science 310:987–991.

Grimmond, C. S. B., and T. R. Oke. 1995. Comparison of heat fluxes from summertime observations in the suburbs of four North American cities. Journal of Applied Meteorology 34:873–889.

Grimmond, C. S. B., and T. R. Oke. 2000. A local-scale urban meteorological pre-processing scheme (LUMPS). Pages 73–83 in M. Schatzmann, J. Brechler, B. Fisher (eds.), Preparation of Meteorological Input Data for Urban Site Studies, Prague, Czech Republic, June 15, 2000.

Grimmond, C. S. B., and T. R. Oke. 2000. Variability of evapotranspiration rates in urban areas. Pages 475–480 in R. J. Dear, J. D. Kalma, T. R. Oke, and A. Auliciems (eds.), Biometeorology and Urban Climatology at the Turn of the Millennium. Selected Papers from the Conference ICB-ICU'99. WMO, Geneva.

Groffman, P. M., D. J. Bain, L. E. Band, K. T. Belt, G. S. Brush, J. M. Grove, R. M. Puouyat, I. C. Yesilonis, and W. C. Zipperer. 2003. Down by the riverside: Urban riparian ecology. Frontiers in Ecology and the Environment 1:315–321.

Groffman, P. M., N. J. Boulware, W. C. Zipperer, R. V. Pouyat, L. E. Band, and M. F. Colosimo. 2002. Soil nitrogen cycle processes in urban riparian zones. Environmental Science and Technology 36(21):4547–4552.

Groffman, P. M., and M. K. Crawford. 2003. Denitrification potential in urban riparian zones. Journal of Environmental Quality 32(3):1144–1149.

Groffman, P. M., A. M. Dorsey, and P. M. Mayer. 2005. N processing within geomorphic structures in urban streams. Journal of the North American Benthological Society 24(3):613–625.

Groffman, P. M., N. L. Law, K. T. Belt, L. E. Band, and G. T. Fisher. 2004. Nitrogen fluxes and retention in urban watershed ecosystems. Ecosystems 7(4):393–403.

Groffman, P. M., R. V. Pouyat, M. L. Cadenasso, W. C. Zipperer, K. Szlavecz, I. D. Yesilonis, L. E. Band, and G. S. Brush. 2006. Land use context and natural soil controls on plant community composition and soil nitrogen and carbon dynamics in urban and rural forests. Forest Ecology and Management 236(2–3):177–192.

Grove, J. M. 1996. The relationship between patterns and processes of social stratification and vegetation of an urban-rural watershed. Published Doctoral Dissertation, Yale University, New Haven, CT.

Grove, J. M., and W. R. Burch. 1997. A social ecology approach to applications of urban ecosystem and landscape analyses: A case study of Baltimore, Maryland. Urban Ecosystems 1:259–275.

Gunderson, L. H., 2000. Ecological resilience In theory and application. Annual Reviews of Ecological Systems 31:425–439.

Gunderson, L. H., and C. S. Holling (eds.), 2002. Panarchy: Understanding Transformations in Human and Natural Systems. Island Press, Washington, DC.

Gustafson, E. 1998. Quantifying landscape spatial patterns: What is the state of art? Ecosystems 1:143–156.

Gustafson, E. J., and G. R. Parker. 1992. Relationships between land cover proportion and indices of landscape spatial pattern. Landscape Ecology 7:101–110.

Haagen-Smit, A. J., C. E. Bradley and M. M. Fox. 1953. Ozone formation in photochemical oxidation of organic substances. Industrial and Engineering Chemistry 45:2086–2089.

Hachmoller, B., R. A. Matthews, and D. F. Brakke. 1991. Effects of riparian community structure, sediment size, and water-quality on the macroinvertebrate communities in a small, suburban stream. Northwest Science 65(3):125–132.

Hall, M. J. 1984. Urban Hydrology. Elsevier Applied Science Publishers, New York.

Hallsworth, E. G. 1978. Benefits and Costs of Land Resource Survey and Evaluation. Commonwealth and State Government Collaborative Soil Conservation Study 1975–77, Report 5, AGPS, Canberra.

Hamm, B. 1982. Hamm, B. 1982. Social area analysis and factorial ecology: A review of substantive findings. Pages 316–337 in A. Theodorson (ed.), Urban Patterns: Studies in Human Ecology. Pennsylvania State University Press, University Park.

Hammer, T. R. 1972. Stream and channel enlargement due to urbanization. Water Resources Research 8:1530–1540.

Hansen, A. J., R. L. Knight, J. M. Marzluff, S. Powell, K. Brown, P. H. Gude, and K. Jones. 2005. Effects of exurban development on biodiversity: Patterns, mechanisms, and research needs. Ecological Applications 15:1893–1905.

Hansen, A. J., R. Rasker, B. Maxwell, J. J. Rotella, A. Wright, U. Langner, W. Cohen, R. Lawrence, and J. Johnson. 2002. Ecology and socio-economics in the New West: A case study from Greater Yellowstone. BioScience 52:151–168.

Hargis, C. D., J. A. Bissonette, and J. L. David. 1998. The behavior of landscape metrics commonly used in the study of habitat fragmentation. Landscape. Ecology 13:167–186.

Harris, B. 1962. Linear programming and the projection of land uses. Paper 20, Penn-Jersey Transportation Study, Philadelphia.

Harris, B. 1994. The real issues concerning Lee's "Requiem." Journal of the American Planning Association 60:31–34.

Harris, L. 1984. The Fragmented Forest: Application of Island Biogeography Principles to Preservation of Biotic Diversity. University of Chicago Press, Chicago.

Harrison, S. 1991. Local extinction in a metapopulation context: An empirical evaluation. Biological Journal of the Linnaean Society. 42:73–88.

Hartley, D. M., C. R. Jackson, and G. Lucchetti. 2001. Discussion: Stream health after urbanization. Journal of the American Water Resources Association 37:751–753.

Hartvigsen, G., A. Kinzig, and G. Peterson. 1998. Use and analysis of complex adaptive systems in ecosystem science: Overview of special section. Ecosystems 1:427–430.

Harvey, D. 1973. Social Justice and the City. Arnold, London.

Haskell, D. G. 2000. Effects of forest roads on macroinvertebrate soil fauna of the Southern Appalachian mountains. Conservation Biology. 14:57–63.

Hauglustaine, D. A., J. Lathiere, S. Szopa, and G. A. Folberth. 2005. Future tropospheric ozone simulated with a climate-chemistry-biosphere model. Geophysical Research Letters 32, L24807, doi:10.1029/2005 GL024031.

He, F., and P. Legendre. 1996. On species–area relations. American Naturalist 148:719–737.

Hemmens, G. 1967. Experiments in urban form and structure. Highway Research Record 207:32–41.

Henderson, J. V., and E. Slade. 1993. Development games in non-monocentric cities. Journal of Urban Economics 34:207–229.

Henein, K., and G. Merriam. 1990. The elements of connectivity where corridor quality is variable. Landscape Ecology 4(2–3):157–170.

Henshaw, P. C., and D. B. Booth. 2000. Natural restabilization of stream channels in urban watersheds. Journal of the American Water Resources Association. 36:1219–1236.

Hepinstall, J. A., M. Alberti, and J. M. Marzluff. In review. Predicting land cover change and avian community responses in rapidly urbanizing environments. Submitted to Landscape Ecology.

Hepinstall, J. A., J. M. Marzluff, and M. Alberti. In Press. Predicting avian community responses to increasing urbanization. In C. A. Lipczyk, and

P. S. Warren (eds.), New Directions in Urban Bird Ecology and Conservation. Studies in Avian Biology.

Herbert, J. D., and B. H. Stevens. 1960. A model for the distribution of residential activity in urban areas. Journal of Regional Science 2:21–36.

Herold, M., H. Coucleis, and K. C. Clarke. 2005. The role of spatial metrics in the analysis and modeling of urban land use change. Computers, Environment and Urban Systems 29:369–399.

Herold, M., Liu, X., and Clarke, K. C. 2003. Spatial metrics and image texture for mapping urban land use. Photogrammetric, Engineering and Remote Sensing, 69(8):991–1001.

Hess, P. M., A. V. Moudon, and M. Logsdon. 2001. Measuring land use patterns. Transportation Research Record 1(780):17–24.

Hewitt, J. E., S. F. Thrush, P. K. Dayton, and E. Bonsdorff. 2007. The effect of spatial and temporal heterogeneity on the design and analysis of empirical studies of scale-dependent systems. The American Naturalist 169(3):398–408.

Hill, K., E. Botsford, and D. Booth. 2002. A rapid land cover classification method for use in urban watershed analysis. University of Washington Department of Civil and Environmental Engineering, Water Resources Series Technical Re-port No. 173.

Hippocrates, Ca. 5th century B. C. E. Airs, waters, places. In Hippocrates Vol 1 The Loeb Classical Library, ed., T E Page. Cambridge, MA Harvard University Press, 1962.

Hoare, R. A. 1984. Nitrogen and phosphorus in Rotorua urban streams. New Zealand Journal of Marine and Freshwater Research 18:451–454.

Hoffmann, M., H. Kelley, and T. P. Evans. 2002. Simulating land-cover change in Indiana: an agent-based model of dereforestation. Pages 218–247 in M. A. Janssen (ed.), Complexity and Ecosystem Management: The Theory and Practice of Multi-Agent Systems. Edward Elgar Publishers.

Hogan, A. W., and M. G. Ferrick. 1998. Observations in nonurban heat islands. Journal of Applied Meteorology 37(2):232–236.

Hogeweg, P., and B. Hesper. 1990. Crowns crowding: an individual oriented model of the Acanthaster phenomenon. Pages 169–188 in R. H. Bradbury (ed.), Acanthaster and the Coral Reef: a Theoretical Perspective. Springer, Berlin.

Hogrefe, C., B. Lynn, K. Civerolo, J.-Y. Ku, J. Rosenthal, C. Rosenzweig, R. Goldberg, S. Gaffin, K. Knowlton, and P. L. Kinney, 2004. Simulating changes in regional air pollution over the eastern United States due to changes in global and regional climate and emissions, Journal of Geophysical Research 109, D22301, doi:10.1029/2004JD004690.

Holling, C. S. 1973. Resilience and stability of ecological systems. Annual Review of Ecology and Systematics 4:1–23.

Holling, C. S. 1978a. Adaptive Environmental Assessment and Management. John Wiley and Sons, London.

Holling, C. S. 1978b. The Spruce-Budworm/forest-management problem. Pages 143–82 in C. S. Holling (ed.), Adaptive Environmental Assessment and Management. International Series on Applied Systems Analysis 3. John Wiley & Sons.

Holling, C. S. 1992. Cross-scale morphology, geometry, and dynamics of ecosystems. Ecological Monographs 62:447–502.

Holling, C. S. 1996. Surprise for science, resilience for ecosystems, and incentives for people. Ecological Applications 6:733–735.

Holling, C. S. 2001. Understanding the complexity of economic, ecological, and social systems. Ecosystems 4:390–405.

Holling, C. S., S. Carpenter, W. Brock, and L. H. Gunderson 2002a. Discoveries for sustainable futures. Pages 395–417 in L. H. Gunderson, and C. S. Holling (eds.), Panarchy: Understanding Transformations in Human and Natural Systems. Island Press, Washington.

Holling, C. S., and L. H. Gunderson. 2002. Resilience and adaptive cycles. Pages 25–62 in L. H. Gunderson, and C. S. Holling (eds.), Panarchy: Understanding Transformations in Human and Natural Systems. Island Press, Washington, DC.

Holling, C. S., L. H. Gunderson, and D. Ludwig. 2002b. In quest of a theory of adaptive change. Pages 3–24 in L. H. Gunderson, and C. S. Holling (eds.), Panarchy: Understanding Transformations in Human And Natural Systems. Island Press,Washington, DC.

Holling, C. S., and G. Orians. 1971. Toward an Urban Ecology. Ecological Society of America Bulletin 52.2–6.

Holling, C. S., D. W. Schindler, B. W. Walker, and J. Roughgarden. 1995. Biodiversity in the functioning of ecosystems: An ecological synthesis. Pages 44–83 in C. Perrings, K.-G. Mäler, C. Folke, C. S. Holling, and B.-O. Jansson (eds.), Biodiversity Loss: Economic and Ecological Issues. Cambridge University Press, New York.

Hollis, G. E. 1975. The effects of urbanization on floods of different recurrence intervals. Water Resources Research 11:431–435.

Holt, R. D., and J. Pickering. 1985. Infectious diseases and species coexistence: A model of Lotka-Volterra form. American Naturalist 726:196–211.

Holzer, M., and G. J. Boer. 2001. Simulated changes in atmospheric transport climate. Journal Climate 14:4398–4420.

Hooper, D. U., F. S. Chapin III, J. J. Ewel, A. Hector, P. Inchausti, S. Lavorel, J. H. Lawton, D. M. Lodge, M. Loreau, S. Naeem, B. Schmid, H. Setälä, A. J. Symstad, J. Vandermeer, and D. A. Wardle. 2005. Effects of biodiversity on ecosystem functioning: A consensus of current knowledge. Ecological Monographs 75:3–35.

Hope, D., C. Gries, W. Zhu, W. F. Fagan, C. L. Redman, N. B. Grimm, A. L. Nelson, C. Martin, and A. Kinzig. 2003. Socioeconomics drive urban plant diversity. Proceedings of the National Academy of Sciences of the United States of America 100:8788–8792.

Horbert, M., H. P. Blume, H. Elvers, and H. Sukkopp. 1982. Ecological contribution to urban planning. Pages 255–275 in R. Bornkamm, J. A. Lee, and M. R. D. Seward (eds.), Urban Ecology: 2nd European Ecological Symposium. Blackwell Scientific Publications, Oxford, UK.

Horner, R. R., D. B. Booth, A. Azous, and C. W. May. 1997. Watershed determinants of ecosystem functioning. Pages 251–274 in C. Roessner (ed.). Effects of Watershed Development and Management on Aquatic Ecosystems. American Society of Civil Engineers, New York.

Horowitz, A. J., M. Meybeck, Z. Idlafkih, and E. Biger. 1999. Variations in trace element geochemistry in the Seine River Basin based on flood-plain deposits and bed sediments. Hydrological Processes 13:1329–1340.

Horwitz, R. J. 1978. Temporal variability patterns and the distributional patterns of stream fishes. Ecological Monographs 11:307–321.

Hough, M. 1984. City Form and Natural Process – Towards a New Urban Vernacular. Croom Helm, London, UK.

Hough, M. 1995. Cities and Natural Processes. Routledge, London.

Howard, E. 1898. Garden Cities of Tomorrow. 2nd ed., 1902. Sonnenschein, London.

Howard, L. 1833. The Climate of London, vols. I–III. Harvey and Darton, London.

Howarth, R. W., G. Billen, D. Swaney, A. Townsend, N. Jaworski, K. Lajtha, A. Downing, R. Elmgreen, N. Caraco, T. Jordan, F. Berendse, J. Freney, V. Kudeyarov, P. Murdoch, and Z. Zhao-liang. 1996. Regional nitrogen budgets and riverine N & P fluxes for the drainages to the North Atlantic Ocean: Natural and human influences. Biogeochemistry 35:181–226.

Hubbell, S. P. 2001. The Unified Neutral Theory of Biodiversity and Biogeography. Princeton University Press, Princeton, NJ.

Hughes, T. P. 1994. Catastrophes, phase shifts, and large-scale degradation of a Caribbean coral reef. Science 265:1547–1551.

Hultman, J. 1991. Miljoproblematikens geografistad, land och kretslopp. Rapporter och Notiser 103. Inst. f. Kulturgeografi och Ekonomisk Geografi, Lund Univ.

Hultman, J. 1993. Approaches and methods in urban ecology. Geografiska Annaler 75B(1):41–49.

Hunsaker, C. T., D. A. Levine, S. P. Timmons, B. L. Jackson, and R. V. O'Neill. 1992. Landscape characterization for assessing regional water quality. Pages 997–1006 in D. H. McKenzie, D. E. Hyatt, and V. J. McDonald (eds.). Ecological Indicators. Elsevier Applied Science, New York.

Husté, A. and T. Boulinier. 2007. Determinants of local extinction and turnover rates in urban bird communities. Ecological Applications 17:168–180.

Huston, M. 1997. Hidden treatments in ecological experiments: Re-evaluating the ecosystem function of biodiversity. Oecologia 110:449–460.

Hutchinson, G. E. 1957. Concluding remarks. Cold Spring Harbor Symposium on Quantitative Biology 22:415–427.

Imhoff, M. L., L. Bounoua, R. S. DeFries, W. T. Lawrence, D. Stutzer, J. T. Compton, and T. Ricketts. 2004. The consequences of urban land transformations on net primary productivity in the United States. Remote Sensing of the Environment 89:434–443.

Imhoff, M. L., L. Bounoua, T. Ricketts, C. Loucks, R. Harriss, and W. T. Lawrence. 2004. Global patterns in human consumption of net primary production. Nature 429:870–873.

Imhoff, M. L., L. Bounoua, T. Tucker, R. DeFries, and T. Ricketts. 2002. How has urbanization altered the carbon cycle in the United States? Poster, AGU, San Francisco, December 6.

Imhoff, M. L., W. T. Lawrence, C. D. Elvidge, T. Paul, E. Levine, M. Prevalsky, and V. Brown. 1997. Using nighttime DMSP/OLS images of city lights to estimate the impact of urban land use on soil resources in the United States. Remote Sensing of the Environment 59:105–117.

Imhoff, M. L., C. J. Tucker, W. T. Lawrence, and D. C. Stutzer. 2000. The use of multisource satellite and geospatial data to study the effect of urbanization on primary productivity in the United States. IEEE Transactions on Geoscience and Remote Sensing 38(6):2549–2556.

Innes, J. E., and D. E. Booher. 1999. Metropolitan development as a complex system: A new approach to sustainability. Economic Development Quarterly 13:141–156.

Inwood, S. E., J. L. Tank, and M. J. Bernot. 2005. Patterns of denitrification associated with land use in 9 midwestern headwater streams. Journal of the North American Benthological Society 24:227–245.

IPCC. 2007. Climate Change 2007: The Physical Science Basis. Contribution of Working Group I to the Fourth Assessment Report of the Inter-governmental Panel on Climate Change. Solomon, S., D. Qin, M. Manning, Z. Chen, M. Marquis, K. B. Averyt, M. Tignor, and H. L. Miller (eds.). Cambridge University Press, Cambridge, United Kingdom and New York, NY, USA, 996 pp.

Ironmonger, D. S., C. K. Aitken, and B. Erbas. 1995. Economies of scale in energy use in adult-only households. Energy Economics 17:301–310.

Ives, A. R., and S. R. Carpenter. 2007. Stability and diversity of ecosystems. Science 317:58–62.

Jackson, J. B. C., M. X. Kirby, W. H. Berger, K. A. Bjorndal, L. W. Botsford, B. J. Bourque, R. H. Bradbury, R. Cooke, J. Erlandson, J. A. Estes, T. P. Hughes, S. Kidwell, C. B. Lange, H. S. Lenihan, J. M. Pandolfi, C. H. Peterson, R. S. Steneck, M. J. Tegner, and R. R. Warner. 2001. Historical overfishing and the recent collapse of coastal ecosystems. Science 293:629–638.

Jacobs, J. 2000. The Nature of Economies. Random House, New York.

Jacobson, R. B., S. R. Femmer, and R. A. McKenney. 2001. Land-use changes and the physical habitat of streams – A review with emphasis on studies within the US Geological Survey Federal-State Cooperative Program. USGS Circular 1175.

Janssen, M. A., and W. Jager. 2000. The human actor in ecological economic models. Ecological Economics 35(3):307–310.

Jansson, A., C. Folke, J. Rockstrom, and L. Gordon. 1999. Linking freshwater flows and ecosystem services appropriated by people: The case of the Baltic Sea drainage basin. Ecosystems 2:351–366.

Jenks, M., E. Burton, and K. Williams (eds.), 1996. The Compact City: A Sustainable Urban Form? E & FN Spon, an imprint of Chapman and Hall, London.

Jo, H.-K., and G. E. McPherson. 1995. Carbon storage and flux in urban residential greenspace. Journal of Environmental Management 45:109–133.

Johnson, D. W., H. Van Miegroet, S. E. Lindberg, D. E. Todd, and R. B. Harison. 1991. Nutrient cycling in red spruce forests of the Great Smoky Mountains. Canandian Journal of Forest Research 21:769–787.

Johnson, K. H., K. A. Vogt, H. J. Clark, O. J. Schmitz, and D. J. Vogt. 1996. Biodiversity and the productivity and stability of ecosystems. Trends in Ecology and Evolution 11:372–377.

Johnson, L. B., C. Richards, G. E. Host, and J. W. Arthur. 1997. Landscape influences on water chemistry in Midwestern stream ecosystems. Freshwater Biology 37:193–208.

Jokimaki, J., and J. Suhonen. 1993. Biological integrity: A long-neglected aspect of water resource management. Ecological Applications 1:66–84.

Jones, C. G., J. H. Lawton, and M. Shachak. 1994. Organisms as ecosystem engineers. Oikos 69:373–386.

Jones, J. A., F. J. Swanson, B. C. Wemple, and K. U. Snyder. 2000. Effects of roads on hydrology, geomorphology, and disturbance patches in stream networks. Conservation Biology 14(1):76–85.

Jones, K., L. Militana, and J. Martini. 1989. Ozone trend analysis for selected urban areas in the continental U.S. Proceedings of the 82nd Annual Meeting of the Air & Waste Management Association, Anaheim, CA.

Jorgensen, S. E. 1997. Integration of Ecosystem Theories: A Pattern. Kluwer Academic Publishers, Boston, USA.

Kaika, M. 2005. City of flows: Modernity, nature, and the city. Routledge, London.

Kalma, J. D., M. Johnson, and K. J. Newcombe. 1978. Energy use and atmospheric environment in Hong Kong. Part II. Waste heat, land use and urban climate. Urban Ecology 3:59–83.

Karr, J. R. 1991. Biological integrity: A long-neglected aspect of water resource management. Ecological Applications 1:66–84.

Karr, J. R. 1995. Clean water is not enough. Illahee 11:51–59.

Karr, J. R. 1998. Rivers as sentinels: Using the biology of rivers to guide landscape management. Pages 502–528 in R. J. Naiman and R. Bilby (eds.), River Ecology and Management: Lessons from the Pacific Coastal Ecoregion. Springer, New York.

Karr, J. R., and I. J. Schlosser. 1978. Water resources and the land-water interface. Science 20:229–234.

Kates, R. W., B. L. Turner II, and W. C. Clark. 1990. The great transformation. Pages 1–17 in B. L. Turner, II, W. C. Clark, R. W. Kates, J. E. Richards,

J. T. Mathews, and W. B. Meyer (eds.), The Earth as Transformed by Human Action. Cambridge University Press, Cambridge, UK.

Kauffman, S. A. 1993. The Origins of Order: Self-Organization and Selection in Evolution. Oxford University Press, New York.

Kaye, J. P., P. M. Groffman, N. B. Grimm, L. A. Baker, and R. V. Pouyat. 2006. A distinct urban biogeochemistry? Trends in Ecology & Evolution 21:192–199.

Kaye, J. P., R. L. McCulley, and I. C. Burke. 2005. Carbon fluxes, nitrogen cycling, and soil microbial communities in adjacent urban, native and agricultural ecosystems. Global Change Biology 11:575–587.

Keitt, T. H., and D. L. Urban. 2005. Scale-specific inference using wavelets. Ecology 86(9):2497–2504.

Keitt, T. H., D. L. Urban, and B. T. Milne. 1997. Detecting critical scales in fragmented landscapes. Conservation Ecology [online] 1:4.

Kelly, N. A., M. A. Ferman, and G. T. Wolff. 1986. The chemical and meteorologcal conditions associated with high and low ozone concentrations in southeastern Michigan and nearby areas of Ontario. Journal of the Air Pollution Control Association 36:150–158.

Kennedy, C., J. Cuddihy, and J. Engel-Yan. 2007. The changing metabolism of cities. Journal of Industrial Ecology 11(2):43.

Keyes, D. L. 1982. Reducing travel and fuel through urban planning. Pages 214–232 in R. W. Burchell, and D. Listokin (eds.). Energy and Land Use. Rutgers University Press, New Brunswick, NJ.

Keyes, D. L., and G. Peterson. 1977. Urban development and energy consumption. WP-5049–1.5. The Urban Land Institute, Washington, DC.

Khamer, M., D. Bouya, and C. Ronneau. 2000. Metallic and organic pollutants associated with urban wastewater in the waters and sediments of a Moroccan river. Water Qual. Research Journal of Canada 35:147–61.

Kidder, S. Q., and O. M. Essenwanger. 1995. The effect of clouds and wind on the difference in nocturnal cooling rates between urban and rural areas. Journal of Applied Meteorology 34:2440–2448.

Kim, S. 2007. Changes in the nature of urban spatial structure in the United States, 1890–2000. Journal of Regional Science 47:2 273.

Kim, T. J. 1989. Integrated Urban System Modeling: Theory and Practice. Martinus Nijhoff, Norwell, MA.

Kitchell, J. F., and S. R. Carpenter. 1993. Variability in lake ecosystems: Complex responses by the apical predator. Pages 111–124 in

M. McDonnell, and S. Pickett (eds.). Humans as Components of Ecosystems. Springer-Verlag, New York.

Klausnitzer, B. 1993. Ökologie der Großstadtfauna. 2nd Edition. Auflage. Gustav Fischer, Stuttgart.

Klein, R. D. 1979. Urbanization and stream quality impairment. Water Resources Bulletin 15:948–963.

Knowlton, K., J. E. Rosenthal, C. Hogrefe, B. Lynn, S. Gaffin, R. Goldberg, C. Rosenzweig, K. Civerolo, J.-Y. Ku, and P. L. Kinney. 2004. Assessing ozone-related health impacts under a changing climate. Environmental Health Perspectives 112(15):1557–1563.

Knutson, K. L., and V. L. Naef. 1997. Management Recommendations for Washington's Priority Habitats: Riparian. Washington Department of Fish and Wildlife, Olympia, WA.

Kolasa, J., and S. T. A. Pickett. 1991. Ecological Heterogeneity. Springer-Verlag, New York.

Konrad, C. P. 2000. The Frequency and Extent of Hydrologic Disturbances in Stream in the Puget Lowland, Washington. Ph.D. Dissertation. Department of Civil Engineering, University of Washington, Seattle.

Konrad, C. P. 2003. Effects of Urban Development on Floods. U. S. Geological Survey Fact Sheet FS-076-03, Tacoma, WA. [online] http://pubs.usgs.gov/fs/fs07603/.

Konrad, C. P., and D. Booth. 2002. Hydrologic trends resulting from urban development in western Washington streams. USGS Water-Resources Investigation Report 02–4040, University of Washington. Seattle.

Konrad, C. P., and D. B. Booth. 2005. Hydrologic changes in urban streams and their ecological significance. American Fisheries Society Symposium 47:157–177.

Konrad, C. P., D. B. Booth, and S. J. Burges. 2005. Effects of urban development in the Puget Lowland, Washington, on interannual stream-flow patterns: Consequences for channel form and streambed disturbance. Water Resources Research 41(7), W07009, doi:10.1029/2005WR004097.

Korzun, V. I., A. A. Sokolov, M. I. Budyko, K. P. Voskresensky, P. Kalinin, A. A. Konoplyantsev, E. S. Korotkevich, and M. I. L'vovitch, eds. 1978. Atlas of world water balance. USSR National Committee for the International Hydrological Decade. English translation. Paris, UNESCO.

Kowarik, I. 1990. Some responses of flora and vegetation to urbanization in Central Europe. Pages 45–74 in H. Sukopp, S. Mejny, and I. Kowarik (eds.), Urban Ecology: Plants and Plant Communities in Urban Environments. SPB Academic, The Hague.

Kowarik, I. 1995. On the role of alien species in urban flora and vegetation. Pages 85–103 in P. Pyšek, K. Prach, M. Remánek, and M .Wade (eds.). Plant Invasions: General Aspects and Special Problems. SPB Academic Publishing, Amsterdam, The Netherlands.

Krugman, P. 1993. The narrow and broad arguments for free trade. American Economic Review 83:362–366.

Krugman, P. 1995. Development, Geography, and Economic Theory. MIT Press, London.

Krugman, P. 1996. The Self-Organizing Economy. Blackwell Publishing Inc., Malden, MA.

Krugman, P. 1998. Space: The final frontier. Journal of Economic Perspectives 12(2):161–174.

Kühn, I., R. Brandl, and S. Klotz. 2004. The flora of German cities is naturally species rich. Evolutionary Ecology Research 6:749–764.

Lamberti, A., and V. H. Resh. 1985. Comparability of introduced tiles and natural substrates for sampling lotic bacteria, algae and macroinvertebrates. Freshwater Biology 15:21–30.

Lammert, M., and D. Allan. 1999. Assessing biotic integrity of streams: Effects of scale in measuring the influence of land use/cover and habitat structure on fish and microinvertebrates. Environmental Management 23:257–270.

Landis, J. D., and M. Zhang. 1998a. The second generation of the California urban futures model. Part I: Model logic and theory. Environment and Planning B: Planning and Design 25(6):657–666.

Landis, J. D., and M. Zhang. 1998b. The second generation of the California urban futures model. Part II: Specification and calibration results of the land-use change submodel. Environment and Planning B: Planning and Design 25(6):795–824.

Landsberg, H. E. 1979. Atmospheric changes in a growing community, Urban Ecology 4:53–82.

Landsberg, H. E. 1981. The Urban Climate. Academic Press, New York.

LaValle, P. D. 1975. Domestic sources of stream phosphates in urban streams. Water Research 9:913–15.

Lavelle, P., E. Blanchart, A. Martin, S. Martin, A. Spain, F. Toutain, I. Barois, and R. Schaefer. 1993. A hierarchical model for the decomposition in terrestrial ecosystems: Application to the soils of the humid tropics. Biotropica 25:130–150.

Lavelle, P., R. Dugdale, and R. Scholes (Lead Authors) 2005. Nutrient Cycling, Millennium Ecosystem Assessment. Chapter 12. Island Press, Washington, DC.

Law, N., L. E. Band, and J. M. Grove. 2004. Nitrogen input from residential lawn care practices in suburban watersheds in Baltimore county, MD. Journal of Environmental Planning and Management 47:737–755.

Lawton, J. H. 1994. What do species do in ecosystems? Oikos 71:367–374.

Lazaro, T. R. 1990. Urban Hydrology, a Multidisciplinary Perspective. Technomic Publishing Company, Inc. Lancaster, PA.

Le Comte, D. M., and H. E. Warren. 1981. Modeling the impact of summer temperatures on national electricity consumption. Journal of Applied Meteorology 20:1415–1419.

Ledig, F. T. 1992. Human impacts on genetic diversity in forest ecosystems. Oikos 63:87−108.

Lefohn, A. S., J. D. Husar, and R. B. Husar. 1999. Estimating historical anthropogenic global sulfur emission patterns for the period 1850–1990. Atmospheric Environment 33:3435–3444.

Lenat, D. R., and J. K. Crawford. 1994. Effects of land use onwater quality and aquatic biota of three North Carolina Piedmont streams. Hydrobiologia 294:185–99.

Leontief, W. 1967. Input-Output Economics. Oxford University Press, New York.

Leopold, L. B. 1968. Hydrology for Urban Planning—A Guidebook on the Hydrologic Effects of Urban Land Use. Circular 554, U.S. Geological Survey, Washington, DC.

Leopold, L. B. 1972. Hydrologic research on instrumented watersheds. Symposium of Wellington: Results of Research on Reporesentative and Experimental Basins, IASH Pub. No. 97:150–162.

Leopold, L. B., M. G. Wolman, and J. P. Miller. 1964. Fluvial Processes in Geomorphology. W.H. Freeman, San Francisco, CA.

Leung, L. R., and W. I. Gustafson, Jr. 2005. Potential regional climate and implications to U.S. air quality. Geophysical Research Letters 32, L16711, doi:10.1029/2005GL022911.

LeVeen, E. P., and W. R. Z. Willey. 1983. A political economic analysis of urban pest management. pp. 19–40 in G. W. Frankie, and C. S. Kohler (eds.). Urban Entomology: Interdisciplinary Perspectives. Praeger, New York.

Levenson, J. 1981. Woodlots as biogeographic islands in Southern Wisconsin. Pages 13–39 in R. Burgess, and D. Sharpe (eds.), Forest Island Dynamics in Man-Dominated Landscapes. Springer-Verlag, New York.

Levin, S. A. 1970. Community equilibria and stability, and an extension of the competitive exclusion principle. American Naturalist 104:413–423.

Levin, S. A. 1992. The problem of pattern and scale in ecology. Ecology 73:1943–1967.

Levin, S. A. 1998. Ecosystems and the biosphere as complex adaptive systems. Ecosystems 1:431–436.

Levin, S. A. 1999. Fragile Dominion: Complexity and the Commons. Perseus, Cambridge, UK.

Levin, S. A., and R. T. Paine. 1974. Disturbance, patch formation and community structure. Proceedings of the National Academy of Science 71:2744–2747.

Levins, R. 1970. Extinction. Lecture Notes in Mathematics 2:75–107.

Li, H., and J. F. Reynolds. 1993. A new contagion index to quantify spatial patterns of landscapes. Landscape Ecology 8:155–168.

Li, H., and J. F. Reynolds. 1994. A simulation experiment to quantify spatial heterogeneity in categorical maps. Ecology 75:2446–2455.

Liebhold, A. M., and J. Gurevitch. 2002. Integrating the statistical analysis of spatial data in ecology. Ecography 25:553–557.

Liebrand, W. B. G., A. Nowak, and R. Hegselmann, eds. 1988. Computer modeling of social processes. Sage Publications, London, UK.

Likens, G. E. 1998. Limitations to intellectual progress in ecosystem science. Pages 247–271 in M. L. Pace, and P. M. Groffman (eds.). Successes, Limitations and Frontiers in Ecosystem Science. Springer-Verlag, Millbrook, NY.

Likens, G. E. 2001. Biogeochemistry, the watershed approach: Some uses and limitations. Marine and Freshwater Research 52(1):5–12.

Limburg, K. E., and R. E. Schmidt. 1990. Patterns of fish spawning in Hudson River tributaries: Response to an urban gradient? Ecology 71:1238–1245.

Lindgren, M., and H. Bandhold. 2003. Scenario Planning. The Link Between Future and Strategy. Palgrave MacMillan, Basingstoke, Hampshire.

Liu, J., T. Dietz, S. Carpenter, M. Alberti, C. Folke, E. Moran, A. Pell, P. Deadman, T. Kratz, J. Lubchenco, E. Ostrom, Z. Ouyang, W. Provencher, C. Redman, S. Schneider, and W. Taylor. 2007. Complexity of coupled human and natural systems. Science 317:1513–1516.

Liu, J. G., G. C. Daily, P. R. Ehrlich, and G. W. Luck. 2003. Effects of household dynamics on resource consumption and biodiversity. Nature 421:530–533.

Logan, J., and H. Molotch. 1987. Urban Fortunes: The Political Economy of Place. University of California Press, Berkeley and Los Angeles.

Loreau, M. 2000. Biodiversity and ecosystem functioning: Recent theoretical advances. Oikos 91:3–17.

Loreau, M. 2001. Biodiverisity and ecosystem functioning: Recent theoretical advances. Oikos 91:3–17.

Loreau, M., S. Naeem, P. Inchausti, J. Bengtsson, J. P. Grime, A. Hector, D. U. Hooper, M. A. Huston, D. Raffaelli, B. Schmid, D. Tilman, and D. A. Wardle. 2001. Biodiversity and ecosystem functioning: Current knowledge and future challenges. Science 294:804–808.

Lovett, G. M., C. G. Jones, M. G. Turner, and K. C. Weathers. 2005. Conceptual frameworks: plans for a half-built house. Pages 463–470 in: G. M. Lovett, C. G. Jones, M. G. Turner and K. C. Weathers, eds. Ecosystem Function in Heterogeneous Landscapes. Springer, New York.

Lovett, G. M., K. C. Weathers, and W. V. Sobczak. 2000. Nitrogen saturation and retention in forested watersheds of the Catskill Mountains, New York. Ecological Applications 10:73–84.

Lowry, I. S. 1964. A Model of Metropolis. RM-4035-RC. The RAND Corporation, Santa Monica, California.

Luck, G. W. 2007. The relationships between net primary productivity, human population density and species conservation. Journal of Biogeography 34: 201–212.

Luck, G. W., T. H. Ricketts, G. C. Daily, and M. Imhoff. 2004. Alleviating spatial conflict between people and biodiversity. Proceedings of the National Academy of Sciences of the United States of America 101:182–186.

Luck, M., and J. Wu. 2002. A gradient analysis of the landscape pattern of urbanization in the Phoenix metropolitan area of USA. Landscape Ecology 17:327–339.

Luck, M. A., G. D. Jenerette, J. Wu, and N. B. Grimm. 2001. The urban funnel model and the spatially heterogeneous ecological footpring. Ecosystems 4(8):782–796.

Ludwig, D., B. H. Walker, and C. S. Holling. 2002. Models and Metaphors of Sustainability, Stability and Resilience, in Resilience and the Behavior of Large-Scale Systems. Island Press, Washington, DC.

Ludwig, D., D. Jones, and C. S. Holling. 1978. Qualitative analysis of insect outbreak systems: The spruce budworm and the forest. Journal of Animal Ecology 47:315–332.

Lyle, J. T. 1985. Design for Human Ecosystem. Van Nostrand Reinhold, New York.

Lynch, K. 1961. The patterns of the metropolis. Daedalus 90:79–98.

Lynch, K. 1981. Good City Form. MIT Press, Cambridge, MA.

Lynch, K., and L. Rodwin. 1958. A Theory of Urban Form. Journal of American Institute of Planners 24, No 4, pp. 201–214.

MacArthur, R. H. 1969. Patterns of communities in the tropics. Biological Journal of the Linnean Society of London 1:19–30.

MacArthur, R. 1970. Species packing and competitive equilibrium for many species. Theoretical Population Biology 1:1–11.

MacArthur, R. H., and J. W. MacArthur. 1961. On bird species diversity. Ecology 42:594–598.

MacArthur, R. H., and E. O. Wilson. 1963. An equilibrium theory of insular zoogeography. Evolution 17:373–387.

MacArthur, R. H., and E. O. Wilson. 1967. The Theory of Island Biogeography. Princeton University Press, Princeton, NJ.

MacKellar, F. L., W. Lutz, C. Prinz, and A. Goujon. 1995. Population, households, and $CO_2$ emissions. Population and Development Review 21:849–865.

Machlis, G. E., J. E. Force, and W. R. Burch, Jr. 1997. The human ecosystem part I: The human ecosystem as an organising concept in ecosystem management. Society and Natural Resources 10(4):347–368.

Mackett, R.L. 1985. Integrated land use-transport models. Transportation Reviews 5:325–343.

Mackett, R. L. 1990. MASTER Model. (Micro-Analytical Simulation of Transport, Employment and Residence). Report SR 237. Transport and Road Research Laboratory. Crowthorne, England.

Mackett, R. L. 1994. The use of land-use transportation models for policy analysis. Paper at the 73rd Annual Transportation Research Board Meetings, January 9–13, Washington, DC.

MacRae, C. R. 1996. Experience from morphological research on Canadian streams: Is control of the two-year frequency runoff event the best basis for stream channel protection? Pages 144–160 in Effects of Watershed Development and Management on Aquatic Ecosystems, Proceedings of the ASCE Engineering Foundation Conference, Snowbird, UT, August 4–9, 1996.

MacRae, C. R., and A. C. Rowney. 1992. The role of moderate flow events and bank structure in the determination of channel response to urbanization. Pages 12.1–12.21 in D. Shrubsole (editor). Resolving Conflicts and Uncertainty in Water Management. Canadian Water Resources Association, Kingston, Ontario.

Mader, H-J., C. Schel, and P. Kornacker. 1990. Linear barriers to arthropod movements in the landscape. Biological Conservation 54:209–222.

Manne, L. L. 2003. Nothing has yet lasted forever: Current and threatened levels of biological diversity. Evolutionary Ecology Research 5:517–527.

Marien, M. 2002. Futures studies in the 21st Century: A reality based view. Futures 34(3–4):261–281.

Marland, G., T. Boden, and R. J. Andres. 2005. Global, Regional and National CO2 Emissions, in Trends: A Compendium of Data on Global Change. Carbon Dioxide Analysis Center, Oak Ridge National Laboratory, U.S. DOE http://cdiac.esd.ornl.gov /trends/emis/meth_reg.htm

Martinez, F. 1992. The bid-choice land use model: An integrated framework. Environment and Planning A 24:871–885.

Marzluff, J. 2005. Island biogeography for an urbanizing world: How extinction and colonization may determine biological diversity in human dominate landscapes. Urban Ecosystems 8:157–177.

Marzluff, J. M. 2001. Worldwide urbanization and its effects on birds. Pages 19–47 in J. M. Marzluff, R. Bowman, and R. Donnelly (eds.). Avian Ecology and Conservation in an Urbanizing World. Kluwer Academic Publishers, Norwell, MA.

Marzluff, J. M., and T. Angell. 2005. In the Company of Crows and Ravens. Yale University Press, New Haven, CT.

Marzluff, J. M., and K. Ewing. 2001. Restoration of fragmented landscapes for the conservation of birds: A general framework and specific recommendations for urbanizing landscapes. Restoration Ecology 9:280–292.

Marzluff, J. M., and K. P. Dial. 1991. Life history correlates of taxonomic diversity. Ecology 72:428–439.

Marzluff, J., F. Gehlbach, and D. Manuwal. 1998. Urban environments: Influences on avifauna and challenges for the avian conservationist. Pages 283–306 in J. Marzluffm, and R. Sallabanks (eds.). Avian Conservation: Research and Management. Island, Washington, DC.

Marzluff, J. M., K. J. McGowan, R. E. Donnelly, and R. L. Knight. 2001. Causes and consequences of expanding American Crow populations. Pages 331–363 in J. M. Marzluff, R. Bowman, and R. E. Donnelly (eds.). Avian Conservation And Ecology In An Urbanizing World. Kluwer: Norwell, MA.

Marzluff, J., J. Millspaugh, P. Hurvitz, and M. Handcock. 2004. Relating resources to a probabilistic measure of space use: Forest fragments and steller's jays. Ecology 85:1411–1427.

Marzluff, J. M., M. G. Raphael, and R. Sallabanks. 2000. Understanding the effects of forest management on avian species. Wildlife Society Bulletin 28:1132–1143.

Mason, R. P., and K. A. Sullivan. 1998. Mercury and methyl-mercury transport through an urban watershed. Water Research 32:321–30.

May, R. M. 1977. Thresholds and breakpoints in ecosystems with a multiplicity of stable states. Nature 269:471–477.

McColl, J. G., and D. S. Bush. 1978. Precipitation and throughfall chemistry in the San Francisco bay area. Journal of Environmental Quality 7:352–357.

McDonnell, M. J., and S. T. A. Pickett. 1990. Ecosystem structure and function along urban-rural gradients: An unexploited opportunity for ecology. Ecology 71:1232–1237.

McDonnell, M. J., and S. Pickett (eds.). 1993. Humans as Components of Ecosystems: The Ecology of Subtle Human Effects and Populated Areas. Springer-Verlag, New York.

McDonnell, M., S. Pickett, P. Groffman, P. Bohlen, R. Pouyat, W. Zipperer, R. Parmelee, M. Carreiro, and K. Medley. 1997. Ecosystem processes along an urban-to-rural gradient. Urban Ecosystems 1:21–36.

McDonnell, M. J., S. T. A. Pickett, and R. V. Pouyat. 1993. The application of the ecological gradient paradigm to the study of urban effects. Pages 175–189 in M. J. McDonnell, and S. T. A. Pickett (eds.), Humans as Components of Ecosystems: Subtle Human Effects and the Ecology of Populated Areas. Springer-Verlag, New York.

McFadden, D. 1973. Conditional logit analysis of qualitative choice behavior. Pages 105–142 in P. Zarembka (ed.). Frontiers in Econometrics. Academic Press, New York.

McFadden, D. 1978. Modelling the choice of residential location. Pages 75–96 in A. Karlqvist, L. Lundqvist, F. Snickars and J.W. Wiebull (eds.), Spatial Interaction Theory and Planning Models. North Holland, Amsterdam.

McGarigal, K., S. A. Cushman, M. C. Neel, and E. Ene. 2002. Spatial Pattern Analysis Program for Categorical Maps, FRAGSTATS 3.1. [Online, URL: www.umass.edu/landeco/research/fragstats/fragstats.html].

McGarigal, K., and B. J. Marks. 1995. FRAGSTATS: Spatial pattern analysis program for quantifying landscape structure. USDA For. Serv. Gen. Tech. Rep. PNW-351.

McGowan, K. 2001. Demographic and behavioral comparisons of suburban and rural American crows. Pages 367–383 in J. M. Marzluff, R. Bowman, and R. Donnely (eds.). Avian Ecology in an Urbanizing World. Kluwer Academic, Norwell, MA.

McGuinness, K. A. 1984. Equations and explanations in the study of species–area curves. Biological Reviews 59:423–440.

McHarg, I. 1964. The place of nature in the city of man. Annals of the American Academy of Political and Social Science 352 (March) 1–12.

McHarg, I. 1969. Design with Nature. Natural History Press, Garden City, NY.

McIntyre, S., and R. Hobbs. 1999. A framework for conceptualizing human effects on landscapes and its relevance to management and research models. Conservation Biology 13(6):1282–1292.

McKinney, M. L. 2002. Urbanization, biodiversity, and conservation. BioScience 52:883–890.

McMahon, G., and T. F. Cuffney. 2000. Quantifying urban intensity in drainage basins for assessing stream ecological conditions. Journal of the American Water Resources Association 36:1247–1261.

McNeely, J. A., M. Gadgil, C. Levèque, C. Padoch, and K. Redford. 1995. Human influences on biodiversity. pp. 711–821 in: V. H. Heywood (ed.). Global Biodiversity Assessment. Cambridge University Press, Cambridge, UK.

McNulty, S. G., J. D. Aber, T. M. McLellan, and S. M. Katt. 1990. Nitrogen cycling in high elevation forests of the northeastern U.S. in relation to nitrogen deposition. Ambio 19:38–40.

McPherson, E. G. 1994. Energy-saving potential of trees in Chicago. Pages 7.1–7.22 in E. G. McPherson, D. J. Nowak, and A. Rountree (eds.). Chicago's Urban Forest Ecosystem: Results of the Chicago Urban Forest Climate Project. Northeastern Forest Experiment Station, Radnor, PA.

McPherson, E. G., D. J. Nowak, and R. A. Rowntree. 1994. Chicago's Urban Forest Ecosystem: Results of the Chicago Urban Forest Climate Project. Radnor, PA: Northeastern Forest Experiment Station.

McPherson, E. G., and J. R. Simpson. 2002. A comparison of municipal forest benefits and costs in Modesto and Santa Monica, California, USA. Urban For. Urban Green. 1:61–74.

McPherson, E. G., and J. R. Simpson. 2003. Potential energy savings in buildings by an urban tree planting program in California. Urban Forestry & Urban Greening 2:73–86.

MEA, 2005. Ecosystems and Human Well-being: Synthesis. Millennium Ecosystem Assessment. Island Press, Washington, DC.

Medley, K. E., M. J. McDonnell, and S. T. A. Pickett. 1995. Forest-landscape structure along an urban-to-rural gradient. The Professional Geographer 47:159–68.

Meentemeyer, V. 1989. Geographical perspectives of space, time, and scale. Landscape Ecology 3:163–173.

Melillo, J. M., C. B. Field, and B. Moldan (eds.). 2003. Interactions of the Major Biogeochemical Cycles: Global Change and Human Impacts. Island Press, Washington, DC.

Melosi, M. 2000. The sanitary city: Urban infrastructure in America from colonial times to the present. Johns Hopkins University Press, Baltimore, MD.

Melles, S., S. Glenn, and K. Martin. 2003. Urban bird diversity and landscape complexity: Species-environment associations along a multiscale habitat gradient. Conservation Ecology 7(1):5. [online] URL: http://www.consecol.org/vol7/iss1/art5/.

Mennechez, G., and P. Clefgeau. 2001. Settlement of breeding European Starlings in urban areas: Importance of lawns vs. anthropogenic wastes. Pages 275–287 in J. M. Marzluff, R. Bowman, and R. Donnelly (eds.). Avian Ecology and Conservation in an Urbanizing World. Klewer Academic Publishers, Boston.

Meybeck, M. 1998. Man and river interface: Multiple impacts on water and particulates chemistry illustrated in the Seine River Basin. Hydrobiologia 373–374: 1–20.

Meyer, J. L., W. H. McDowell, T. L. Bott, J. W. Elwood, C. Ishizaki, J. M. Melack, B. L. Peckarsky, B. J. Peterson, and P. A. Rublee. 1988. Elemental dynamics in streams. Journal of the North American Benthological Society 7:410–432.

Meyer, J. L., M. J. Paul, and W. K. Taulbee. 2005. Stream ecosystem function in urbanizing landscapes. Journal of the North American Benthological Society 24(3):602–612.

Meyer, J. L., and J. B. Wallace. 2001. Lost linkages and lotic ecology: rediscovering small streams. Pages 295–317 in M. C. Press, N. J. Huntly, and S. Levin (eds.). Ecology: Achievement and Challenge. Blackwell Science, Malden, MA.

Mickley, L. J., D. J. Jacob, B. D. Field, and D. Rind. 2004. Effects of future climate change on regional air pollution episodes in the United States. Geophysical Research Letters 31, L24103, doi:10.1029/2004GL021216.

Mielke, H. W., C. R. Gonzales, M. K. Smith, and P. W. Mielke. 2000. Quantities and associations of lead, zinc, cadmium, manganese, chromium, nickel, vanadium, and copper in fresh Mississippi Delta alluvium and New Orleans alluvial soils. Science of the Total Environment 246:249–259.

Miess, M. 1979. The Climate of Cities. In I. C. Laurie (ed.), Nature in Cities. J. Wiley, and Sons: Chichester, UK.

Mills, G. S., J. B. Dunning, Jr., and J. M. Bates. 1989. Effects of urbanization on breeding bird community structure in southwestern desert habitats. Condor 91:416–428.

Mittelbach, G. G., C. F. Steiner, S. M. Scheiner, K. L. Gross, H. L. Reynolds, R. B. Waide, M. R. Willig, S. I. Dodson, and L. Gough. 2001. What is the observed relationship between species richness and productivity? Ecology 82:2381–2396.

Monticino, M., M. Acevedo, J. B. Callicott, T. Cogdill, and C. Lindquist. 2007. Coupled human and natural systems: A multi-agent-based approach. Environmental Modelling and Software 22:656–663.

Moore, A. A., and M. A. Palmer. 2005. Invertebrate biodiversity in agricultural and urban headwater streams: Implications for conservation and management. Ecological Applications 15:1169–1177.

Moore, J. W., D. E. Schindler, M. D. Scheuerell, D. Smith, and J. Frodge. 2003. Lake eutrophication at the urban fringe, Seattle region, USA. Ambio 32(1):13–18.

Moring, J. B., and D. R. Rose. 1997. Occurrence and concentrations of polycyclic aromatic hydrocarbons in semipermeable membrane devices and clams in three urban streams of the Dallas-FortWorth Metropolitan Area, Texas. Chemosphere 34:551–66.

Morris, C. J. G., I. Simmonds, and N. Plummer. 2001. Quantification of the influences of wind and cloud on the nocturnal urban heat island of a large city. Journal of Applied Meteorology 40:169–182.

Morris, R., M. Gery, M. Liu, G. Moore, C. Daly, and S. Greenfield. 1989. Sensitivity of a regional oxidant model to variations in climate parameters. In The potential effects of global climate changes on the United States. Appendix F: Air Quality. U.S. EPA. Washington, DC.

Morse, D. R., J. H. Lawton, M. M. Dodson, and M. H. Williamson. 1985. Fractal dimension of vegetation and the distribution of arthropod body lengths. Nature 314:731–733.

Moudon, A. V. 1997. Urban morphology as an emerging interdisciplinary field. Urban Morphology 1:3–10.

Moudon, A. V., S. Kavage, J. E. Mabry and D. W. Sohn. 2005. Transportation-efficient land use mapping index. Transportation Research Record 1902:134–144.

Mulder, C. P. H., J. Koricheva, K. Huss-Danell, P. Högberg, and J. Joshi. 1999. Insects affect relationships between plant species richness and ecosystem processes. Ecology Letters 2237–246.

Mumford, L. 1925. Regions to live in. Survey Graphic 7(May): 151–152. New York, NY: Regional Planning Association of America.

Mumford, L. 1961. The City in History. New York. Harcourt Brace JovanovichMumford, L. 1956. The Transformations of Man. Harper & Row, New York.

Murcia, C. 1995. Edge effects in fragmented forests: Implications for conservation. Trends in Ecology and Evolution 10:58–62.

Muschak, W. 1990. Pollution of street runoff by traffic and local conditions. Science of the Total Environment 93:419–431.

Naeem, S. 2006. Expanding scales in biodiversity-based research: Challenges and solutions for marine systems. Marine Ecology Progress Series 311:273–283.

Naeem, S., J. M. H. Knops, D. Tilman, K. M. Howe, T. Kennedy, and S. Gale. 2000. Plant diversity increases resistance to invasion in the absence of covarying extrinsic factors. Oikos 91:97–108.

Naeem, S., M. Loreau, and P. Inchausti. 2002. Biodiversity and ecosystem functioning: the emergence of a synthetic ecological framework. Pages 3–11 in M. Loreau, S. Naeem, and P. Inchausti, editors. Biodiversity and Ecosystem Functioning: Synthesis and Perspectives. Oxford University Press, Oxford.

NAHB, 2005. Housing Facts, Figures and Trends. National Association of Home Builders, Washington, DC.

Naiman, R. J. 1992. New perspectives for watershed management: balancing long-term sustainability with cumulative environmental change. Pages 3–1 1 in R. J. Naiman (ed.). Watershed Management: Balancing Sustainability and Environmental Change. Springer-Verlag, New York.

Naiman, R. J., and H. Decamps (eds.). 1990. The ecology and management of aquatic ecotones: Man and the Biosphere Series, Volume 4. UNESCO, Paris.

Naiman, R. J., S. R. Elliott, J. M. Helfield, and T. C. O'Keefe. 1999. Biophysical interactions and the structure and dynamics of riverine ecosystems: The importance of biotic feedbacks. Hydrobiologia 410:79–86.

Naiman, R. J., G. Pinay, C. S. Johnston, and J. Pastor. 1994. Beaver influences on the long-term biogeochemical characteristics of boreal forest drainage networks. Ecology 75:905–921.

Nassauer, J. I. 1995. Culture and changing landscape structure. Landscape Ecology 10(4):229–237.

National Research Council. 1991. Rethinking the Ozone Problem in Urban and Regional Air Pollution, National Academy Press, Washington, DC.

Neal, C., and A. J. Robson. 2000. A summary of river water quality data collected within the Land-Ocean Interaction Study: Core data for eastern UK rivers draining to the North Sea. Science of the Total Environment 251/252:585–665.

Neil, D., and B. Yu. 1999. A method of analyzing stream channel response to environmental change: Gauge data for the Tully River. Australian Geographer 30:239–252.

Newbury, R. W. 1988. Hydrologic determinants of aquatic insect habitats. Pages 323–357 in V. H. Resh, and D. M. Rosenburg (eds.), The Ecology of Aquatic Insects. Praeger Scientific. New York.

Newcombe, K. 1977. Nutrient flow in a major urban settlement: Hong Kong. Human Ecology 5:179–208.

Newcombe, K., J. Kalma, and A. Aston. 1978. The metabolism of a city: The case of Hong Kong. Ambio 7:3–15.

Newman, A. 1995. Water pollution sources still significant in urban areas. Environmental Science and Technology 29:114.

Newman, P., and J. Kenworthy. 1989a. Cities and Automobile Dependence: An International Sourcebook. Gower, Aldershot, UK.

Newman, P., and J. R. Kenworthy. 1989b. Gasoline consumption and cities: A comparison of U.S. cities with a global survey. Journal of the American Planning Association 55:24–37.

Newmark, W. 1987. A land-bridge is land perspective on mammalian extinction in western North American parks. Nature 325:430–432.

Newton, I. 1998. Population Limitation in Birds. London: Academic.

Nicolis G. and I. Prigogine. 1977. Self-Organization in Non-Equilibrium Systems. John Wiley & Sons, New York.

Nicolis, G., and I. Prigogine. 1989. Exploring Complexity: An Introduction. W. H. Freeman, New York.

Niemela, J. 1999a. Ecology and urban planning. Biodiversity and Conservation 8:119–131.

Niemela, J. 1999b. Is there a need for a theory of urban ecology? Urban Ecosystems 3:57–65.

Niemelä, J., and J. R. Spence. 1991. Distribution and abundance of an exotic ground-beetle (Carabidae): A test of community impact. Oikos 62:351–359.

Nilon, C. H., and R. C. Pais. 1997. Terrestrial vertebrates in urban ecosystems: Developing hypotheses for the Gwynns Falls Watershed in Baltimore, Maryland. Urban Ecosystems 1:247–257.

Nilsson, J. 1995. A phosphorus budget for a Swedish municipality. Journal of Environmental Management 45(3):243–253.

Nixon, S. W., J. W. Ammerman, L. P. Atkinson, V. M. Berounsky, G. Bilen, W. C. Boicourt, W. R. Boynton, T. M. Church, D. M. DiToro, R. Elmgren, J. Garber, A. E. Giblin, R. A. Jahnke, N. J. P. Owens, M. E. Q. Pilson, and S. P. Seitzinger. 1996. The fate of nitrogen and phosphorus at the land-sea margin of the North Atlantic Ocean. Biogeochemistry 35:141–180.

Nkemdirim, L. C. 1976. Dynamics of an urban temperature field: A case study. Journal of Applied Meteorology 15:818–828.

Norberg, J. 2004. Biodiversity and ecosystem functioning: A complex adaptive systems approach. Limnology and Oceanography. 49 (4 part 2):1269–1277.

Noss, R. F. 1989. Longleaf pine and wiregrass: Keystone components of an endangered ecosystem. Natural Areas Journal 9:211–213.

Nowak, D. J. 1993. Atmospheric carbon-reduction by urban trees. Journal of Environmental Management 37(3):207–217.

Nowak, D. J., and D. E. Crane. 2002. Carbon storage and sequestration by urban trees in the USA. Environmental Pollution 116:381–389.

NRC (National Research Council). 1986. Global Change in the Geosphere-Biosphere. Initial Priorities for an IGBP. Page 91 in U.S. Committee for an International Geosphere-Biosphere Program (J. A. Eddy, Chair). National Academy Press, Washington, DC.

Nuhn, T. P., and C. G. Wright. 1979. An ecological survey of ants in a landscaped suburban habitat. American Midland Naturalist 102:353–362.

Nystrom, M., and C. Folke. 2001. Spatial resilience of coral reefs. Ecosystems 4:406–417.

Odum, P. E. 1953. Fundamentals of Ecology. W. B. Saunders Company, Philadelphia. USA.

Odum, E. P. 1963. Ecology. Holt, Rinehart and Winston, New York.

Odum, E. P. 1975. Ecology: The Link between the Natural and Social Sciences. Second edition. Holt, Rinehart and Winston, New York, USA.

Odum, E. P. 1997. Ecology: A Bridge between Science and Society. Sinauer, Sunderland, MA.

Office of Financial Management, Washington State, 2005. Washington State Data Book, OFM Forecasting Division. Olympia, Washington.

Ogilvy, J.A., and P. Schwartz. 1998. Plotting your scenarios. Pages 57–80 in L. Fahey and R. Randall (eds.). Learning from the Future: Competitive Foresight Scenarios. John Wiley & Sons, New York.

Oke, T. R. 1973. City size and urban heat island. Atmospheric Environment 7:769–779.

Oke, T. R. 1976. The distinction between canopy and boundary-layer urban heat islands. Atmosphere 14:268–277.

Oke, T. R. 1982. The energetic basis of the urban heat island. Quarterly Journal of the Royal Meteorological Society 108:1–24.

Oke, T. R. 1987. Boundary Layer Climates. Methuen, London.

Oke, T. R. 1995. The heat island of the urban boundary layer: Characteristics, causes and effects. Pages. 81–107 in J. E. Cermak, A. G. Davenport, E. J. Plate, and D. X. Viegas (eds.). Wind Climate in Cities. Kluwer Academic Publishers, the Netherlands.

Oke, T. R. 1997. Urban environments. Pages 303–327 in W. G. Bailey, T. R. Oke, and W. R. Rouse (eds.). The Surface Climates of Canada. McGill/Queens University Press, Montreal, Canada.

Oke, T. R. 2004. Initial guidance to obtain representative meteorological observations at urban sites. Instruments and Observing Methods Report No. 81, WMO/TD. No. 1250. World Meteorological Organization, Geneva.

Oke, T. R. 2006. Towards better scientific communication in urban climate. Theoretical and Applied Climatology 84:179–190.

Oke, T. R., and C. East. 1971. The urban boundary layer. Boundary Layer Meteorology 1:411–437.

Oke, T. R., C. S. B. Grimmond, and R. Spronken-Smith. 1998. On the confounding role of rural wetness in assessing urban effects on climate. Pages 59–62 in Preprints of the AMS Second Urban Environment Symposium.

Oke, T., and S. Kanae. 2006. Global hydrological cycles and world water resources. Science 313:1068–1072.

Olden, J. D., and N. L. Poff. 2003. Toward a mechanistic understanding and prediction of biotic homogenization. American Naturalist 162: 442–460.

Olden, J. D., and N. L. Poff. 2004. Ecological processes driving biotic homogenization: Testing a mechanistic model using fish faunas. Ecology 85:1867–1875.

Omernik, J. M. 1976. The influence of land use on stream nutrient levels. EPA-600/3–76–014. Washington, DC: U.S. Environmental Protection Agency.

Omernik, J. M. 1987. Ecoregions of the conterminous United States. Annals of the Association of American Geographers 77(1):118–125.

O'Neill, R., D. L. DeAngelis, J. Waide, and T. Allen. 1986. A Hierarchical Concept of Ecosystems. Princeton University Press, Princeton, NJ.

O'Neill, R. V., J. R. Krummel, R. H. Gardner, G. Sugihara, B. Jackson, D. L. DeAngelis, B. T. Milne, M. G. Turner, B. Zygmunt, S. W. Christensen, V. H. Dale, and R. L. Graham. 1988. Indices of landscape pattern. Landscape Ecology 1:153–162.

O'Neill, R. V., S. J. Turner, V. I. Cullinan, D. P. Coffin, T. Cook, W. Conley, J. Brunt, J. M. Thomas, M. R. Conley, and J. Gosz. 1991. Multiple landscape scales: An intersite comparison. Landscape Ecology 5:137–144.

Onisto, L. K., E. Krause, and M. Wackernagel. 1998. How Big is Toronto's Ecological Footprint? Centre for Sustainability Studies and the City of Toronto, Toronto, Canada.

Opdam, P., F. Foppen, and C. Vos. 2002. Bridging the gap between ecology and spatial planning in landscape ecology. Landscape Ecology 16:767–779.

Openshaw, S. 1995. Human systems modelling as a new grand challenge area in science: What has happened to the science in the social science: Commentary. Environment and Planning A 27:159–164.

Ostrom E, 1991. Governing the Commons. Cambridge University Press, Cambridge, UK.

Ottensmann, J. R. 1977. Urban sprawl, land values and the density of development. Land Economics 53:389–400.

Owens, P. 1993. Neighbourhood form and pedestrian life: Taking a closer look. Landscape and Urban Planning, 26:115–135.

Owens, S. 1984. Energy demand and spatial structure. Pages 215–240 in D. Cope, P. Hills, and P. James (eds.). Energy Policy and Land Use Planning. Pergamon, Oxford, UK.

Owens, S. 1986. Energy, Planning and Urban Form. London: Pion.

Owens, S., and P. Rickaby. 1992. Settlements and energy revisited. Built Environment, 18:247–252.

Paine, R. T. 1966. Food web complexity and species diversity. American Naturalist 100:65–75.

Paine, R. T. 1984. Ecological determinism in the competition for space. Ecology 65:1339–1348.

Palumbi, S. R. 2001. Humans as the world's greatest evolutionary force. Science. 293:1786–1790.

Park, H.-S. 1986. Features of the heat island in Seoul and its surrounding cities. Atmospheric Environment 20:1859–1866.

Park, R. E., E. W. Burges, and R. D. McKenzie (eds.). 1925. The City. University of Chicago Press, Chicago.

Parker, D. C., T. Berger, and S. M. Manson. 2001. Agent-Based Models Of Land-Use and Land-Cover Change. Proceedings of an International Workshop October 4–7, 2001, Irvine, CA.

Parker, D. C., A. Manson, M. Janessen, M. Hoffmann, and P. Deadman. 2003. Multi-agent systems for the simulation of land-use and land-cover change: a review. Annals of the Association of American Geographers 93(2):314–337.

Parris, K. M., and D. L. Hazell. 2005. Biotic effects of climate change in urban environments: The case of the grey-headed flying fox (*Pteropus poliocephalus*) in Melbourne, Australia. Biological Conservation 124:267–276.

Parrott, L. 2002. Complexity and the limits of ecological engineering. Transactions of the American Society of Agricultural Engineers (ASAE) 45:1697–1702.

Pataki, D. E., R. J. Alig, A. S. Fung, N. E. Golubiewski, C. A. Kennedy, E. G. McPherson, D. J. Nowak, R. V. Pouyat, and P. Romero Lankao. 2006. Urban ecosystems and the North American carbon cycle. Global Change Biology 12(11):2092–2102.

Patten, C. B. 1995. Network integration of ecological extremal principles: exergy, emergy, power, ascendency, and indirect effects. Ecological Modelling 79(1–3):75–84.

Paul, M. J., and J. L. Meyer. 2001. Streams in the urban landcape. Annual Review of Ecological Systematics 32:333–365.

Pautasso, M. 2007. Scale dependence of the correlation between human population presence and vertebrate and plant species richness. Ecology Letters 10:16–24.

Peterson, G. D. 2000. Political ecology and ecological resilience: an integration of human and ecological dynamics. Ecological Economics 35:323–336.

Peterson, G. D. 2004. A simple Model of Resilience http://www.geog.mc gill.ca/faculty/peterson/susfut/resilience/rLandscape.html

Peterson, G. D., C. R. Allen, and C. S. Holling. 1998. Ecological resilience, biodiversity, and scale. Ecosystems 1:6–18.

Peterson, A. G., J. T. Ball, Y. Luo, C. B. Field, P. B. Reich, P. S. Curtis, K. L. Griffin, C. A. Gunderson, R. J. Norby, D. T. Tissue, M. Forstreuter, A. A. Rey, C. S. Vogel, and CMEAL participants. 1999. The photosynthesis-leaf nitrogen relationship at ambient and elevated atmospheric carbon dioxide: A meta-analysis. Global Change Biology 5:331–346.

Peterson, G. D., G. S. Cumming, and S. R. Carpenter. 2003. Scenario planning: A tool for conservation in an uncertain world. Conservation Biology 17:358–366.

Petren, K., and T. J. Case. 1996. An experimental demonstration of exploitation competition in an ongoing invasion. Ecology 77:118–132.

Philandras, C. M., D. A. Metaxas, and P. T. Nastos. 1999. Climate variability and urbanization in Athens. Theoretical and Applied Climatology 63:65–72.

Phillips, J. D. 1999. Divergence, Convergence, and Self-Organization in Landscapes. Annals of the Association of American Geographers 89:466–488.

Pickett, S. T. A., and F. A. Bazzaz. 1976. Divergence of two co-occurring successional annuals on a soil moisture gradient. Ecology 57(1):169–176.

Pickett, S. T. A., W. R. Burch, S. E. Dalton, T. W. Foresman, J. M. Grove, and R. Rowntree. 1997. A conceptual framework for the study of human ecosystems in urban areas. Urban Ecosystems 1:185–199.

Pickett, S. T. A., I. C. Burke, V. H. Dale, J. R. Gosz, R. G. Lee, S. W. Pacala, and M. Shachak. 1994. Integrated models in forested regions. Pages 120–141 in P. M. Groffman, and G. E. Likens (eds.). Integrated Regional Models. Chapman & Hall, New York.

Pickett, S., and M. Cadenasso. 1995. Landscape ecology: Spatial heterogeneity in ecological systems. Science 269:331–334.

Pickett, S. T. A., and M. L. Cadenasso. 2002. The ecosystem as a multidimensional concept: Meaning, model, and metaphor. Ecosystems 5:1–10.

Pickett, S. T. A., and M. L. Cadenasso. 2006. Advancing urban ecological studies: Frameworks, concepts, and results from the Baltimore Ecosystem Study. Austral Ecology 31:114–125.

Pickett, S. T. A., M. L. Cadenasso, J. M. Grove, C. H. Nilon, R. V. Pouyat, W. C. Zipperer, and R. Costanza. 2001. Urban ecological systems: Linking terrestrial ecological, physical, and socioeconomic components of metropolitan areas. Annual Review of Ecology and Systematics 32:127–157.

Pickett, S. T. A., M. L. Cadenasso, and C. G. Jones. 2000. Generation of heterogeneity by organisms: Creation, maintenance, and transformation. Pages 33–52 in M. Hutchings, L. John, and A. Stewart (eds.). Ecological Consequences of Habitat Heterogeneity. Blackwell, New York.

Pickett, S. T. A., C. Jones, and J. Kolasa. 1994. Ecological Understanding: The Nature of Theory and the Theory of Nature. Academic Press, San Diego, California.

Pickett, S. T. A., and R. S. Ostfeld. 1994. The shifting paradigm in ecology. Ecological Environment 3:151–159.

Pickett, S. T. A., V. T. Parker, and P. Fiedler. 1992. The new paradigm in ecology: Implications for conservation biology above the species level. Pages 65–88 in P. L. Fiedler, and S. K. Jain (eds.). Conservation Biology: The Theory and Practice of Nature Conservation, Preservation, and Management. Chapman and Hall, New York.

Pickett, S. T. A., and K. Rogers. 1997. Patch dynamics: The transformation of landscape structure and function. Pages 101–127 in J. Bissonnette (ed.). Wildlife and Landscape Ecology. New York: Springer-Verlag.

Pickett, S. T. A., and P. S. White. 1985. Patch dynamics: A synthesis. Pages 371–384 in S. T. A. Pickett, and P. S. White (eds.). The Ecology of Natural Disturbance and Patch Dynamics. Academic Press, New York.

Pickett, S. T. A., and R. White (eds.). 1985. The Ecology of Natural Disturbance and Patch Dynamics. Academic Press, Orlando, FL.

Pickett, S. T. A., J. Wu, and M. L. Cadenasso. 1999. Patch dynamics and the ecology of disturbed ground: A framework for synthesis. Pages 707–722 in L. R. Walker (ed.). Ecosystems of the World: Ecosystems of Disturbed Ground. Elsevier Science, Amsterdam.

Pimm, S. L. 1991. The Balance of Nature? Ecological Issues in the Conservation of Species and Communities. University of Chicago Press, Chicago.

Pimm, S. L., M. P. Moulton, and L. J. Justice. 1994. Bird extinctions in the central Pacific. Philosophical Transactions of the Royal Society of London, B. 344:27–33.

Pimm, S. L., and P. H. Raven. 2000. Biodiversity: Extinction by numbers. Nature 403:843–845.

Pisarski, A. E. 1991. Overview. Pages 3–10 in Transportation, Urban Form, and the Environment. Transportation Research Board Special Report 231. National Research Council, Washington DC.

Pitt, R., R. Field, M. Lalor, and M. Brown. 1995. Urban stormwater toxic pollutants: Assessment, sources, and treatability. Water Environment Research 67(3):260–275.

Plotnick, R. E., R. H. Gardner, W. W. Hargrove, K. Prestegaard, and M. Perlmutter. 1996. Lacunarity analysis: A general technique for the analysis of spatial patterns. Physical Review E 53(5):5461–5468.

Plotnick, R. E., R. H. Gardner, and R. V. O'Neill. 1993. Lacunarity indices as measures of landscape texture. Landscape Ecology 8(3):201–211.

Poff, N. L., and J. V. Ward. 1989. Implications of streamflow variability and predictability for lotic community structure: A regional analysis of streamflow patterns. Canadian Journal of Fisheries and Aquatic Sciences 46:1805–1818.

Poff, N. L., and J. V. Ward. 1990. The physical habitat template of lotic systems: Recovery in the context of historical pattern of spatio-temporal heterogeneity. Environmental Management 14:629–646.

Poff, N. L., J. D. Allan, M. B. Bain, J. R. Karr, K. L. Prestegaard, B. D. Richter, R. E. Sparks, and J. C. Stromberg. 1997. The natural flow regime: A paradigm for river conservation and restoration. BioScience 47:769–784.

Porter, et al. 2003, Porter, E. E., B. R. Forschner, and R. B. Blair. 2001. Woody vegetation and canopy fragmentation along a fort-to-urban gradient. Urban Ecosystems 5, 131–151.

Portugali, J. 2000. Self-Organization and the City. Springer, Berlin.

Postel, S. L., G. C. Daily, and P. R. Ehrlich. 1996. Human appropriation of renewable fresh water. Science 271–785–788.

Pouyat, R. V., P. Bohlen, V. Eviner, M. M. Carreiro, and P. M. Groffman. 1996. Short and long term effects of earthworms on N dynamics in forest soils. Supplemental Bulletin of the Ecological Society of America 77:359.

Pouyat, R. V., and M. J. McDonnell. 1991. Heavy metal accumulations in forest soils along an urban-rural gradient in Southeastern New York, USA. Water, Air & Soil Pollution 57–58(1):797–807.

Pouyat, R. V., M. J. McDonnell, and S. T. A. Pickett. 1995. Soil characteristics of oak stands along an urban-rural land-use gradient. Journal of Environmental Quality 24:516–526.

Pouyat, R. V., M. J. McDonnell, and S. T. A. Pickett. 1997. Litter decomposition and nitrogen mineralization in oak stands along an urban-rural land use gradient. Urban Ecosystems 1:117–131.

Pouyat, R. V., R. W. Parmelee, and M. M. Carreiro. 1994. Environmental effects of forest soil-invertebrate and fungal densities in oak stands along an urban-rural land use gradient. Pedobiologia 38:385–399.

Pouyat, R. V., I. D. Yesilonis, and D. J. Nowak. 2006. Carbon storage by urban soils in the United States. Journal of Environmental Quality 35(4):1566–1575.

Power, M. E., D. Tilman, J. A. Estes, B. A. Menge, W. J. Bond, L. S. Mills, D. Gretchen, J. C. Castilla, J. Lubchenco, and R. T. Paine. 1996. Challenges in the quest for keystones. BioScience 46:609–620.

Power, M. E., R. J. Stout, C. E. Cushing, P. P. Harper, F. R. Hauer, W. J. Matthews, P. B. Moyle, B. Stazner, and I. R. Wais De Bagden. 1988. Biotic and abiotic controls in river and stream communities. Journal of the North American Benthological Society. 7:456–479.

Power, S., T. Casey, C. Folland, A. Colman, & V. Mehta, (1999) Interdecadal modulation of the impact of ENSO in Australia. Climate Dynamics, 15, 319–324.

Prastacos, P. 1986. An integrated land-use-transportation model for the San Francisco Region: 1. Design and mathematical structure. Environment and Planning A 18:307–322.

Prather, M., D. Ehhalt, F. Dentener, R. Derwent, E. Dlugokencky, E. Holland, I. Isaksen, J. Katima, V. Kirchhoff, P. Matson, P. Midgley, and M. Wang. 2001. Atmospheric chemistry and greenhouse gases. Pages 239–287 in J. T. Houghton, Y. Ding, D. J. Griggs, M. Noguer, P. J. Van der Linden, X. Dai, K. Maskell, and C. A. Johnson (eds.). Climate Change 2001: The Scientific Basis. Contribution of Working 5 Group I to the Third Assessment Report of the Intergovernmental Panel on Climate Change. Cambridge University Press, Cambridge, UK.

Pratt, J. M., R. A. Coler, and P. J. Godfrey. 1981. Ecological effects of urban stormwater runoff on benthic macroinvertebrates inhabiting the Green River, Massachusetts. Hydrobiologia 83:29–42.

Preston, F. W. 1960. Time and space and the variation of species. Ecology 41:611–627.

Preston, F. W. 1962. The canonical distribution of commonness and rarity. Ecology 43:185–215 (part I), 410–432 (part II).

Prins, H. H. T., and H. P. Van der Jeud. 1993. Herbivore population crashes and woodland structure in East Africa. Journal of Ecology 81:305–14.

PRISM (Puget Sound Regional Integrated Synthesis Model) 2000. Human dimension of PRISM. Alberti M. and P. Waddell, Presentation. University of Washington Seattle, WA.

PSRC. 2005. Regional Demographics and Growth Trends. Puget Sound Regional Council, Seattle, WA.

Pucher, J., and C. Lefevre. 1996. Urban Transport Crisis in Europe and North America. Macmillan Press, London, England.

Pumain, D. 2000. Settlement systems in the evolution. Geografiska Annaler 82 B(2):73–87.

Putman, S. H. 1979. Urban Residential Location Models. Martinus Nijhoff Publishing, Norwell, MA.

Putman, S.H. 1983. Integrated Urban Models. Pion Limited, London. England.

Putman, S. H. 1991. Integrated Urban Models 2. Pion Limited, London, England.

Putman, S.H. 1995. EMPAL and DRAM Location and Land use Models: An Overview. In G. A. Shunk, (ed.). LandUse Modeling Conference Proceedings. February 19–21, 1995. Report DOT-T-96–09. US Department of Transportation, Washington, DC, 1995.

Racey, G. D., and D. L. Euler. 1982. Small mammal and habitat response to shoreline cottage development in central Ontario, Canada. Canadian Journal of Zoology 60:865–880.

Ralph, S. C., G. C. Poole, L. L. Conquest, and R. J. Naiman. 1994. Stream channel morphology and woody debris in logged and unlogged basins of western Washington. Canadian Journal of Fisheries and Aquatic Science 51:37–51.

Ranney, J., M. Bruner, and J. Levenson. 1981. The importance of edge in the structure and dynamics of forest lands.Pages 67–95 in R. Burgess and D. Sharpe (eds.). Forest Island Dynamics in Man-Dominated Landscapes. Springer-Verlag, New York.

Rao, T., E. Zalewsky, and I. Zurbenko. 1995. Determining temporal and spatial variations in ozone air quality. Journal of the Air and Waste Management Association 45:57–61.

Rapoport, A. 1977. Human aspects of urban form. Oxford: Pergamon Press.

Rapoport, E. H. 1993. The process of plant colonization in small settlements and large cities. Pages 190–207 in M. J. McDonnell, and S. T. A. Pickett (eds.), Humans as Components of Ecosystems. Springer-Verlag, New York.

Rapport, D. J., H. A. Regier, and T. C. Hutchinson. 1985. Ecosystem behavior under stress. American Naturalist 125:617–640.

Real, R., A. M. Barbosa, D. Porras, M. C. Kin, A. L. Marquez, J. C. Guerrero, L. J. Palomo, E. R. Justo, and J. M. Vargas. 2003. Relative importance of environment, human activity and spatial situation in determining the distribution of terrestrial mammal diversity in Argentina. Journal of Biogeography 30: 939–947.

Rebele, F. 1994. Urban ecology and special features of urban ecosystems. Global Ecology and Biogeography Letters. 4:173–187.

Redman, C. L. 2005. The urban ecology of metropolitan Phoenix: a laboratory for interdisciplinary study. Pages 163–192 in B. Entwistle and P. C. Stern, editors. Population, Land Use, and Environment: Research Directions. National Research Council, The National Academies Press, Washington, DC.

Redman, C., J. M. Grove, and L. Kuby. 2000. Toward a Unified Understanding of Human Ecosystems: Integrating Social Sciences into Long-Term Ecological Research. White Paper of the Social Science Committee of the LTER Network. Online: http://intranet.lternet.edu/archives/documents/Publications/sosciwhtppr/.

Redman, C.L., M.J. Grove, and L.H. Kuby. 2000. Toward a Unified Understanding of Human Ecosystems: Integrating Social Science into Long-Term Ecological Research http://intranet.lternet.edu/archives/documents/ Publications/sosciwhtppr/

Rees, W. E. 1995. Achieving sustainability: Reform or transformation. Journal of Planning Literature 9:343–361.

Rees, W. E. 1996. Revisiting carrying capacity: Area-based indicators of sustainability. Population and Environment 17:195–215.

Rees, W., and M. Wackernagel. 1994. Ecological footprints and appropriated carrying capacity: Measuring the natural capital requirements of the human economy. Pages 362–390 in A.-M. Jansson, M. Hammer, C. Folke, and R. Costanza (eds.), Investing in Natural Capital: The Ecological Economics Approach to Sustainability. Island Press, Washington, DC.

Rees, W., and M. Wackernagel, 1996. Urban ecological footprints: Why cities cannot be sustainable and why they are a key to sustainability. Environmental Impact Assessment Review 16:223–248.

Resh, V. H., A. V. Brown, A. P. Coovich, M. E. Gurtz, H. W. Li, G. W. Minshal, S. R. Reice, A. L. Sheldon, J. B. Wallace, and R. Wissmar. 1988. The role of disturbance in stream ecology. Journal of the North American Benthological Society 7:433–455.

Resnick, M. R. 1994. Turtles, Termites, and Traffic Jams. MIT Press, Cambridge, MA.

Reynolds, C. S. 1995. The intermediate disturbance hypothesis and its applicability to planktonic communities. Comments on the views expressed in Padisak—v—Wilson. New Zealand Journal of Ecology 19:219–225.

Reynolds, J., and J. Wu. 1999. Do landscape structural and functional units exist? Pages 273–296 in J. Tenhunen, and P. Kabat (eds.), Integrating Hydrology, Ecosystem Dynamics, and Biogeochemistry in Complex Landscapes. Wiley, Chichester.

Richards, C., L. B. Johnson, and G. E. Host. 1996. Landscape-scale influences on stream habitats and biota. Canadian Journal Fisheries and Aquatic Sciences 53:295–311.

Ricketts, T. H. 2001. The matrix matters: Effective isolation in fragmented landscapes. American Naturalist 158:87–99.

Riebsame, E. W., W. B. Meyer, and B. L. Turner II. 1994. Modeling land use and cover as part of global environmental change. Special Issue, Climatic Change 28:45–65.

Rind, D., J. Lerner, and C. McLinden. 2001. Changes of tracer distribution in the doubled $CO_2$ climate. Journal of Geophysical Research 106:28,061–28,080.

Risser, P. 1995. Biodiversity and ecosystem function. Conservation Biology 9:742–746.

Robbins, P. and J. Sharp. 2003. Producing and consuming chemicals: the moral economy of the American lawn. Economic Geography 79:425–451.

Robinson, S. K., and D. S. Wilcove. 1994. Forest fragmentation in the temperate zone and its effects on migratory songbirds. Pages 233–249 in K. Young, and M. A. Ramos (eds.), The Conservation of Migratory Birds in the Neotropics. Bird Conservation International, Washington, DC.

Rojstaczer, S., S. M. Sterling, and N. J. Moore. 2001. Human appropriation of photosynthesis products. Science 294:2549–2552.

Rolando, A., G. Pulcher, and A. Giuso. 1997. Avian community structure along an urbanization gradient. Italian Journal of Zoology 64:341–349.

Rosenfeld, A. H., H. Akbari, J. J. Romm, and M. Pomerantz. 1998. Cool communities: Strategies for heat island mitigation and smog reduction. Energy and Buildings 28:51–62.

Rosenfeld, J. S. 2002. Functional redundancy in ecology and conservation. Oikos 98(1):156–162.

Rosenzweig, M. L., and Z. Abramsky. 1993. How are diversity and productivity related? Pages 52–65 in R. E. Ricklefs, and D. Schluter (eds.), Species Diversity in Ecological Comminutes: Historical and Geographical Perspectives. University of Chicago Press, Chicago.

Rosser, J. B. In press. Dynamic discontinuities in ecologic-economic systems in Cross-Scale Structure and Discontinuities. In Craig R. Allen and C.S. Holling (eds.), Cross-Scale Structure and Discontinuities in Ecosystems and Other Complex Systems. Columbia University Press, New York, forthcoming.

Roth, M., and T. R. Oke. 1995. Relative efficiencies of turbulent transfer of heat, mass and momentum over a patchy urban surface. Journal of Atmospheric Science 52:1864–1874.

Roth, N. E., J. D. Allan, and D. E. Erickson. 1996. Landscape influences on stream biotic integrity assessed at multiple spatial scales. Landscape Ecology 11:141–156.

Roughgarden, J., R. M. May, and S. A. Levin (eds.), 1989 . Perspectives in Ecological Theory. Princeton University Press, N. J. Princeton.

Roy, A. H., M. C. Freeman, B. J. Freeman, S. J.Wenger, W. E. Ensign, and J. L. Meyer. 2005. Investigating hydrological alteration as a mechanism of fish assemblage shifts in urbanizing streams. Journal of the North American Benthological Society 24:656–678.

Rozoff, C. M., W. R. Cotton, and J. O. Adegoke. 2003. Simulation of St. Louis, Missouri, land use impacts on thunderstorms. Journal of Applied Meteorology 42:716–738.

Rudnicky, J. L., and M. J. McDonnell. 1989. Forty-eight years of canopy change in a hardwood-hemlock forest in New York City. Bulletin of the Torrey Botanical Club 116:52–64.

Rusk, D. 1999. Inside Game, Outside Game: Winning Strategies for Saving Urban America. Brookings Institution, Washington, DC.

Sala, O. E., F. S. Chapin 3rd, J. J. Armesto, E. Berlow, J. Bloomfield, R. Dirzo, E. Huber-Sanwald, L. F. Huenneke, R. Jackson, A. Kinzig, R. Leemans, D. M. Lodge, H. A. Mooney, M. Oesterheld, N. L. Poff,

M. T. Sykes, B. H. Walker, M. Walker, and D. H. Wall. 2000. Global bio-diversity scenarios for the year 2100. Science 287:1770–1774.

Sarkar, S. 2006. Ecological diversity and biodiversity as concepts for conservation planning: Comments on Ricotta. Acta Biotheoretica 54:133–140.

Sartor, J. D., G. B. Boyd, and F. J. Agardy. 1974. Water pollution aspects of street surface contaminants. Journal Water Pollution Control Federation 46:458–467.

Savard, J.-P.L., and G. Falls. 2001. Survey techniques and habitat relationships of breeding birds in residential areas of Toronto, Canada. Pages 543–568 in J. M. Marzluff, R. Bowman, and R. Donnelly (eds.), Avian Conservation and Ecology in an Urbanizing World. Kluwer Academic, Boston.

Sax, D. F., and S. D. Gaines. 2003. Species diversity: From global decreases to local increases. Trends in Ecology and Evolution 11:561–566.

Scheffer, M. 1998. Ecology of Shallow Lakes. London: Chapman and Hall.

Scheffer, M., S. Carpenter, J. A. Foley, C. Folkes, and B. Walker. 2001. Catastrophic shifts in ecosystems. Nature 413:591–596.

Scheffer, M., and S. R. Carpenter. 2003. Catastrophic regime shifts in ecosystems: linking theory to observation. Trends in Ecology and Evolution 18(12): 648–656.

Schimel, D., J. Melillo, H. Tian, A. D. McGuire, D. Kicklighter, T. Kittel, N. Rosenbloom, S. Running, P. Thornton, D. Ojima, R. Kelly, M. Sykes, R. Neilson, and B. Rizzo. 2000. Contribution of increasing $CO_2$ and climate to carbon storage by ecosystems of the United States. Science 287:2004–2006.

Schindler, D. W. 1990. Experimental perturbations of whole lakes as tests of hypotheses concerning ecosystem structure and function. OIKOS 57:25–41.

Schippers, P., J. Verboom, J. P. Knaapen, and R. C. vanApeldoorn. 1996. Dispersal and habitat connectivity in complex heterogeneous landscapes: An analysis with a GIS-based random walk model. Ecography 19(2):97–106.

Schläpfer, F. and B. Schmid. 1999. Ecosystem effects of biodiversity: a classification of hypotheses and exploration of empirical results. Ecological Applications 9: 893–912.

Schlosser, I. 1991. Stream fish ecology: A landscape perspective. BioScience 41:704–712.

Schlosser, I. J. 1985. Flow regime, juvenile abundance, and the assemblage structure of stream fishes. Ecology 66:1484–1490.

Schmidlin, T. W. 1989. The urban heat island at Toledo, Ohio. Ohio Journal of Science 89:38–41.

Schneider, D. C. 1994. Quantitative Ecology: Spatial and Temporal Scaling. Academic Press, San Diego, CA.

Schneider, D. C., R. Walters, S. F. Thrush, and P. K. Dayton. 1997. Scale-up of ecological experiments: density variation in the mobile bivalve *Macomona liliana Iredale*. Journal of Experimental Marine Biology and Ecology 216:129–152.

Schoener, T. W. 1983. Field experiments on interspecific competition. American Naturalist 122:240–285.

Schueler, T. R. 1994. The importance of imperviousness. Watershed Protect. Techn. 1:100–111.

Schueler, T. R., and J. Galli. 1992. Environmental impacts of stormwater ponds. Pages 159–180 in P. Kumble, and T. Schueler (eds.), Watershed Restoration Source Book. Metropolitan Washington Council of Governments, Washington, DC.

Schumaker, N. H. 1996. Using landscape indices to predict habitat connectivity. Ecology 77:1210–1225.

Schwartz, P. 1991. The Art of the Long View: Paths to Strategic Insight for Yourself and Your Company. Doubleday, New York.

Schwartz, P., and J. Ogilvy. 2004. Plotting Your Scenarios: An Introduction to the Art and Process of Scenario Planning. Global Business Network, Emeryville, CA.

Scott, M. C., and G. S. Helfman. 2001. Native invasions, homogenization, and the mismeasure of integrity of fish assemblages. Fisheries 26:6–15.

Seinfeld, J. H., and S. N. Pandis. 2006. Atmospheric Chemistry and Physics: From Air Pollution to Climate Change. Third edition. John Wiley, New York.

Senge, P. 1990. The Fifth Discipline: The Art and Practice of the Learning Organization, Currency Doubleday, New York.

Shaeffer, J. R., K. Wright, W. Taggart, and R. Wright. 1982. Urban storm drainage management. New York Marcel Dekker.

Shared Strategy Development Committee. 2005.

Shelford, V. E., and S. Eddy. 1929. Methods for the study of stream communities. Ecology 10:382–391.

Shen, W., G. D. Jenerette, J. Wu, and R. H. Gardner. 2004. Evaluating empirical scaling relations of pattern metrics with simulated landscapes. Ecography 27:459–469.

Shepherd, J. M., H. Pierce, and A. Negri. 2002. Rainfall modification by major urban areas from spaceborne rain RADAR on the TRMM Satellite. Journal of Applied Meteorology 41:689–701.

Shevky, E., and W. Bell. 1955. Social Area Analysis: Theory, Illustrative Application, and Computational Procedures. Stanford University Press.

Shiklomanov, I. A., and J. C. Rodda. 2003. World Water Resources at the Beginning of the 21st Century. International Hydrology Series by IHP (International Hydrological Program) – UNESCO – in collaboration with Cambridge Press, UK, 435 pp.

Shochat, E. 2004. Credit or debit? Resource input changes population dynamics of city-slicker birds. Oikos 106:622–626.

Shochat, E., S. B. Lerman, M. Katti, and D. B. Lewis. 2004. Linking optimal foraging behavior to bird community structure in an urban–desert landscape: Field experiments with artificial food patches. American Naturalist 164:232–243.

Shochat, E., W. L. Stefanov, M. E. A. Whitehouse, and S. H. Faeth. 2004. Urbanization and spider diversity: Influences of human modification of habitat structure and productivity. Ecological Applications 14:268–280.

Shochat, E., P. S. Warren, S. H. Faeth, N. E. McIntyre, and D. Hope. 2006. From patterns to emerging processes in mechanistic urban ecology. Trends in Ecology & Evolution 21:186–191.

Short, J. R. 2006. Urban Theory. A Critical Assessment. Palgrave Macmillan, New York, NY.

Shutes, R. B. E., D. M. Revitt, A. S. Mungur, and L. N. L. Scholes. 1984. The design of wetland systems for the treatment of urban runoff. Water Science and Technology 35(5):19.

Sklar, F. H., and R. Costanza. 1991. The development of dynamic spatial models for landscape ecology: a review and prognosis. Pages 239–288 in M. G. Turner and R. Gardner (eds.), Quantitative Methods in Landscape Ecology. Springer, New York.

Simon, H. A. 1973. The organization of complex systems. Pages 1–27 in H. H. Pattee (ed.), Hierarchy Theory: The Challenge of Complex Systems. George Braziller, New York.

Simon, H. A. 1962. The architecture of complexity. Proceedings of the American Philosophical Society 106(6):467–482.

Slabbekoorn, H., and M. Peet. 2003. Birds sing at a higher pitch in urban noise – great tits hit the high notes to ensure that their mating calls are heard above the city's din. Nature 424:267.

Small, K. A., and S. Song. 1994. Population and employment densities: Structure and change. Journal of Urban Economics 36: 292–313.

Smil, V. 2000. Phosphorus in the environment: Natural flows and human interferences. Annual Review of Energy and the Environment 25(1): 53–88.

Smith, V. H. 1992. Effects of nitrogen: phosphorus supply ratios on nitrogen fixation in agricultural and pastoral ecosystems. Biogeochemistry 18(1):19–35.

Smith, J. B. and D. A. Tirpak (eds.). 1989. The Potential Effects of Global Climate Change on the United States. EPA-230-05-89, Office of Policy, Planning and Evaluation, U.S. Environmental Protection Agency, Washington, DC.

Sole, R. V., Manrubia, S. C., Benton, M. et al. 1999. Criticality and scaling in evolutionary ecology. Trends in Ecology and Evolution 14:156–160.

Solomon, S., D. Qin, M. Manning, R.B. Alley, T. Berntsen, N.L. Bindoff, Z. Chen, A. Chidthaisong, J.M. Gregory, G.C. Hegerl, M. Heimann, B. Hewitson, B.J. Hoskins, F. Joos, J. Jouzel, V. Kattsov, U. Lohmann, T. Matsuno, M. Molina, N. Nicholls, J. Overpeck, G. Raga, V. Ramaswamy, J. Ren, M. Rusticucci, R. Somerville, T.F. Stocker, P. Whetton, R.A. Wood and D. Wratt, 2007. Technical Summary. In: Climate Change 2007: The Physical Science Basis. Contribution of Working Group I to the Fourth Assessment Report of the Intergovernmental Panel on Climate Change [Solomon, S., D. Qin, M. Manning, Z. Chen, M. Marquis, K.B. Averyt, M. Tignor and H.L. Miller (eds.)]. Cambridge University Press, Cambridge, United Kingdom and New York, NY, USA. online: http://ipccwg1.ucar. edu/wg1/Report/AR4WG1_Pub_TS.pdf .

Song, S. 1992. Monocentric and Polycentric Density Functions and Their Required Commutes. Working Paper UCTC No. 198. University of California Transportation Center, Berkeley, CA.

Song, Y., and G. J. Knaap. 2004. Measuring urban form: Is Portland winning the war on sprawl? Journal of American Planning Association 70(2):210–225.

Souch, C., and S. Grimmond. 2006. Applied Climatology: Urban Climate, Progress in Physical Geography 30(2):270–279.

Soulé, M., D. Bolger, A. Albert, J. Wright, M. Sorice, and S. Hill. 1988. Reconstructed dynamics of rapid extinctions of chaparral-requiring birds in urban habitat islands. Conservation Biology 2:75–92.

Spence, J. R., and D. H. Spence. 1988. Of groundbeetles and men: Introduced species and the synanthropic fauna of western Canada. Memoirs of the Entomological Society of Canada 144:151–168.

Spirn, A. W. 1984. The Granite Garden: Urban Nature and Human Design. Basic Books, New York.

Spirn, A. 1998. The language of landscape. Yale University Press, New Haven Conn.

Srivastava, D. S., and M. Vellend. 2005. Biodiversity-ecosystem function research: is it relevant to conservation? Annual Review of Ecology, Evolution and Systematics 36: 267–294.

Starrett, D. A. 1974. Principles of Optimal Location in a Large Homogeneous Area," Journal of Economic Theory 9(4):418–48.

Steedman, R. J. 1988. Modification and assessment of an index of biotic integrity to quantify stream quality in southern Ontario. Canadian Journal of Fisheries and Aquatic Sciences 45:492–501.

Steinberg, D. A., R. V. Pouyat, R. W. Parmelee, and P. M. Groffman. 1997. Earthworm abundance and nitrogen mineralization rates along an urban-rural land use gradient. Soil Biology and Biochemistry 29:427–430.

Steiner, F. 1991. Landscape planning: a method applied to a growth management example. Environmental Management, 15:519–529.

Steiner, F. 2000. The living landscape: an ecological approach to landscape planning. 2nd edn. McGraw-Hill, New York.

Steiner, F. 2004. Urban human ecology. Urban Ecosyst 7:179–197.

Stern, F.W. T. Montag. 1974. The Urban Ecosystem. The holistic Approach. Dowde. Hutchinson and Ross, Stroudsburg. Pennsilvania.

Stewart, J. W. B., and R. W. Howarth. 1992. Introduction. In R. W. Howarth, J. W. B. Stewart, and M. V. Ivanov (eds.), Sulphur Cycling on the Continents: Wetlands, Terrestrial Ecosystems, and Associated Water Bodies. Wiley, U.K.

Stone, B. 2005. Urban heat and air pollution: An emerging role for planners in the climate change debate. Journal of the American Planning Association 71:13–25.

Stone, P. A. 1973. The Structure, Size, and Costs of Urban Settlements. Cambridge University Press, Cambridge, UK.

Sukopp, H. 1990. Urban ecology and its application in Europe. Pages 1–22 in H. Sukopp, and S. Hejny (eds.), Urban Ecology: Plants and Plant Communities in Urban Environments. SPB Publishing, The Hague.

Sukopp, H. 1998. On the study of anthropogenic plant migrations in Central Europe. Pages 43–56 in Plant Invasions: Ecological Mechanisms and Human Responses. Backhuys Publishers, Leiden, the Netherlands.

Sukopp, H., H. P. Blume, and W. Kunick. 1979. The soil, flora and vegetation of Berlin's wastelands. Pages 115–132 in I. C. Laurie (ed.), Nature in Cities. John Wiley, Chichester.

Sukopp, H., and M. Numata 1995. Foreword. Page vii in H. Sukopp, M. Numata, and A. Huber (eds.), Urban Ecology as the Basis for Urban Planning. SPB Academic, The Hague.

Sukopp, H., M. Numata, and A. Huber (eds.), 1995. Urban Ecology as the Basis for Urban Planning. SPB Academic, The Hague.

Sukopp, H., and U. Starfinger. 1999. Disturbance in urban ecosystems. Pages 397–412 in L. R. Walker (ed.), Ecosystems of the World 16: Ecosystems of Disturbed Ground. Elsevier, Amsterdam.

Sukopp, H. and Werner, P. 1982. Nature in Cities. Nature and Environment Series, No. 36. Council of Europe, Strasbourg.

Sustainable Seattle. 1999. Indicators of Sustainable Community. Sustainable Seattle, Seattle.

Sweeney, B. W. 1984. Factors influencing life history patterns of aquatic insects. Pages. 56–100 in V. H. Resh, and D. M. Rosenberg (eds.), The Ecology of Aquatic Insects. Praeger. New York.

Taylor, P. D., L. Fahrig, K. Henein, and G. Merriam. 1993. Connectivity is a vital element of landscape structure. Oikos 68:571–572.

Thomson, J. R., P. S. Lake, and B. J. Downes. 2002. The effect of hydrological disturbance on the impact of a benthic invertebrate predator. Ecology 83: 628–642.

Thompson, W. 1975. Economic processes and employment problems in declining metropolitan areas. Pages 187–196 in G. Sternlieb, and J. W. Hughes (eds.), Post-industrial America: Metropolitan Decline and Inter-regional Job Shifts. Rutgers, New Brunswick, NJ.

Thorne, R. S. J., W. P. Williams, and C. Gordon. 2000. The macroinvertebrates of a polluted stream in Ghana. Journal of Freshwater Ecology 15:209–217.

Tilghman, N. 1987. Characteristics of urban woodlands affecting breeding bird diversity and abundance. Landscape & Urban Planning Journal 14:481–495.

Tilman, D. 1996. Biodiversity: Population versus ecosystem stability. Ecology 77:350–363.

Tilman, D. 1999. The ecological consequences of changes in biodiversity: A search for general principles. Ecology 80:1455–1474.

Tilman, D., and S. Pacala. 1993. The maintenance of species richness in plant communities. Pages 13–25 in R. E. Ricklefs, and D. Schluter (eds.), Species Diversity in Ecological Communities: Historical and Geographical Perspectives. University of Chicago Press, Chicago.

Tilman, D., D. Wedin, and J. Knops. 1996. Productivity and sustainability influenced by biodiversity in grassland ecosystems. Nature 379:718–720.

Tischendorf, L., and L. Fahrig. 2000. On the usage and measurement of landscape connectivity. Oikos 90:7–19.

Tobler, W. R. 1979. Smooth pycnophylactic interpolation for geographic regions. Journal of the American Statistical Association 74:519–530.

Torrens, P. 2003. Automata-based models of urban systems. In: P. Longley and M. Batty (Eds.), Advanced Spatial Analysis. ESRI Press, Redlands, CA.

Torrens, P. M. & M. Alberti. 2000. Measuring sprawl. CASA Working Paper 27. University College London, Centre for Advanced Spatial Analysis.

Townsend, C. R. 1989. The patch dynamics concept of stream community ecology. Journal of the North American Benthological Society. 8:36–50.

Tress, G., B. Tress, G. Fry, (2004) Clarifying integrative research concepts in landscape ecology. Landsc Ecol 20:479–493.

Trepl, L. 1994. Holism and reductionism in ecology: Technical, political, and ideological implications. Capit. Nat. Soc. 5:13–31.

Trepl, L. 1995. Towards a theory of urban biocoenoses. Pages 3–21 in H. Sukopp, M. Numata, and A. Huber (eds.), Urban Ecology as the Basis for Urban Planning. SPB Academic, The Hague.

Trombulak, S. C., and C. A. Frissell. 2000. Review of ecological effects of roads on terrestrial and aquatic communities. Conservation Biology 14(1):18–30.

Turner, B. L. II., W. Clark, R. Kates, J. Richards, J. Mathew, and W. Meyer (eds.), 1990. The Earth as Transformed by Human Action: Global and Regional Changes in the Biosphere Over the Past 300 Years. Cambridge University Press, Cambridge, UK.

Turner, M. G. 1989. Landscape ecology: the effect of pattern on process. Annual Review of Ecology and Systematics 20:171–197.

Turner, M. G., G. J. Arthaud, R. T. Engstrom, S. J. Hejl, J. Liu, S. Loeb, and K. McKelvey. 1995. Usefulness of spatially explicit population models in land management. Ecological Applications 5:12–16.

Turner, M. G., S. R. Carpenter, and E. J. Gustafson. 1998. Land use. Pages 37–61 in M. J. Mac, P. A. Opler, and C. E. P. Haecker (eds.), Status and

Trends of The Nation's Biological Resources, Volume 1. U.S. Department of the Interior, Geological Survey, Washington, DC.

Turner, M. G., and F. S. I. Chapin. 2005. Causes and consequences of spatial heterogeneity in ecosystem function. Pages 9–30 in G. M. Lovett, C. G. Jones, M. G. Turner, and K. C. Weathers (eds.), Ecosystem Function in Heterogeneous Landscapes. Springer, New York.

Turner, M. G., R. Costanza, and F. H. Sklar. 1989. Methods to evaluate the performance of spatial simulation models. Ecological Modelling 48:1–18.

Turner, M., and R. Gardner (eds.), 1991. Quantitative Methods in Landscape Ecology. Ecological Studies. Springer, New York.

Turner, M. G., W. H. Romme, R. H. Gardner, R. V. O'Neill, and T. K. Kratz. 1993. A revised concept of landscape equilibrium: Disturbance and stability across scaled landscapes. Landscape Ecology 8(3): 213–227.

Turner, M. G., D. N. Wear, and R. O. Flamm. 1996. Land ownership and land-cover change in the southern Appalachian highlands and the Olympic peninsula. Ecological Applications 6:1150–1172.

UERL 2005a http://www.urbaneco.washington.edu/landcover.htm.

UERL 2005b. Alberti, M. (PI) Weeks, R. and C. Russel. Puget Sound Landcover 2002. PRISM, University of Washington, Seattle. WA.

UERL 2005c. Marina Alberti (PI), Jeff Hepinstall, Michal Russo, Bekkah Coburn. 2005. CTED Landscape Benchmarks Project. Trade and Community Development, WA.

UERL 2006. Marina Alberti (PI), Paul Waddell, John Marzluff, Mark Handcock. 2006. Modeling Interactions Among Urban Development, Land-Cover Change, and Bird Diversity. National Science Foundation Biocomplexity project, 2001–2006 (BCS-0120024).

UNESCO-IHP (UN Educational, Scientific and Cultural Organization - International Hydrological Programme), 2004. The World-wide Hydrogeological Mapping and Assessment Programme (WHYMAP), A consortium mapping project of International Hydrological Program, International Geological Correlation Programme, International Association of Hydrogeologists, Commission for the Geological Map of the World, International Agricultural Exchange Association, Bundesanstalt fur Geowissenschaften und Rohstoffe—Germany). Available at www.bgr.de/b1hydro/index.html?/b1hydro/fachbeitraege/a200401/e_whymap.htm.

Unger, N., D. T. Shindell, D. M. Koch, M. Amann, J. Cofala, and D. G. Streets. 2006. Influences of man-made emissions and climate changes on tropospheric ozone, methane, and sulfate at 2030 from a broad range of possible futures. Journal of Geophysical Research 111, D12313, doi:10.1029/2005JD006518.

United Nations, 1999. The State of World Population 1999–6 Billion: A Time for Choices. New York: United Nations Population Fund.

United Nations, 2006. World Urbanization Prospects: The 2005 Revision. United Nations, Department of Economic and Social Affairs, Population Division, Working Paper No. ESA/P/WP/200.

Upmanis, H., I. Eliasson, and S. Lindqvist. 1998. The influence of green areas on nocturnal temperatures in a high latitude city (Göteborg, Sweden). International Journal of Climatology 18:681–700.

Urban, D. L., R. V. O'Neill, and H. H. Shugart Jr. 1987. Landscape ecology — a hierarchical perspective can help scientists understand spatial patterns. BioScience 37:119–127.

US Census. 2000. U.S. Census Bureau, 2000 Census of Population and Housing, Summary Population and Housing Characteristics, PHC-1–1, United States. Washington, DC.

US EPA. 2004. The Ozone Report: Measuring Progress Through 2003. EPA 454/K-04-001. U.S. Environmental Protection Agency, Office of Air Quality Planning and Standards, Research Triangle Park, NC.

US EPA. 2005. National Management Measures to Control Nonpoint Source Pollution from Urban Areas. EPA-841-B-05-004. US Environmental Protection Agency, Office of Water, Washington, DC.

USGS. 1999. The Quality of Our Nation's Waters – Nutrients and Pesticides. U.S. Geological Survey Circular 1225. U.S. Geological Survey, U.S. Department of the Interior, Washington, DC. <http://water.usgs.gov/pubs/circ/circ1225/index.html>

Vermeulen, H. J. W. 1994. Corridor function of a road verge for dispersal of stenotopic heathland ground beetles Carabidae. Biological Conservation 69:339–349.

Vickers, A. 2001. Handbook of Water Use and Conservation. WaterPlow Press, Amherst, MA.

Vitousek, P. M. 1990. Biological invasions and ecosystem processes: Towards an integration of population biology and ecosystem studies. Oikos 57:7–13.

Vitousek, P. M., K. Cassman, C. Cleveland, T. Crews, C. B. Field, N. B. Grimm, R. W. Howarth, R. Marino, L. Martinelli, E. B. Rastetter, and J. I. Sprent. 2002. Towards an ecological understanding of biological nitrogen fixation. Biogeochemistry 57/58:1–45.

Vitousek, P. M., C. M. D'Antonio, L. L. Loope, M. Rejmanek, and R. Westbrooks. 1997a. Introduced species: A significant component of human-caused global change. New Zealand Journal of Ecology 21:1–16.

Vitousek, P. M., P. R. Ehrlich, A. H. Ehrlich, and P. A. Matson. 1986. Human appropriation of the products of photosynthesis. BioScience 36:368–373.

Vitousek, P. M, and D. U. Hooper. 1993. Biological diversity and terrestrial ecosystem biogeochemistry. Pages 3–14 in E.-D. Schulze and H. A. Mooney (eds.), Biodiversity and Ecosystem Function. Springer-Verlag, Berlin, Germany.

Vitousek, P. M., and P. A. Matson. 1990. Gradient analysis of ecosystems. Pages 287–298 in J. J. Cole, G. M. Lovett, S. E. G. Findlay (eds.), Comparative Analysis of Ecosystems: Patterns, Mechanisms and Theories. Springer Verlag, New York.

Vitousek, P. M., H. A. Mooney, J. Lubchenco, and J. Melillo. 1997b. Human domination of Earth's ecosystems. Science 277:494–499.

Vitruvius. ca. 1st century B.C.E. In The Ten Books on Architecture Cambridge, MA Harvard University Press, 1914.

Waddell, P. A. 1998. UrbanSim: The Oregon prototype metropolitan land use model, Proceedings of the 1998 ASCE Conference on Land Use, Transportation and Air Quality: Making the Connection, Portland, Oregon.

Waddell, P. 2000. A behavioral simulation model for metropolitan policy analysis and planning: residential location and housing market components of UrbanSim. Environment and Planning B: Planning and Design 27(2):247–263.

Waddell, P. 2002. UrbanSim: Modeling urban development for land use, transportation, and environmental planning. APA Journal 68(3): 297–314.

Waddell, P. A. and M. Alberti, 1998. Integration of an Urban Simulation Model and an Urban Ecosystems Model, Proceedings of the International Conference on Modeling Geographical and Environmental Systems with Geographical Information Systems, Hong Kong.

Waddell, P. A., A. Borning, M. North, N. Freier, M. Becke, and G. Ulfarsson. 2003. Microsimulation of urban development and location choices: design and implementation of UrbanSim. Networks and Spatial Economics 3: 43–67.

Waddell, P., G.F. Ulfarsson, J. Franklin and J. Lobb. 2007. Incorporating Land Use in Metropolitan Transportation Planning, Transportation Research Part A: Policy and Practice 41:382–410.

Wagener, F. O. 2003. Skiba points and heteroclinic bifurcations with applications to the shallow lake system. Journal of Economic Dynamics and Control 27, 1533–1561.

Wagner, H. H., and M-J. Fortin. 2005. Spatial analysis of landscapes: Concepts and statistics. Ecology 86:1975–1985.

Waide, R. B., M. R. Willig, C. F. Steiner, G. Mittelbach, L. Gough, S. I. Dodson, G. P. Juday, and R. Parmenter. 1999. The relationship between productivity and species richness. Annual Review of Ecology and Systematics 30:257–301.

Wakim, P. G. 1989. Temperature-adjusted Ozone Trends for Houston, New York, and Washington, 1981–1987. Paper 89–35.1. Presented at the 82nd Annual Meeting and Exhibition of the Air and Waste Management Association, Anaheim, CA, June 25–30, 1989. Air and Waste Management Association, Pittsburgh, PA.

Walbridge, M. R. 1997. Urban ecosystems. Urban Ecosystems 1:1–2.

Wali, A., G. Darlow, C. Fialkowski, M. Tudor, H. del Campo, and D. Stotz. 2003. New methodologies for interdisciplinary research and action in an urban ecosystem in Chicago. Conservation Ecology 7(3):2.

Walker, B., and J. A. Meyers. 2004. Thresholds in ecological and social-ecological systems: a developing database. Ecology and Society 9(2):3, [online] URL: http://www.ecologyandsociety.org/vol9/iss2/art3/.

Walker, B. H. 1992. Biodiversity and ecological redundancy. Conservation Biology 6:18–23.

Walker, B. H., L. H. Gunderson, A. Kinzig, C. Folke, S. R. Carpenter, and L. Schultz. 2006. A handful of heuristics and some propositions for understanding resilience in social-ecological systems. Ecology and Society 11:13.

Walker, B., C. S. Holling, S. R. Carpenter, and A. Kinzig. 2004. Resilience, adaptability and transformability in social-ecological systems. Ecology and Society 9(2): 5. [online] URL: http://www.ecologyandsociety.org/vol9/iss2/art5/.

Walsh, C. J., T. D. Fletcher, and A. R. Ladson. 2005. Stream restoration in urban catchments through redesigning stormwater systems: Looking to the catchment to save the stream. Journal of the North American Benthological Society. 24:690–705.

Walters, S. 2007. Modeling scale-dependent landscape pattern, dispersal, and connectivity from the perspective of the organism. Landscape Ecology 22:867–881.

Wandeler, P., S. M. Funk, and C. R. Largiadèr, S. Gloor, and U. Breitenmoser. 2003. The city-fox phenomenon: Genetic consequences of a recent colonization of urban habitat. Molecular Ecology 12(3):647–656.

Wang, L., J. Lyons, P. Kanehl, and R. Gatti. 1997. Influences of watershed land use on habitat quality and biotic integrity in Wisconsin streams. Fisheries 22:6–12.

Warren-Rhodes, K., and A. Koenig, 2001. Ecosystem appropriation by Hong Kong and its implications for sustainable development. Ecological Economics, 39:347–359.

Waschbusch, R. J., W. R. Selbig, and R. T. Bannerman. 1999. Sources of phosphorus in stormwater and street dirt from two urban residential basins in Madison, Wisconsin. USGS Water Resources Investigative Report 99–4021.

Watkins, R., J. Palmer, M. Kolokotroni, and P. Littlefair. 2002. The London Heat Island, Results from a summertime monitoring, Building Service Engineering Research Technology 23(2):97–106.

Wear, D. N., and P. Bolstad. 1998. Land-use changes in Southern Appalachian landscapes: Spatial analysis and forecast evaluation. Ecosystems 1(6):575–594.

Wear, D. N., M. G. Turner, and R. J. Naiman. 1998. Land cover along an urban-rural gradient: implications for water quality. Ecological Applications 8(3):619–630.

Wegener, M. 1983. Description of the Dortmund Region Model. Working Paper 8. Dortmund: Institute für Raumplanung.

Wegener, M. 1994. Operational urban models: state of the art. Journal of the American Planning Association 60(1):17–30.

Wegener, M. 1995. Current and Future Land Use Models. In G. A. Shunk (ed.) LandUse Modeling Conference Proceedings, February 19–21, 1995. Report DOT-T-96-09. US Department of Transportation, Washington, DC, 1995, pages 13–40.

Wegener, M., P. A. Waddell, and I. Salomon. 1999. Sustainable lifestyles? Microsimulation of household formation, housing choice and travel behaviour. In: Proceedings of the National Science Foundation–European Science Foundation Conference on Social Change and Sustainable Transportation, Berkeley, California.

Webster, J. R., and E. F. Benfield. 1986. Vascular plant breakdown in freshwater ecosystems. Annual Review of Ecology and Systematics 17:567–94.

Weigmann, G. 1982. The colonization of ruderal biotopes in the city of Berlin by arthropods. Pages 75–82 in R. Bornkamm, J. A. Lee and

M. R. D. Seaward (eds.), Urban Ecology, The Second European Ecological Symposium. Blackwell Scientific Publications, Oxford.

Werner, B. T. 1999. Complexity in natural landform patterns. Science 284:102–104.

Wernick, B. G., K. E. Cook, and H. Schreier. 1998. Land use and streamwater nitrate-N dynamics in an urban-rural fringe watershed. Journal of the American Water Resources Association 34:639–650.

Western, D. 2001. Human-modified ecosystems and future evolution. Proceedings of the National Academy of Sciences 98(10):5458–5465.

Westley, F., S. R. Carpenter, W. Brock, C. S. Holling, and L. H. Gunderson. 2002. Why systems of people and nature are not just social and ecological systems. Pages 103–120 in L. H. Gunderson, and C. S. Holling (eds.), Panarchy: Understanding Transformations In Human and Natural Systems. Island Press, Washington, D. C.

Wetterer, J. K. 1997. Urban ecology: Encyclopedia of environmental sciences. New York: Chapman & Hall.

Wheaton, W. C. 1974. Linear Programming and Locational Equilibrium: The Herbert-Stevens Model Revisited. Journal of Urban Economics 1:278–287.

Whipple, W. Jr., and J. V. Hunter. 1979. Petroleum hydrocarbons in urban runoff. Water Resources Bulletin 15:1096–104.

White, R., and G. Engelen. 1993. Cellular automata and fractal urban form: a cellular modelling approach to the evaluation of urban land-use patterns. Environment and Planning A 25:1175–1199.

White, R., and G. Engelen. 1997. Cellular automata as the basis of integrated dynamic modeling. Environment and Planning B 24:235–246.

White, R., G. Engelen, and I. Uljee. 1997. The use of constrained cellular automata for high-resolution modeling of urban land use dynamics. Environment and Planning B 24:323–343.

Whiting, E. R., and H. F. Clifford. 1983. Invertebrates and urban runoff in a small northern stream, Edmonton, Alberta, Canada. Hydrobiologia 102:73–80.

White, P. S., and S. T. A. Pickett. 1985. Natural disturbance and patch dynamics: An introduction. Chapter 1 in S. T. A. Pickett, and P. S. White (eds.), The Ecology of Natural Disturbance and Patch Dynamics. Academic Press, Orlando, FL.

White, R., and J. Whitney. 1992. Cities and the environment: An overview. Pages 8–51 in R. Stren, R. White, and J. Whitney (eds.), Sustainable

Cities: Urbanization and the Environment in International Perspective. Westview, Oxford.

Whittaker, R. H. 1967. Gradient analysis of vegetation. Biological Reviews. 42:207–264.

Wiegand, T., F. Jeltsch, I. Hanski, and V. Grimm. 2003. Using pattern-oriented modeling for revealing hidden information: A key for reconciling ecological theory and conservation practice. Oikos 100: 209–222.

Wiens, J. A. 1989. Spatial scaling in ecology. Functional Ecology 3: 385–397.

Wilcove, D. S. 1985. Nest predation in forest tracts and the decline of migratory songbirds. Ecology 66:1211–1214.

Williams, C. B. 1964. Patterns in the Balance of Nature. Academic Press, London.

Williams, C. B. 1943. Area and number of species. Nature 152:264–267.

Wilsey, B. J., and C. Potvin. 2000. Biodiversity and ecosystem functioning: Importance of species evenness in an old field. Ecology 81:887–892.

Wilson, A. G., J. D. Coelho, S. M. Macgill, and H. C. W. L. Williams. 1981. Optimisation in Locational and Transport Analysis. John Wiley, Chichester.

Wilson, A. G., P. H. Rees, and C. M. Leigh. 1977. Models of Cities and Regions. John Wiley, Chichester.

Wilson, E. O. 1992. The Diversity of Life. Belknap Press of Harvard University Press, Cambridge, MA.

Wilson, J. B. 1990. Mechanisms of species coexistence: Twelve explanations for Hutchinson's "paradox of the plankton": Evidence from New Zealand plant communities. New Zealand Journal of Ecology 13:17–42.

Wilson, J. B. 1994. The intermediate disturbance hypothesis of species coexistence is based on patch dynamics. New Zealand Journal of Ecology 18:176–181.

Wingo, L. 1961. Transportation and Urban Land. The Johns Hopkins University Press: Baltimore, MD.

Wirth, L. 1925. A bibliography of the urban community. Pages 161–228 in R. E. Park, E. W. Burgess, and R. D. McKenzie (eds.), The City. University of Chicago Press, Chicago.

With, K. A., and T. O. Crist. 1995. Critical Thresholds in Species Responses to Landscape Structure. Ecology 76:2446–2459.

With, K. A., R. H. Gardner, and M. G. Turner. 1997. Landscape connectivity and population distributions in heterogeneous environments. Oikos 78:151–169.

With, K. A., and A. W. King. 1999. Dispersal success on fractal landscapes: a consequence of lacunarity thresholds. Landscape Ecology 14:73–82.

Wollheim, W. M., B. A. Pellerin, C. J. Vörösmarty, and C. S. Hopkinson. 2005. N retention in urbanizing headwater catchments. Ecosystems 8:871–884.

Wolman, A. 1965. The Metabolism of Cities. Scientific American 213: 179–188.

Wolman, M. G. 1967. A cycle of sedimentation and erosion in urban river channels. Geografiska Annaler, Series A: Physical Geography 49: 385–395.

Wolman, M. G., and J. P. Miller. 1960. Magnitude and frequency of forces in geomorphic processes. Journal of Geology 68:54–74.

Wolman, M. G., and A. P. Shick. 1967. Effects of construction on fluvial sediment, urban and suburban areas of Maryland. Water Resources Research 3:451–464.

Woodbury, P. B., J. E. Smith, and L. S. Heath. 2007. Carbon sequestration in the U.S. forest sector from 1990 to 2010. Forest Ecology and Management 241(1–3):14–27.

Wood, W. E., and S. M. Yezerinac. 2006. Song Sparrow (Melospiza melodia) song varies with urban noise. Auk 123:650–659.

WRI. 2005. Climate and Atmosphere Data Tables. World Resources Institute, Washington, DC. online: http://earthtrends.wri.org /datatables/ index.cfm?theme=3.

Wu, F. 1998a. An experiment on the generic polycentricity of urban growth in a cellular automata city. Environment and Planning B 25:731–752.

Wu, F. 1998b. Simulating urban encroachment on rural land with fuzzy-logic controlled cellular automata in a geographical information system. Journal of Environmental Management 53:293 308.

Wu, F. 1999. A simulation approach to urban changes: Experiments and observations on fluctuations in cellular automata. Pages 20 in P. Rizzi, and F. Angeli (eds.), 6th International Conference on Computers in Urban Planning and Management. Stratema Istituto Universitario di Architettura di Venezia,Venezia.

Wu, F. and C. J. Webster. 1998. Simulation of land development through the integration of cellular automata and multicriteria evaluation. Environment and Planning B 25:103–126.

Wu, J. 1999. Hierarchy and scaling: Extrapolating information along a scaling ladder. Canadian Journal of Remote Sensing 25(4):367–380.

Wu, J. G. 2004. Effects of changing scale on landscape pattern analysis: scaling relations. Landscape Ecology 19:125–138.

Wu, J., and J. David. 2002. A spatially explicit hierarchical approach to modeling complex ecological systems: Theory and applications. Ecological Modelling 153:7–26.

Wu, J., and R. Hobbs. 2002. Key issues and research priorities in landscape ecology: An idiosyncratic synthesis. Landscape Ecology 17:355–365.

Wu, J. G., D. E. Jelinski, M. Luck, and P. T. Tueller. 2000. Multiscale analysis of landscape heterogeneity: Scale variance and pattern metrics. Geographic Information Sciences 6:6–19.

Wu, J., J. Huang, X. Han, Z. Xie, and X. Gao. 2003. Three-Gorges Dam: Experiment in Habitat Fragmentation? Science 300:1239–1240.

Wu, J., and S. A. Levin. 1994. A spatial patch dynamic modeling approach to pattern and process in an annual grassland. Ecological Monographs 64:447–464.

Wu, J., and S. A. Levin. 1997. A patch-based spatial modeling approach: Conceptual framework and simulation scheme. Ecological Modelling 101:325–346.

Wu, J., and O. L. Loucks. 1995. From balance of nature to hierarchical patch dynamics: A paradigm shift in ecology. The Quarterly Review of Biology 70:439–466.

Wu, J., and Y. Qi. 2000. Dealing with scale in landscape analysis: An overview. Geographic Information Sciences 6:1–5.

Wu, J., W. Shen, W. Sun, and P. T. Tueller. 2002. Empirical patterns of the effects of changing scale on landscape metrics. Landscape Ecology 17:761–782.

Xie, Y., M. Yu, Y. Bai, and X. Xing. 2006. Ecological analysis of an emerging urban landscape pattern—desakota: A case study in Suzhou, China. Landscape Ecology 21:1297–1309.

Yamashita, S., K. Sekine, M. Shoda, K. Yamashita, and Y. Hara. 1986. On relationships between heat island and sky view factor in the cities of Tama River Basin, Japan. Atmospheric Environment 20:681–686.

Yeh, P. J. 2004. Rapid evolution of a sexually selected trait following population establishment in a novel habitat. Evolution 58:166–174.

Yeh, P. J., and T. D. Price. 2004. Adaptive phenotypic plasticity and the successful colonization of a novel environment. American Naturalist 164:531–542.

Yoder, C. O., R. J. Miltner, and D. White. 1999. Assessing the status of aquatic life designated uses in urban and suburban watersheds. In Retrofit Opportunities for Water Resource Protection in Urban Environment: A National Conference, Chicago, IL, 16–28. EPA/625/R-99/002.

Young, I. M., E. Blanchart, C. Chenu, M. Dangerfields, C. Fragoso, M. Grimaldi, J. Ingram, and L. Jocteur. 1998. The interaction of soil biota and soil structure under global change. Global Change Biology 4: 703–712.

Zeng, G., and J. A. Pyle. 2003. Changes in tropospheric ozone between 2000 and 2100 modeled in a chemistry-climate model. Geophysical Research Letters 30(7), 1392, doi:10.1029/2002GL016708.

Zhu, W., and M. M. Carreiro. 1999. Chemoautotrophic nitrification in acidic forest soils along an urban-to-rural transect. Soil Biology and Biochemistry:1091–1100.

Zhu, W., and M. Carreiro. 2004. Temporal and spatial variations in nitrogen transformations in deciduous forest ecosystems along an urban - rural gradient. Soil Biology and Biochemistry 36(2):267–278.

Zhu, W. X., N. D. Dillard, and N. B. Grimm. 2004. Urban nitrogen biogeochemistry: Status and processes in green retention basins. Biogeochemistry 71:177–196.

Zielinski, K. 1979. Experimental analysis of eleven models of urban population density. Environment and Planning A 11(6):629–641.

Ziemer, R. R., and T. E. Lisle. 1998. Chapter 3: Hydrology. Pages 43–68 in R. J. Naiman, and R. E. Bilby (eds.), River Ecology and Management: Lessons from the Pacific Coastal Ecoregion. Springer-Verlag, New York.

Zimmerer, K. 1994. Human geography and the 'new ecology': the prospect and promise of integration. Ann Assoc Am Geogr 84:108–125.

Zimov, S. A., V. I. Chuprynin, A. P. Oreshko, F. S. Chapin III, J. F. Reynolds, and M. C. Chapin. 1995. Steppe-tundra transition: A herbivore-driven biome shift at the end of the Pleistocene. American Naturalist 146:765–794.

Zipperer, W. C., J. Wu, R. V. Pouyat, and S. T. A. Pickett. 2000. The application of ecological principles to urban and urbanizing landscapes. Ecological Applications 10(3):685–688.

Zipf, G. K. 1949. Human Behavior and the Principle of Least Effort. Addison-Wesley, Cambridge, MA.

Ziska, L. H., J. A. Bunce, and E. W. Goins. 2004. Characterization of an urban-rural $CO_2$/temperature gradient and associated changes in initial plant productivity during secondary succession. Oecologia 139: 454–458.

# INDEX